Cosmetic Lipids and the Skin Barrier

COSMETIC SCIENCE AND TECHNOLOGY

Series Editor

ERIC JUNGERMANN

Jungermann Associates, Inc.
Phoenix, Arizona

1. Cosmetic and Drug Preservation: Principles and Practice, *edited by Jon J. Kabara*
2. The Cosmetic Industry: Scientific and Regulatory Foundations, *edited by Norman F. Estrin*
3. Cosmetic Product Testing: A Modern Psychophysical Approach, *Howard R. Moskowitz*
4. Cosmetic Analysis: Selective Methods and Techniques, *edited by P. Boré*
5. Cosmetic Safety: A Primer for Cosmetic Scientists, *edited by James H. Whittam*
6. Oral Hygiene Products and Practice, *Morton Pader*
7. Antiperspirants and Deodorants, *edited by Karl Laden and Carl B. Felger*
8. Clinical Safety and Efficacy Testing of Cosmetics, *edited by William C. Waggoner*
9. Methods for Cutaneous Investigation, *edited by Robert L. Rietschel and Thomas S. Spencer*
10. Sunscreens: Development, Evaluation, and Regulatory Aspects, *edited by Nicholas J. Lowe and Nadim A. Shaath*
11. Glycerine: A Key Cosmetic Ingredient, *edited by Eric Jungermann and Norman O. V. Sonntag*
12. Handbook of Cosmetic Microbiology, *Donald S. Orth*
13. Rheological Properties of Cosmetics and Toiletries, *edited by Dennis Laba*
14. Consumer Testing and Evaluation of Personal Care Products, *Howard R. Moskowitz*
15. Sunscreens: Development, Evaluation, and Regulatory Aspects. Second Edition, Revised and Expanded, *edited by Nicholas J. Lowe, Nadim A. Shaath, and Madhu A. Pathak*
16. Preservative-Free and Self-Preserving Cosmetics and Drugs: Principles and Practice, *edited by Jon J. Kabara and Donald S. Orth*
17. Hair and Hair Care, *edited by Dale H. Johnson*
18. Cosmetic Claims Substantiation, *edited by Louise B. Aust*
19. Novel Cosmetic Delivery Systems, *edited by Shlomo Magdassi and Elka Touitou*
20. Antiperspirants and Deodorants: Second Edition, Revised and Expanded, *edited by Karl Laden*
21. Conditioning Agents for Hair and Skin, *edited by Randy Schueller and Perry Romanowski*

22. Principles of Polymer Science and Technology in Cosmetics and Personal Care, *edited by E. Desmond Goddard and James V. Gruber*
23. Cosmeceuticals: Drugs vs. Cosmetics, *edited by Peter Elsner and Howard I. Maibach*
24. Cosmetic Lipids and the Skin Barrier, *edited by Thomas Förster*

ADDITIONAL VOLUMES IN PREPARATION

Botanicals in Cosmetics, *edited by Larry Smith*
Skin Moisturization, *edited by James J. Leyden and Anthony V. Rawlings*

Cosmetic Lipids and the Skin Barrier

edited by
Thomas Förster
Henkel KGaA
Düsseldorf, Germany

MARCEL DEKKER, INC. NEW YORK · BASEL

ISBN: 0-8247-0664-1

This book is printed on acid-free paper.

Headquarters
Marcel Dekker, Inc.
270 Madison Avenue, New York, NY 10016
tel: 212-696-9000; fax: 212-685-4540

Eastern Hemisphere Distribution
Marcel Dekker AG
Hutgasse 4, Postfach 812, CH-4001 Basel, Switzerland
tel: 41-61-261-8482; fax: 41-61-261-8896

World Wide Web
http://www.dekker.com

The publisher offers discounts on this book when ordered in bulk quantities. For more information, write to Special Sales/Professional Marketing at the headquarters address above.

Copyright © 2002 by Marcel Dekker, Inc. All Rights Reserved.

Neither this book nor any part may be reproduced or transmitted in any form or by any means, electronic or mechanical, including photocopying, microfilming, and recording, or by any information storage and retrieval system, without permission in writing from the publisher.

Current printing (last digit):
10 9 8 7 6 5 4 3 2 1

PRINTED IN THE UNITED STATES OF AMERICA

Series Introduction

The Cosmetic Science and Technology series was conceived to permit discussion of a broad range of current theories and knowledge in cosmetic science and technology. Authorities from industry, academia, and government agencies from around the world are participating in writing these books. Topics are drawn from a wide spectrum of disciplines ranging from chemistry, physics, biochemistry, and analytical and consumer evaluations to safety, efficacy, toxicity, and regulatory questions. Organic, inorganic, physical and polymer chemistry, emulsion and lipid technology, microbiology, dermatology, and toxicology all play a role in cosmetic science.

There is little commonality among the scientific methods, processes, or formulations required for the wide variety of cosmetics and toiletries on the market.

Cosmetics and toiletries represent a highly diversified field involving many subsections of science and ''art''. Even in these days of high technology, ''art'' and intuition continue to play an important part in the development of formulations, their evaluation and the selection of raw materials. The move toward the application of more sophisticated scientific methodologies that gained momentum in the 1980s has grown in such areas as claim substantiation, safety testing, product testing, and chemical analyses and has led to a better understanding of the properties of skin and hair. Molecular modeling techniques are beginning to be applied to data obtained in skin sensory studies.

The Cosmetic Science and Technology series reports the current status of cosmetic technology and science, changing regulatory climates, and historical reviews. The series has grown to over 20 books dealing with the constantly changing technologies and trends in the cosmetic industry, including globalization. Several books have been translated into Japanese and Chinese. Contributions range from highly sophisticated and scientific treatises to primers, practical applications, and pragmatic presentations. Contributors are encouraged to present their own concepts as well as established theories, and not to shy away from fields that are in a state of transition or to hesitate to present detailed discussions of their own work. Our intention is to develop the series into a collection of critical surveys and ideas covering diverse phases of the cosmetic industry.

Cosmetic Lipids and the Skin Barrier, the twenty-fourth book published in the series, represents a global effort. Ten chapters are contributed by leading authorities from European universities and industrial research centers, including one chapter from Japan. The book covers the chemistry, structure, and biological functions of skin lipids and their interaction with the lipids formulated into skin and hair care preparations. Cosmetic emulsions contain high levels of lipids, often making them the second major component after water. Two chapters discuss body washes, a relatively new class of products that have appeared on the U.S. market since the mid-1990s. These are sophisticated surfactant-based formulations containing a relatively high oil load that provide cleaning as well as moisturization and skin conditioning.

I would like to thank the contributors for participating in this project, particularly the editor, Dr. Thomas Förster, for conceiving, organizing, and coordinating this book. Special thanks are extended to Sandra Beberman and the editorial and production staff at Marcel Dekker, Inc. Finally, I thank my wife, Eva, without whose constant support and editorial help I would not have undertaken this project.

Eric Jungermann

Preface

The benefits of lipids in cosmetic treatment of the skin were already known when Cleopatra took her bath in donkey's milk. Since ancient times, waxes, oils, and amphiphilic lipidic materials have been used in various cosmetic formulations such as creams, rouge, and lipstick to make the skin smooth and supple and to improve the application properties of cosmetic products. Despite extensive traditional knowledge about lipids in cosmetic formulations and over 40 years of intensive research in skin lipid science, a quantitative understanding of the interaction of topically applied lipids with skin barrier lipids has been lacking.

The primary objective of this book is to describe the most important developments in understanding the interaction between cosmetic lipids and the skin barrier over the past 10 years. Emphasis has been placed on connecting the latest scientific findings and methods with possible applications in cosmetic products. Each chapter presents the relevant scientific background in a particular area and an in-depth and up-to-date account of recent research findings.

The first three chapters introduce the fundamentals of lipid chemistry and recent concepts about the structural arrangement of lipids in the skin barrier. Chapter 1 describes the chemistry of natural human skin lipids and presents synthetic skin barrier lipid analogs frequently used in skin care creams. Chapter 2 discusses the fate of lipid liposomes applied to the skin. Molecular dynamics offers a new way to scrutinize the molecular interaction of lipids in lamellar layers. Chapter 3 presents the state of the art in computer modeling of skin barrier lipid layers.

The next two chapters deal with the biological function and activity of skin lipids as studied in vivo and in vitro. Chapter 4 focuses on the role of lipids in stratum corneum barrier function and barrier repair. Chapter 5 discusses the expression of skin barrier lipids in in vitro skin models by different ingredients.

The following three chapters review various methods for analyzing the interaction with the skin of topically applied lipids. Chapter 6 scrutinizes the composition of skin lipids found in humans of different skin types. Chapter 7 reviews several biophysical and analytical methods for assessing lipid status and the improvement of skin condition. Chapter 8 introduces infrared and Raman spectroscopy as powerful tools to detect subtle changes in skin surface lipids after various cosmetic treatments.

The last three chapters are concerned with lipids in cosmetic products and how they interact with the skin. Chapter 9 gives an overview of different lipid types used in skin care formulations and their function in the cosmetic product and on the skin. Chapter 10 discusses the evidence for skin care effects obtained by lipidic refatters in personal cleansing products. Finally, Chapter 11 relates the physicochemical properties of various lipid types with the sensorial perception of these lipids in different leave-on and rinse-off products by consumers.

The contributors represent a cross section of researchers from universities and industry who are specialists in their particular field. The book provides an exhaustive, yet concise view of the interaction of cosmetic lipids with the skin barrier. Students and researchers at universities or in industry who are new to the field will find this book a valuable introduction. At the other end of the spectrum, specialists will profit from the current and comprehensive information in each chapter.

Thomas Förster

Contents

Series Introduction Eric Jungermann   *iii*
Preface   *v*
Contributors   *ix*

PART I. CHEMISTRY AND STRUCTURE OF SKIN LIPIDS

1. The Chemistry of Natural and Synthetic Skin Barrier Lipids   1
   Hinrich Möller

2. Structure of Stratum Corneum Lipid Layers and Interactions with Lipid Liposomes   37
   Joke Bouwstra

3. Computer Modeling of Skin Barrier Lipid Layers by Molecular Dynamics   75
   Monika Höltje

PART II. BIOLOGICAL FUNCTION OF SKIN LIPIDS

4. Role of Lipids in Skin Barrier Function   97
   Mitsuhiro Denda

5. Investigating Human Skin Barrier Lipids with In Vitro Skin Models   121
   Annie Black, Odile Damour, and Kordula Schlotmann

PART III. ANALYSIS OF SKIN LIPIDS AND THEIR EFFECTS ON SKIN

6. Analytical Techniques for Skin Lipids — 149
   Kristien De Paepe and Vera Rogiers

7. Biophysical Methods for Stratum Corneum Characterization — 185
   Hans Lambers and Hans Pronk

8. Detection of Cosmetic Changes in Skin Surface Lipids by Infrared and Raman Spectroscopy — 227
   Thomas Prasch and Thomas Förster

PART IV. LIPIDS IN COSMETIC PRODUCTS

9. Lipidic Ingredients in Skin Care Formulations — 255
   Ghita Lanzendörfer

10. Benefits of Lipidic Refatters in Surfactant Products — 299
    Ulrich Issberner

11. Sensory Assessment of Lipids in Leave-On and Rinse-Off Products — 319
    Thomas Gassenmeier and Peter Busch

Index — *353*

Contributors

Annie Black, Ph.D. Laboratoire des Substituts Cutanés, Hôpital Edouard Herriot, Lyon, France

Joke Bouwstra, Ph.D. Center for Drug Research, Leiden University, Leiden, The Netherlands

Peter Busch, Ph.D. Department of Biophysics and Sensorics/Product Performance, Cognis Deutschland GmbH, Düsseldorf, Germany

Odile Damour, Ph.D. Director of Laboratoire des Substituts Cutanés, Hôpital Edouard Herriot, Lyon, France

Kristien De Paepe, Ph.D. Department of Toxicology, Vrije Universiteit Brussel, Brussels, Belgium

Mitsuhiro Denda, Ph.D. Skin Biology Research Laboratory, Shiseido Life Science Research Center, Yokohama, Japan

Thomas Förster, Ph.D. Department of Cosmetics, Henkel KgaA, Düsseldorf, Germany

Thomas Gassenmeier, M.D. Department of Skin Biochemistry, Henkel KgaA, Düsseldorf, Germany

Monika Höltje, Ph.D. Department of Pharmaceutical Chemistry, Institute of Pharmaceutical Chemistry, Heinrich-Heine University, Düsseldorf, Germany

Ulrich Issberner, M.D.   Department of Care Chemicals, Cognis Deutschland GmbH, Düsseldorf, Germany

Hans Lambers, Ph.D.   Department of Innovation Cosmetics, Sara Lee Household and Body Care Research, The Hague, The Netherlands

Ghita Lanzendörfer, Ph.D.   Advanced Development, Decorative Cosmetics Department, Beiersdorf AG, Hamburg, Germany

Hinrich Möller, Ph.D.   Department of Cosmetics, Henkel KgaA, Düsseldorf, Germany

Thomas Prasch, Ph.D.   Department of Analytics of Chemical Products, Henkel KgaA, Düsseldorf, Germany

Hans Pronk, M.Sc.   Department of Biophysical Evaluation, Sara Lee Household and Body Care Research, The Hague, The Netherlands

Vera Rogiers, Ph.D.   Department of Toxicology, Vrije Universiteit Brussel, Brussels, Belgium

Kordula Schlotmann, M.D.   Department of Skin Biochemistry, Henkel KgaA, Düsseldorf, Germany

1
The Chemistry of Natural and Synthetic Skin Barrier Lipids

Hinrich Möller
Henkel KGaA, Düsseldorf, Germany

I. BACKGROUND

As long ago as the mid-1970s, Elias et al. [1] discovered that the intercellular lipidlike substance of the stratum corneum, which acts as a sort of cement between the corneocytes, has a lamellar structure. The compound system formed by the dead cells and the lamellae is the skin's actual barrier against the environment. Some years later, the same work group [2] became aware of the relatively high proportion of ceramides besides cholesterol, free fatty acids, and triacylglycerol in the intercellular lipid layer. The ceramides are credited with playing an important role in the lamellar lipid structure and in the barrier function of the stratum corneum.

II. CHEMICAL STRUCTURE

The unusual structure of the ceramides should be explained at this point. They are derived from sphinganine (D-erythro-2-amino-octadecane-1,3-diol) (see Fig. 1).

This yields sphingosine, an unsaturated compound with a double bond at the fourth carbon atom. Sphingosine provides the basic structure for the ceramides, which are formed from them by N-acylation. Glycosylation of the primary OH group produces cerebrosides, which are present, mainly in the form of galactosides, in the cell membranes of brain tissue. The final ceramide derivatives are the sphingomyelins, in which the phosphorylated choline group substitutes the terminal OH group. They are found in the myelin sheaths of the nerves.

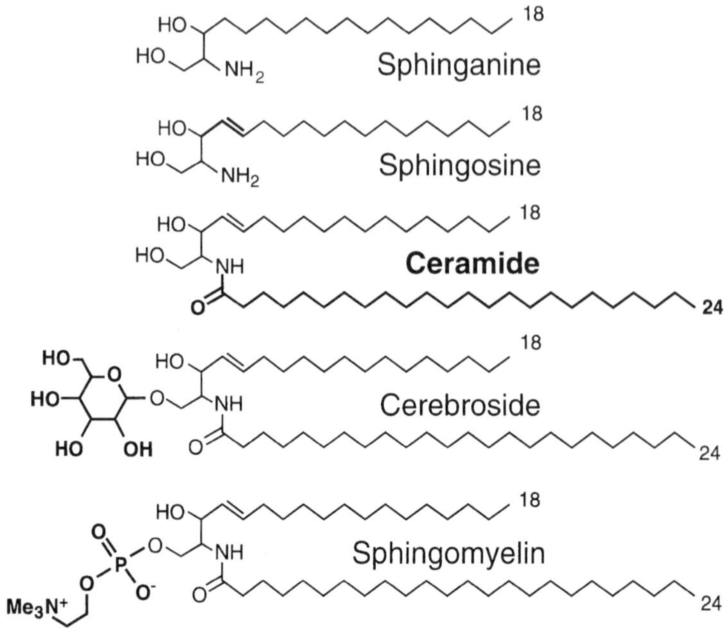

Figure 1 Sphinganine derivatives.

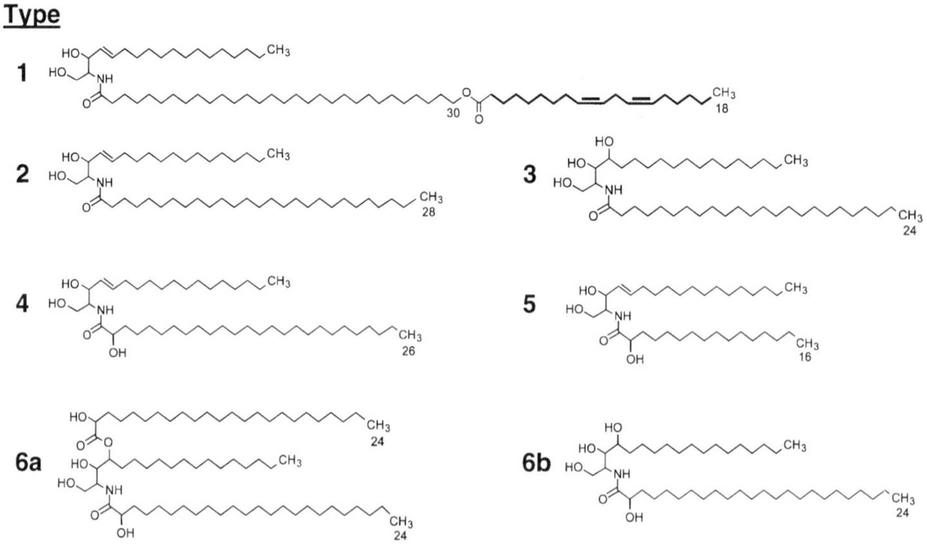

Figure 2 Ceramide structures.

Chemistry of Natural and Synthetic Skin Barrier Lipids

The most important ceramides found in the stratum corneum lipids have been assigned numbers in the order of their polarity and are listed in Figure 2 [3].

Structural variation is achieved by changing not only the chain length of the N-acyl group but also its character, by incorporating 2-hydroxy or ω-acyloxyacyl groups, as in ceramides of types 1, 4, 5, and 6. On the other hand, the double bond from the sphingosine group can be hydrated to increase hydrophilicity, as in ceramides of types 3 and 6a and b. The type 1 ceramide is notable for still having a linolyloxy group in the ω-position of the C_{30}-acyl group. The structure of type 6a is the only one that contains three long alkyl chains, but four hydroxy groups like 6b. It was formerly assigned to a different structure that was derived from type 1, with the linolyl group being exchanged by the 2-hydroxytetracosanoyl group [4]. The cerebrosides are classified similarly.

III. LOCALIZATION AND CREATION

Ceramides, together with other epidermal lipids and water, form intercellular multilamellar liquid crystalline gel structures. These structures were made visible in pig epidermis with the help of electron microscopy (Fig. 3) [5].

The individual lipid bilayers between the corneocytes are clearly separate from each other (a). The effect of a solvent extraction can be seen in part b, where only residues of these layers are recognizable. Figure 4 shows a model of the epidermal barrier.

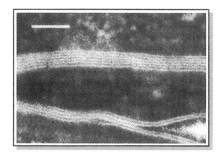
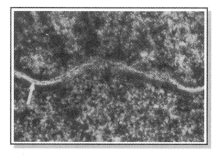

Transmission electron micrograph
(x 240.000, Bar: 50 nm)
before extraction

Transmission electron micrograph
(x 85.000, Bar: 100 nm)
after extraction with
chloroform/methanol (2:1)

Figure 3 Multiple intercellular lamellae of pig stratum corneum [5].

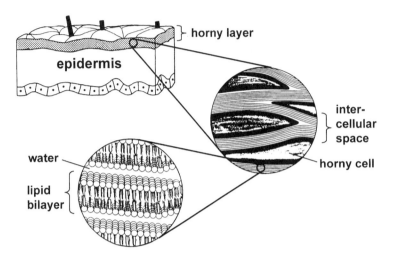

Figure 4 Barrier model of intercellular horny layer lipids.

The very simplified section of the epidermis shows as the outermost layer the stratum corneum (horny layer). The enlargement of this section shows the lamellar structure in the intercellular space between the corneocytes. Further enlargement reveals the molecular structure of the lipid bilayer in the liquid crystalline lipid/water system. The spheres represent the hydrophilic "heads" of the polar lipid molecules, which are aligned toward the aqueous phase; the lipophilic hydrocarbon chains are aligned toward the interlayer gap of the lipid bilayers.

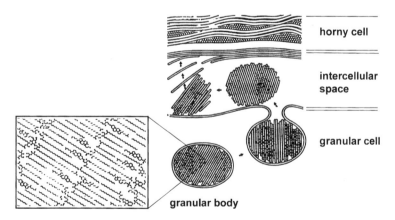

Figure 5 Generation of multiple intercellular lipid lamellae in the stratum corneum [4].

These lamellae are directly related to the lamellar bodies in the stratum granulosum, from which the final epidermal layer, the stratum corneum, is formed as corneocytes are produced (Fig. 5) [3,4].

These lipid bilayers are preformed as closely packed disks in the lamellar bodies of the stratum granulosum. Their molecular structure is very similar to that of the stratum corneum and is shown in model form in the enlargement at the bottom of the illustration. One major difference is that most of the ceramides are glycosylated (acyl glycosyl ceramides). In the course of further cytodifferentiation, these lipid disks are relinquished into the intercellular space at the boundary to the stratum corneum and are absorbed into the extensive lamellar barrier structures of the stratum corneum while undergoing deglycosylation.

IV. COMPOSITION OF THE EPIDERMAL LIPIDS

Before going into detail of the composition of the epidermal lipids coming up to the skin surface from the intercellular space after desquamation, it may be of interest to deal with the second source of lipids on the skin: the sebum. These lipids are produced by the sebaceous glands and contribute the main part of the human superficial fat and are nearly uninvolved in barrier function of the skin. The composition of sebum differs considerably from that of the epidermal origin. An example is shown in Table 1 [6]. There are found no ceramides, little cholesterol, but a substantial proportion of squalene that may serve as a marker for sebum, whereas the free cholesterol and the ceramide content is characteristic of the stratum corneum lipids.

Table 1 Composition of Lipids from Human Sebaceous Glands (Scalp) [6]

Component	Content [weight %] (S.D.)
Squalene	13.6 (0.6)
Wax esters	19.0 (1.6)
Triacyl glycerols	63.6 (1.1)
	94.0
Free fatty acids	1.2 (0.5)
Cholesterol esters	2.0 (0.5)
Cholesterol	0.6 (0.1)
Phospholipids	about 6.0
• Phosphatidyl cholin	• 3.6

S.D. = Standard deviation

Triacylglycerols are the main component of the sebaceous lipids—about 60%. After spread to the skin surface, most of the triacylglycerols are split to free fatty acids and glycerol or to mono- or diacylglycerols. The fatty acids consist mainly of the C_{16}-, followed by the C_{18}- and C_{14}-, homologues and are saturated as well as unsaturated. The double bonds in the 6- and 8-position of the monoenoic alkanoic acids are unusual. Normally, the double bonds are found in the 9- or in the 9,12-position. It is noteworthy, too, that a considerable portion of the fatty acids have odd-numbered chain lengths and methyl branching at the end of the chain [7]. The fairly high squalene content means that in the sebaceous glands only a small amount is transformed to sterols. Phospholipids, mainly phosphatidylcholin, are also present and belong to the polar fraction of lipids. Recently, other authors [8] found varying lipid contents in the sebaceous glands separated from the chest. They argued that the lipid composition might strongly depend on the skin area. The origin of the lipids of Table 1 is the scalp.

On the other hand, the lipids of the stratum corneum contain not only the ceramides but also some other essential components. The functionality of the above-described lamellar, liquid crystalline lipid structures depends on the presence of a range of other lipid components as well as the ceramides. Table 2 shows two typical examples of the composition of human stratum corneum lipids (abdomen [9], lower leg [10]).

In the literature, the results concerning the measurements of the different components obtained by numerous authors diverge to some degree because it is very difficult to separate the lipids only of the stratum corneum from the skin without contamination by lipids from the deeper layer or of the skin surface.

Table 2 Composition of Human Stratum Corneum Lipids [9,10]

	Content [weight%]	
Component	Abdomen	Lower leg
Polar lipids	4.9	—
Cholesteryl sulfate	1.5	—
Neutral lipids	74.8	68.1
• Free sterols	• 14.0	• 23.7
• Sterol esters	• —	• 10.2
• Free fatty acids	• 19.3	• 25.3
• Triacyl glycerols	• 25.2	• 8.9
• Squalene	• 4.8	• —
• n-Alkanes	• 6.1	• —
Ceramides	18.1	41.8

Chemistry of Natural and Synthetic Skin Barrier Lipids 7

These latter lipids originate mainly from the sebaceous glands. The skin area also plays an important role in this respect.

Nevertheless it can be seen that the main components of the neutral lipids are free sterols (predominantly cholesterol), free fatty acids, and triacylglycerols, together with a significant proportion of ceramides. Free fatty acids, cholesterol, and the ceramides that occur mainly in equimolar ratios are especially important for the formation of stable lamellar structures [11].

At this point, the structures of some major lipidic components of the horny layer other than ceramides are compiled in order to demonstrate their chemical features and possible spatial arrangements (Fig. 6).

The different ceramide types in humans have also been studied and analyzed. The same types were found as in pigs. Table 3 shows a typical composition [12]. The most abundant ceramide was type 2, making up more than 20%. The combined values for types 4 and 5, which could not be separated by thin-layer chromatography, and for types 6a and 6b give a percentage content similar to type 2. The content of types 1 and 3 is about 10% each.

The research done by Yardley [13] is of interest. He examined changes in lipid composition during the differentiation of the epidermal cells on their journey from the basal layer through the stratum granulosum to the stratum corneum.

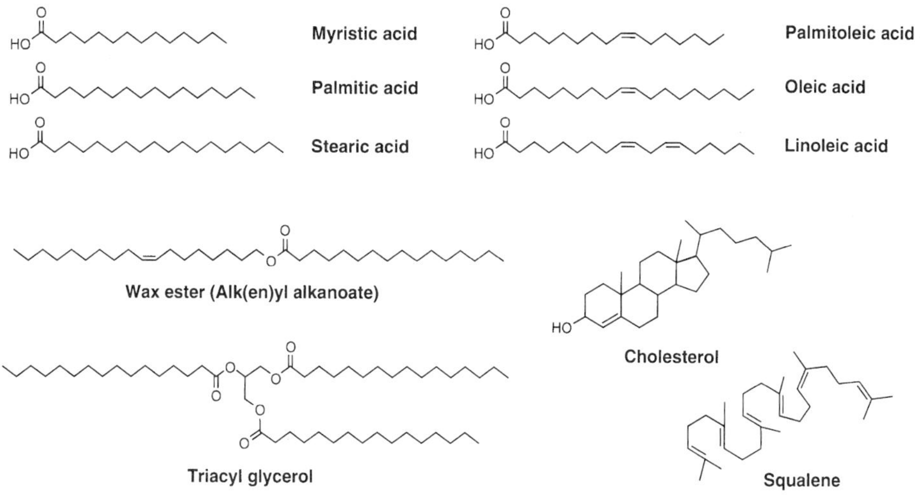

Figure 6 Chemical structures of some major lipidic components of the stratum corneum, without ceramides.

Table 3 Composition of Ceramides Found in Human Stratum Corneum [12]

Ceramide (type)	Content [weight %]
1	7.0 (3.2)[a]
2	21.0 (4.9)
3	13.4 (4.3)
4/5	22.2 (4.5)
6a	9.8 (1.1)
6b	13.6 (4.5)

[a] Values in brackets: standard deviation.

Figure 7 very clearly shows the components that undergo the greatest change: the phospholipids initially increase but are completely absent from the horny layer. The same applies to the glycosylceramides. They are apparently of importance for the formation of the lamellar bodies, but their more hydrophilic character is reversed to a more hydrophobic one before incorporation into the multilamellar structures between the corneocytes.

The marked increase in the content of the hydrophobic ceramides and fatty acids is also very noticeable, especially in the final differentiation step. The cho-

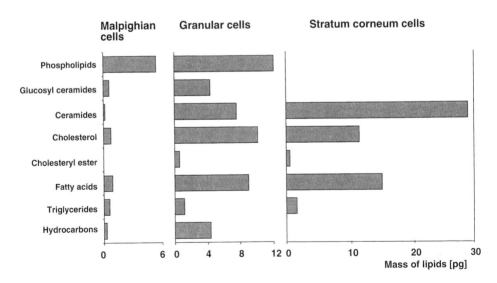

Figure 7 Change in lipid composition during epidermal cell differentiation [13].

Chemistry of Natural and Synthetic Skin Barrier Lipids

lesterol content also increases strongly. The concentration of the other components changes much less, so that this illustration clearly reveals the considerable significance of the three components—ceramides, cholesterol, and free fatty acids—for the generation of the lamellar lipid structure in the stratum corneum that is a basic requirement for the highly efficient barrier function against water permeability.

V. MODEL STUDIES

On the basis of the above-mentioned lipid analyses, Friberg et al. [14] created a model horny layer lipid and studied the influence of different individual components on the lamellar structure (Table 4). The first major surprise occurred when they mixed this model lipid with water and were unable to obtain a lamellar dispersion. Only after the pH of the mixture was adjusted to that of the skin (pH 4.5–6.0) were lamellar structures observed. Moreover the ceramides did not yield lamellar liquid crystals when mixed only with water, although it had long been assumed that they would, in view of their structural similarity to the phospholipids. On the other hand, it was found that a mixture of fatty acids and their salts, together with 30% water, form stable lamellar systems. This occurs most readily

Table 4 Model for Stratum Corneum Lipid [14]

Component	Content [weight %]
Free fatty acids	17
• Myristic	• 10
• Linolic	• 15
• Oleic	• 55
• Palmitic	• 5 — 80 (unsaturated fatty acids)
• Palmitoleic	• 10
• Stearic	• 5
Phosphatidyl ethanolamine	5
Cholesteryl sulfate	4
Cholesterol	17
Triolein	22
Palmityl oleate	5
Squalene	5
Pristane	4
Ceramides	21

at a pH of 4.6, when about 40% of the fatty acids are present as soaps. This correlation is illustrated in a phase diagram (Fig. 8). The trapezoid in the medium pH range is the zone of existence of lamellar liquid crystals at between 20% and 70% water content. The other zones contain an aqueous solution, liquid crystals, crystals, or no structure at all.

Friberg and his colleagues then turned to small angle x-ray scattering (SAXS) to research the special function of the ceramides. This enabled them to measure the interlayer distance in lamellar dispersions [16]. Reflections occur only at polar electron-rich layers, i.e., in the aqueous phase (Fig. 9). The distance d between the layers can be determined with the help of the Bragg equation. This yields the sum of the interlayer distances in the aqueous and lipid phases.

On the basis of the above-mentioned basic lamellar system consisting of a mixture of free fatty acids and their soaps, Friberg and his colleagues then carried out a systematic study of the influence of the individual components of the model lipid (Fig. 10) [17]. The interlayer distances were plotted as a function of the water content. The lowest straight line shows the basic lamellar lipid system, which consists of the unsaturated fatty acids of the model lipid system (Table 4). The gradient of the straight line increases slightly as the water content increases. The thickness of the lipid bilayer is obtained by extrapolating to water content of 0%. If the three saturated fatty acids are now added in sequence, three

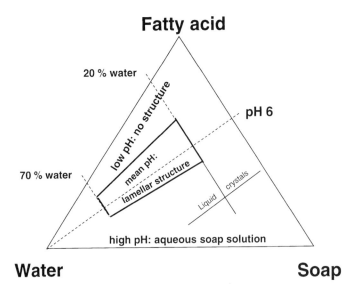

Figure 8 Phase diagram of the system: fatty acid–soap–water. (From Ref. 15.)

Chemistry of Natural and Synthetic Skin Barrier Lipids

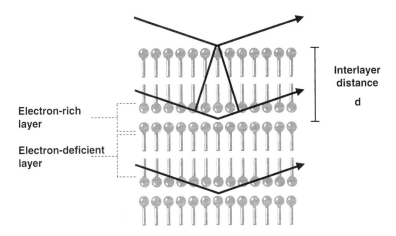

Figure 9 Scheme of x-ray reflection off lipid bilayers.

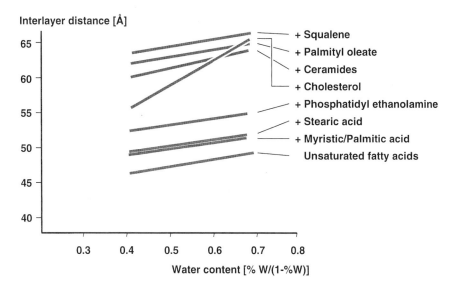

Figure 10 Interlayer distance as a function of the water content in the step by step composed model lipid. (From Ref. 17.)

straight lines are obtained above the original one, each with a similar slope. The interlayer distances are clearly larger, i.e., incorporation into the lipid bilayer occurs not so much parallel to the alkyl chains but rather transversal to them in the zone where the methyl groups of the antipodal alkyl chains meet.

The further addition of phosphatidyl ethanolamine has a similar effect, also indicating that incorporation occurs between the antipodal monolayers. When cholesterol is then added, the gradient increases steeply. This variant behavior is attributed to the lowered water-bonding capacity of the hydrated hydrophilic heads in the lipid bilayers when cholesterol is added. Some of the added water is no longer anchored in the lipid bilayer but remains in the aqueous layer, thus causing a widening of the layers. Extrapolation to a water-free system gives an interlayer distance that corresponds to the basic system. In this case all of the components are aligned parallel to the alkyl chains. This unusual effect of the cholesterol is reversed when ceramide is added, with the gradient of the system again approximating that of the basic system. The thickness of the layer increases again. When, finally, the neutral lipids, wax esters, and squalene are added, only an increase in the interlayer distances is again observed.

Other authors [18] studied a reduced three-component system containing hexadecanoic acid, ceramide, and cholesterol, which was brought into equilibrium with the help of an aqueous buffer solution. They measured the different effects of ceramides 2 and 5, which differ only by a hydroxy group. Separate domains for ceramides and fatty acids were detected in the structural pattern of the lipid bilayers. These mosaic structures, whose presence was recently also postulated by another research group [19], certainly play a crucial role in regulating the water permeability of the epidermis.

VI. EFFECTS OF CERAMIDES

The different effects of the ceramides are shown in Table 5.

A. Biophysical Effects

The overriding property of the ceramides revealed by the in vitro measurement results is their capacity to modulate the lamellar crystalline gel structure of the horny layer lipids. The studies of Friberg and colleagues and the group around Bouwstra [20–23] with model lipids show that the ceramides and cholesterol act as antagonists. This holds true especially in respect to influence on the interlayer distance of the stratum corneum lamellar structure as demonstrated before. Both together stabilize the lamellar system, also making it more compatible with other lipid components. Bouwstra and colleagues found that ceramide 1 and also cera-

Table 5 Activity of Ceramides

Modulation of lamellar gel crystal of the horny layer lipids
- Opponent of cholesterol concerning interlayer distance
- Solution aid for cholesterol
- Augmentation of crystalline parts
- Decrease of membrane fluidity

Influence on the stratum corneum properties
- Decrease of TEWL
- Increase of water content
- Decrease of skin scalines

Biologic activity
- Control of epidermal proliferation
- Induction of cell differentiation
- Stimulation of apoptosis of cancer cells
- Second messenger function

mide 2 possess a pronounced role in building lamellar structures as well as the molar ratio of cholesterol to ceramides and the composition of the latter.

Our own, not yet published studies have shown that the addition of cholesterol clearly increases the extent of the layer structure. Addition of ceramides, on the other hand, partly counteracts this effect and increases the solubility of cholesterol [24]. The literature also contains reports of a lowering of the membrane fluidity by cholesterol [25], as a result of which the lipid bilayer can be set to an optimal viscosity.

The barrier properties of the whole stratum corneum are also significantly influenced by the ceramides and cholesterol and are optimal in the close combination of horny cells and the complete mixture of stratum corneum lipids. In vitro and in vivo methods were used to measure the effects. The reduction in transepidermal water loss (TEWL) and the increase in the water-binding capacity demonstrates the powerful effects of the stratum corneum lipids on the barrier and moisturizing function of the epidermis [16,26–28].

This effect is also impressively confirmed in a barrier repair study [29]. The barrier of hairless mice was damaged by washing it with acetone alone and with acetone containing a substance that inhibits ceramide synthesis, respectively. In the second case, it was shown that regeneration of the barrier took much longer and was accompanied by the (also delayed) restoration of the previous ceramide content.

The in vivo restoration of the barrier of volunteers, whose skin had been damaged with a surfactant solution or by treatment with Scotch tape stripping, was significantly better with a synthetic type-2 ceramide than a placebo [30].

Ceramides also help to increase skin moisture as measured on volunteers with a corneometer [31,32] (for details, see Chapter 4).

A reduction in scaly skin (skin roughness) was also observed when a short-chain hydroxyacylated phytosphingosine (with another hydroxy group in position 4) was applied to human skin [33].

B. Biological Effects

In addition, biological effects were observed with certain ceramides. For example, it was demonstrated that ceramides inhibit the rate of epidermal cell division, whereas their precursors, the glycosylceramides, which are more highly concentrated in the lower layers of the epidermis and do not occur at all in the stratum corneum, accelerate it [34].

Moreover, it was shown that certain ceramides, such as the cell-permeant ceramide analogues N-cetyl-(C2-Cer) and N-thioacetylsphingosine, induce cell differentiation and programmed cell death (apoptosis) in the human epidermis [35]. C2-Cer, as a second messenger of tumor necrosis factor–alpha (TNF-α) in follicles of ovaries, is a stimulant of apoptosis [36], initiating the apoptosis of ovarian cancer cells [37]. Another ceramide analogue (L-threo-1-phenyl-2-decanoyl-3-morpholino-1-propanol) stimulates cortical neurons and ameliorates memory deficits in ischemic rats [38]. Sweeny et al. [39] recently succeeded in demonstrating that in comparison with nonmethylated sphingosine, N,N-dimethylsphingosine has a powerful antitumor effect on various human cancer cell lines. On the other hand, sphingosine is effective as an anti-inflammatory agent in the skin via inhibition of protein kinase C, which plays a key role in inflammation [40]. Also, the antimicrobial activity of a variety of related sphingosines has been observed, and because free sphingosines occur in the stratum corneum and other epidermal layers, they may contribute to the antimicrobial barrier of the skin [41].

Other authors [42] are studying a second messenger function of the ceramides in connection with the epidermal growth factor. Glycosphingolipids (glycosylceramides) with different numbers of monosaccharide units exhibit important activities in cell-cell recognition, cellular differentiation, and growth control [43]. In 1997, Orfanos and coworkers [44] gave an overview of the signaling role of sphingolipids in human epidermis.

In view of cosmetic applications, interesting recent studies describe utilization of *ceramide precursors* to influence epidermal barrier function. Rawlings and coworkers [45] found out that the ceramide production in vitro as well as in vivo could be increased significantly by the application of L-lactic acid. At the same time, the barrier function of the stratum corneum was ameliorated. The D-form was nearly ineffective.

In another example, it was demonstrated that the topical application of the sphingosine derivative tetra-acetylphytosphingosine produced an elevation of ceramide concentration in the stratum corneum and the barrier function was improved too [46]. Precursors for the biosynthesis of ceramide 1 are linoleic acid or its derivatives. The known decreased level of this special ceramide in the stratum corneum causes skin xerosis in winter and could be normalized by topical application of formulations containing linoleic acid–rich triacylglycerol [47]. Using human keratinocytes in culture, it was possible to stimulate ceramide biosynthesis in stratum corneum considerably by the action of N-acetyl cysteine. Other thiols were less active [48].

Additional references concerning biological actions on the barrier function of the stratum corneum are given in a recently published review article [49].

VII. NONNATURAL CERAMIDES: NATURE-IDENTICAL CERAMIDES/CERAMIDE ANALOGUES/ PSEUDOCERAMIDES

Now that the special significance of the ceramides for the functioning and appearance of our skin has been explained in detail, it is not difficult to appreciate why cosmetic chemists in particular show increased interest in this class of substances. Natural ceramides obtained from the brains of cattle were initially used. They were, however, very expensive (more than US $1000 per kilogram) and as products isolated from cattle they became increasingly involved in the bovine spongiform encephalitis (BSE) debate in recent years. Therefore, it was a logical step to consider synthetic products with an identical or similar structure.

The term *nature-identical ceramides* covers nature-identical ceramides and their racemates, whereas *ceramide analogues* are usually compounds based on stereoisomerically pure sphinganine with a different acyl chain length or with a very similar skeletal structure, and interest is focused on their biological effects. *Pseudoceramides* generally have structures that differ more clearly from the basic structure and are often designed for cosmetic applications. The so-called synthetic barrier lipids (SBL) are pseudoceramides.

A. Nature-Identical Ceramides and Ceramide Analogues

It is noteworthy, first, that, in the meantime, methods have been developed to produce nature-identical ceramides biotechnologically. One of these ceramides is 'Ceramide III' from Jan Dekker, obtained by using yeast. But it is not available at a low price. On the other hand, the synthesis of *nature-identical ceramides* has been described [50], but it inevitably involves a large number of steps and the resulting products are therefore very expensive. A more recent, 7-step synthesis starts from L-serine and yields D-erythro-sphingosine [51]. Figure 11 shows the synthetic pathway.

This synthesis is also very costly and is difficult to translate onto a commercial scale. This is a major reason why synthesis methods have also been developed for producing D/L-erythro/threo-sphingosine/sphinganine which contains all possible stereoisomers. It can subsequently be acylated with suitable fatty acid derivatives, thus completing a process for synthesizing ceramides that closely resemble the natural substances on which they are modeled. As an example, the L'Oreal method [52] for producing 2-oleoylamino-octadecane-1,3-diol is described here. The synthetic pathway, which is also suitable for commercial-scale production, is shown in Figure 12.

This method requires four relatively simple reaction steps leading to the formation of racemic D/L-erythro/threo-sphinganine, which is subsequently acylated to the synthetic ceramide. The fact that five different steps have to be carried out inevitably means that this method, too, is economically unattractive.

Syntheses of ceramide analogues, which are mainly of interest for their biological effects, often start with natural (commercially available) sphingosine. To produce C2-, C6-, and C8-ceramides, for example, sphingosine is first completely O- and N-acylated, and the O-acyl groups are then selectively split by hydrolysis [53].

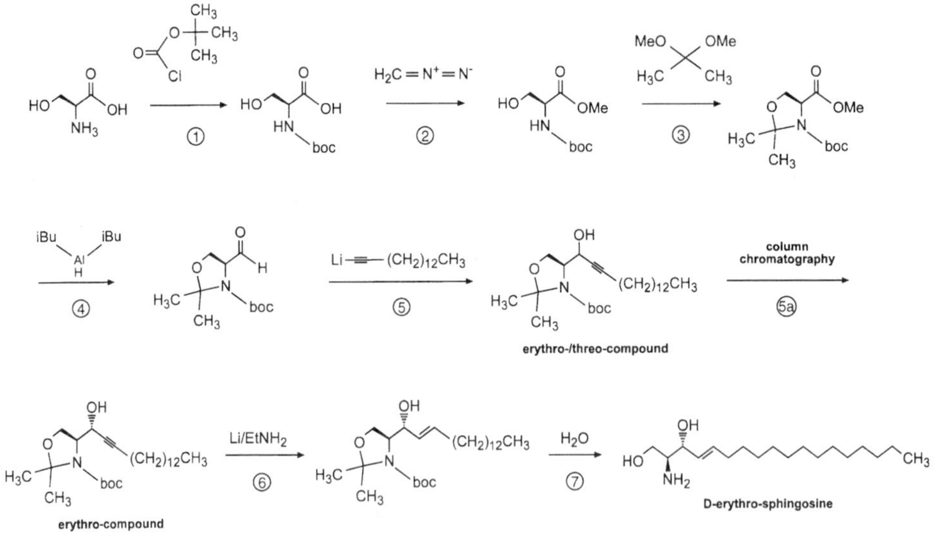

Figure 11 Synthesis of D-erythro-sphingosine. (From Ref. 51.)

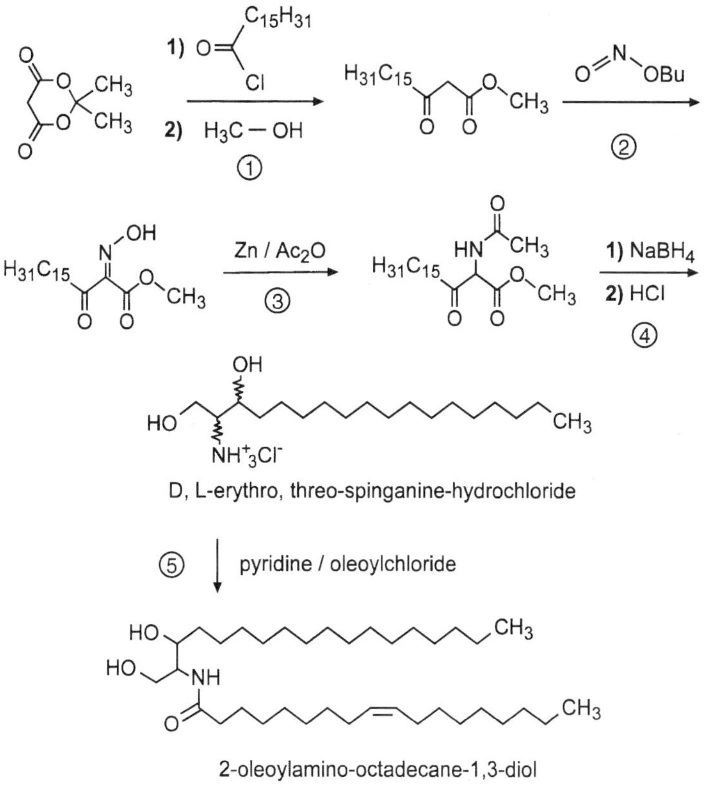

Figure 12 Synthesis of 2-oleoylamino-octadedecane-1,3-diol. (From Ref. 52.)

B. Pseudoceramides

1. General Remarks

There has been made every effort to synthesize ceramide analogues, pseudoceramides, and synthetic barrier lipids (SBL) by simple chemical pathways. There are now a number of patents in this area, although this is not the place to list them all in detail. A summary account of new *pseudoceramides* and special syntheses was recently published [54]. By way of illustration, Figure 13 shows a number of commercially available pseudoceramides with structures that exhibit clear similarities to the natural ceramide structure.

The first listed product in Figure 13 is the above-mentioned Ceramide III from Jan Dekker, biotechnologically obtained. The second product from Kao (SLE) is one of the first pseudoceramides that have been studied in depth [31].

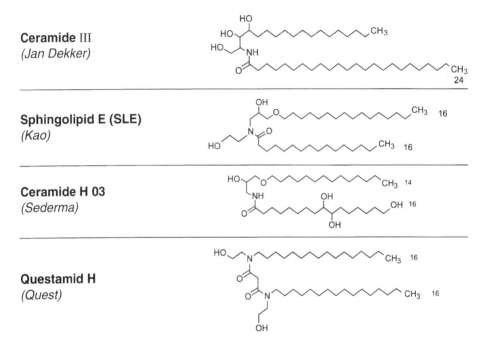

Figure 13 Commercially available pseudoceramides/ceramides.

The C,C double bond of the sphingosine was substituted by an ether group, and the sphingosine's primary OH group was substituted by the N-hydroxyethyl group. This principle is also recognizable in the underlying structure, the ceramide H03, although there is no further hydroxy group in the head of the molecule, but instead several hydroxy groups in the alkyl chain of the acyl group. The fourth pseudoceramide from Quest [55,56] is similar to the second product insofar as its hydrophilicity is also due to the N-hydroxyethyl group. Two fatty alkyl groups and two OH groups are introduced into the molecule through the dicarboxylic acid diamide linkage.

2. Structural and Synthetic Features

The general structure of the above-mentioned pseudoceramides is very similar to that of the natural type 3 ceramide that was for a long time the only commercially available one of high purity for purposes of comparison. The fundamental structure elements for the synthesis of new substances with potential ceramide effects are shown in Figure 14.

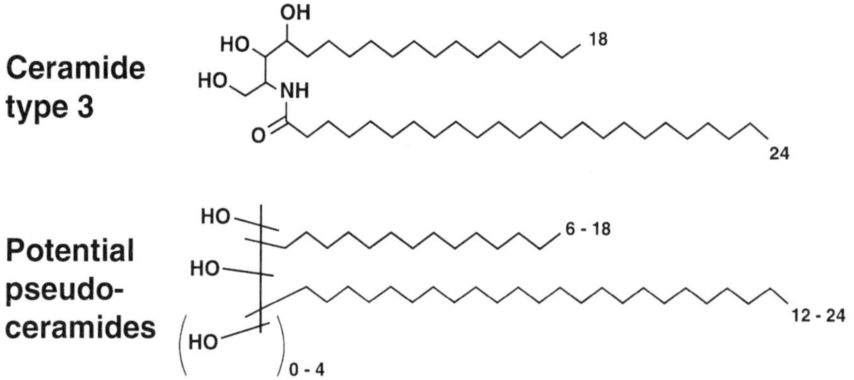

Ceramide type 3

[Structure: Ceramide type 3 with OH, HO, HO, NH, O groups; alkyl chain labeled 18; acyl chain labeled 24]

Potential pseudo-ceramides

[Structure showing HO groups (0-4), with alkyl chains labeled 6-18 and 12-24]

Nonionic compounds containing 2 to 6 hydroxy groups and 2 saturated or unsaturated alkyl chains of the same or different length

Figure 14 Fundamental structure elements of pseudoceramides.

Two saturated or unsaturated alkyl chains of equal or unequal length are linked to an oligohydroxy compound to form a nonionic structure. The number of hydroxy groups should be equal to or greater than two. A poly(ethlenoxy) group can also represent the hydrophilic part of the molecule. A number of possible starting materials are shown in Table 6. The individual elements are linked through C—C, C—N, ester, amide, and ether bonds.

A long-chain alkylamine with one or more hydroxy groups is often produced as pseudosphingosine/sphinganine, which is then acylated with different fatty acid derivatives at the nitrogen atom. One or more hydroxy groups can also be introduced through the fatty acyl group. Methyl esters, acid chlorides, anhydrides, mixed anhydrides, azolides, and even combinations of carboxylic acid with carbodi-imides are employed as reactive carboxylic acid derivatives.

3. Syntheses of a Number of Selected Pseudoceramides

a. Pseudoceramides Based on 1-Amino-3-alkoxy-2-propanols. The synthesis of the most frequently studied pseudoceramide, SLE from Kao (see Fig. 13), is shown in Figure 15 [57,58].

In the first step, hexadecanol is reacted with 3-chloro-1,2-epoxypropane in the presence of sodium hydroxide solution and the phase transfer catalyst tetrabutylammonium bromide (TBAB) to obtain 2,3-epoxypropylhexadecyl ether. This reacts quantitatively with 2-aminoethanol to form a secondary amine. In the third

Table 6 Starting Materials

1 Hydrophiles
    Glucose
    Alkylglucosides
    Hydroxyalkylamines, Oligohydroxyalkylamines
    Sorbitylamine
    N-Methylsorbitylamine
    D-Glucosamine hydrochloride
    D-Gluconic acid δ-lactone
    Mucic acid
    Oligomeric ethylene glycols
2 Lipophiles
    Fatty acid chlorides, -methylesters
    Alkylsuccinic acid anhydrides, -monoesters
    Amino acid esters, amides
    Guerbet acid derivatives
    Fatty alcohols
    Fatty alkylamines
    Guerbet-alcohols
    Malonic acid derivatives
    Hydroxyalkane diacid derivatives
    Difattyacyl glycerols

step, the amine undergoes N-acylation with methyl hexadecanoate acid, yielding the end product SLE.

Unilever has claimed, within a patent application, structures that are based on the same synthesis principle [59]. These include the 2-hydroxy analogue of SLE, N-(3-hexadecyloxy-2-hydroxypropyl-N-(2-hydroxyethyl)-2-hydroxyhexadecanoic acid amide. In this variant, acylation is carried out with the methyl ester of 2-hydroxyalkanoic acid in the final step.

b. Pseudoceramides Based on N-Hydroxy/Dihydroxyalkylaminoalkanes. The structure of the pseudoceramide SLE is more strongly modified in another Unilever patent [60]. As can be seen in Figure 16, the hydrophilic part of the molecule also has two hydroxy groups, which are, however, arranged vicinally. The chain length in the example is shorter. The synthesis is even simpler: dodecylamine reacts with 2,3-epoxypropanol to form 2,3-dihydroxypropyldodecylamine, which is acylated in the usual way with methyl hexadecanoate to generate a pseudoceramide. This patent contains another modification of the structure that is claimed in another patent by Elisabeth Arden [61] as an active ingredient of a cosmetic product for treating aging skin (at the bottom of Fig. 16). This pseudo-

Figure 15 Synthesis of the pseudoceramide SLE (Kao). (From Ref. 57.)

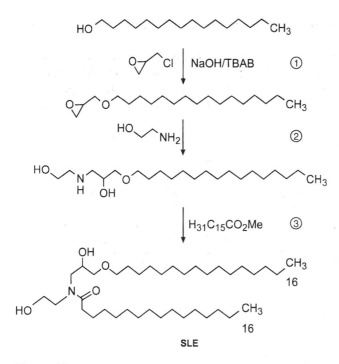

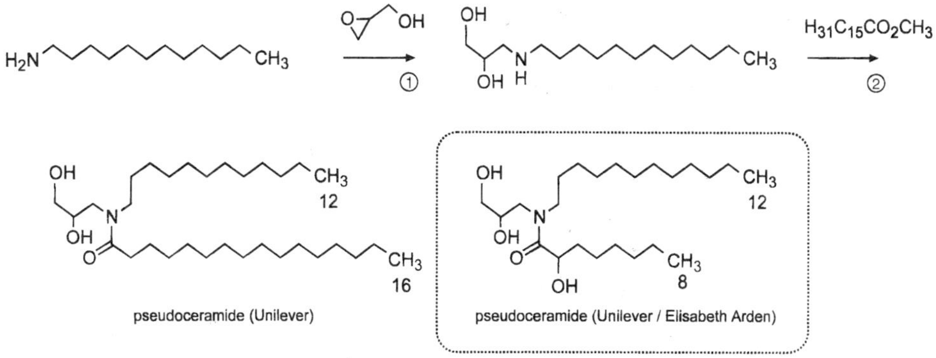

Figure 16 Synthesis of N-(2,3-dihydroxypropyl)-N-dodecyl-hexadecanoic acid amide. (From Ref. 60.)

ceramide contains a shortened and hydroxylated acyl group to increase its hydrophilicity.

A malonic acid bis-(N-2-hydroxyethyl-N-hexadecylamide), Questamide H, listed among the commercially available pseudoceramides, also falls into this category and can be obtained in only two reaction steps: hydroxyethylation of hexadecylamine with ethylene oxide and subsequent reaction with diethyl malonate [56].

c. *Pseudoceramides Based on Polyethylene Oxide Adducts.* SBL KE 3039 from Henkel [62] is an example of this group of pseudoceramides. The reaction pathway is shown in Figure 17.

In the first step, hexadecenylsuccinic acid anhydride is obtained in an ene reaction by addition of 1-hexadecene to maleic acid anhydride. After selective catalytic hydrogenation, the cyclic anhydride is reacted with behenyl alcohol (docosanol) to give the semiester. In the fourth step, after ethoxylation with 7.5 Mol ethylene oxide, hexadecylsuccinic acid monobehenyl/polyethyleneglycol ester is

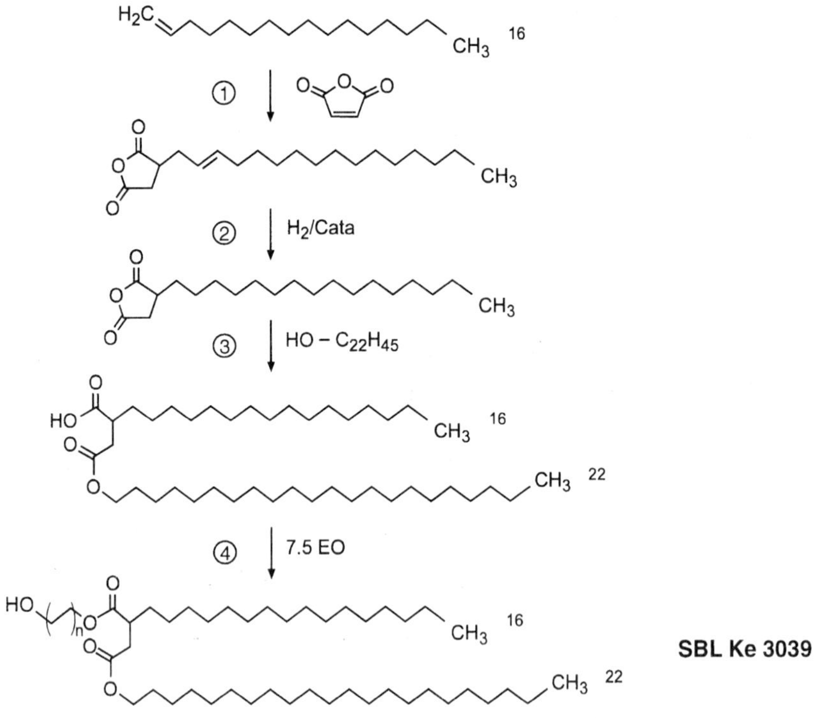

Figure 17 Synthesis of SBL Ke 3039 (Henkel). (From Ref. 62.)

obtained. In this reaction part, the ethylene oxide is additionally inserted into the already existing ester group.

   d. Pseudoceramides Based on Simple Sugar Derivatives. Much work has recently been carried out on a synthesis concept for pseudoceramides that makes use of easily accessible sugar derivatives as hydrophilic components and oleochemical derivatives obtained from natural raw materials as lipophilic ones [54].

Only glucose and some of its derivatives, all of which are easily accessible, are shown in Table 6 as examples of oligohydroxy compounds. N-fatty alkyl glucosylamines, for example, are obtained in a simple reaction by condensing fatty amines with glucose. The corresponding sorbityl amines are obtained from them by catalytic hydrogenation. The other substances are commercially available. The lipophilic substances are mainly fatty acids or other carboxylic acid derivatives, which have a longer alkyl chain, and dicarboxylic acid derivatives or fatty alcohols and fatty amines, all of which are also available.

   e. Pseudoceramides Based on Amino Acid Derivatives. One general synthesis concept is based on the use of various commercially obtainable amino acids. One advantage is that at least two different functional groups are available. Of relevance in this context are, for example, glycine, alanine, sarcosine, β-alanine, lysine, serine, aspartic acid, and glutamic acid. The pseudoceramide derived from aspartic acid is obtainable by a two-step synthesis, as shown in Figure 18 [63].

First, the aspartic acid is reacted with cetaryl($= C_{16/18}$) alcohol to obtain the dialkyl ester of aspartic acid. In the second step, this is acylated in a smooth reaction with D-gluconolactone to give asparagine SBL.

   f. Pseudoceramides Based on Succinic Acid Derivatives. Figure 19 shows suggested structures based on succinic acid derivatives. These molecular structures represent all succinic acid monoamides whose amine component is N-methylsorbitylamine. The two long alkyl chains that are needed are introduced on the one hand by C-substitution of the succinic acid and on the other by esterification of the second carboxyl group [64].

Unsubstituted succinic acid was used in the next three structures, so the two fatty alkyl groups had to be linked to the hydrophilic part of the molecule together with the ester group. In these cases—in contrast to the first structure—the selective introduction of two chains of different length into the target molecule is impossible or can be achieved only with considerable additional effort [65–67].

The ester components of the final three compounds are formed with the following hydroxy compounds: $C_{32/36}$-Guerbet alcohol; 1,3-distearoylglycerol; and malic acid distearyl ester. Because hexadecylsuccinic acid from the above-

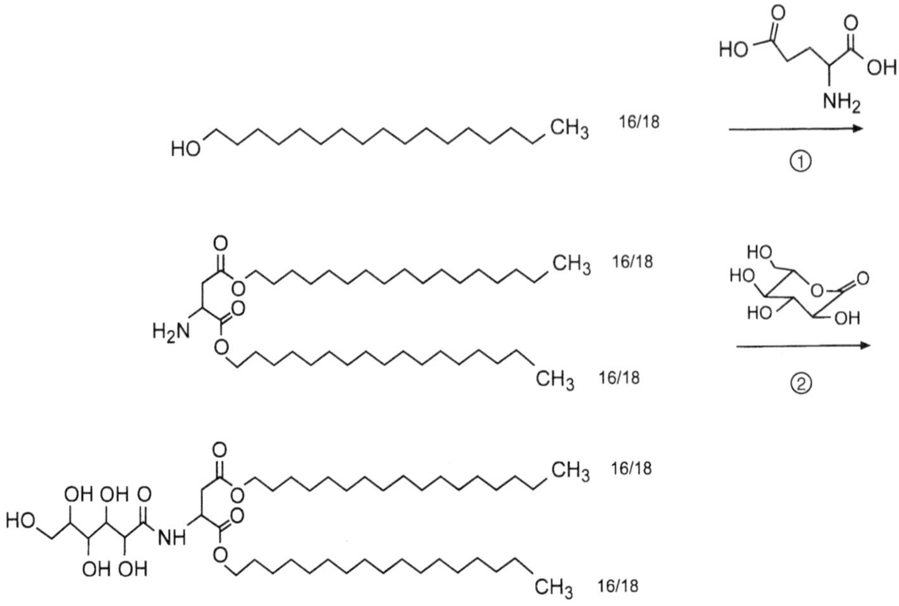

Figure 18 Synthesis of N-gluconoylaspartic acid di-$C_{16/18}$-alkylester. (From Ref. 63.)

mentioned Henkel pseudoceramide can be used for the first target compound and different chain lengths can also be selected here, the synthesis of this compound is an interesting EO-free way of obtaining the above-mentioned pseudoceramide. The other compounds that are shown are synthesized similarly.

The reaction pathway of the synthesis of C_{16}-alkylsuccinic acid monobehenylester N-methylsorbitylamide is shown in Figure 20. The production of hexadecylsuccinic acid anhydride has already been shown in Figure 17. The acid anhydride is then, as also described above, neatly reacted with behenyl alcohol to obtain the semiester, which is converted to the corresponding acid chloride by reaction with phosphorus trichloride. The acid chloride and an aqueous N-methylsorbitylamine solution then are reacted under Schotten-Baumann conditions, forming the end product glucamide SBL (Ke 3193). The yield is 70% to 80%, and this synthesis is suitable for production on a commercial scale.

g. Pseudoceramides Based on Different Sugar and Sugar Acid Derivatives. Four typical representatives of this group are shown in Figure 21. The top two structures are long-chain acylated N-dodecylglucosylamide and N-dodecylsorbitylamide, which have already been described in the literature in other

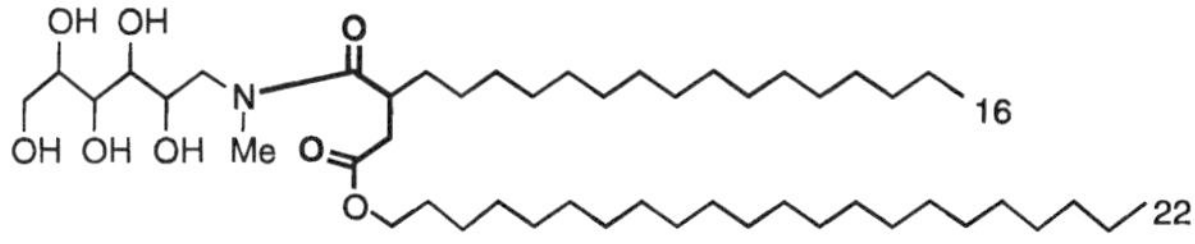

Hexadecylsucc.-monobehenylester N-methylsorbitylamide

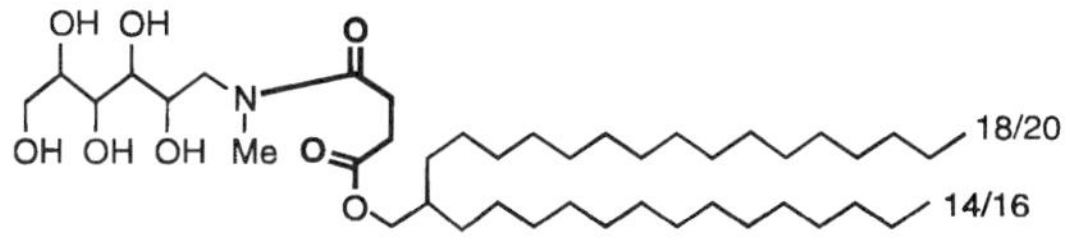

Succ.-mono-$C_{32/36}$-Guerbet-ester N-methylsorbitylamide

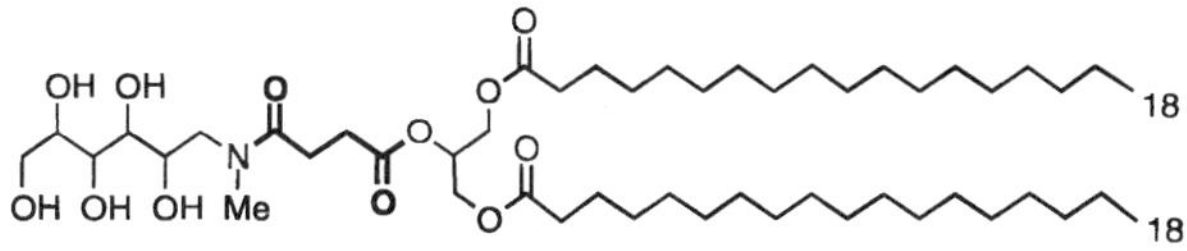

Succ.-mono-[1,3-di-(C_{18}-acyloxy)-2-propylester] N-methylsorbitylamide

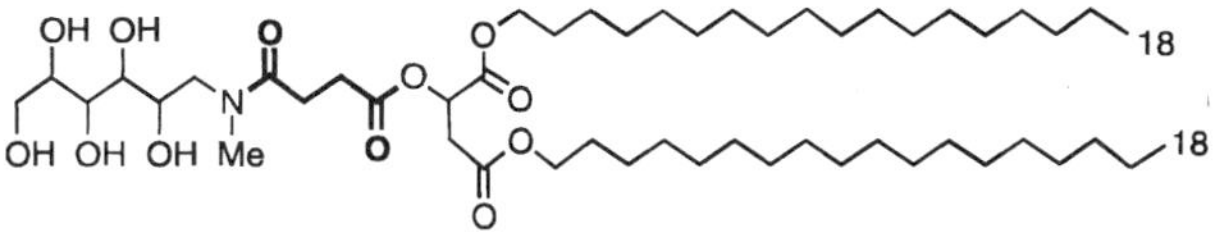

Succ.-mono-[1,2-di-(C_{18}-alkoxycarbonyl)-ethylester] N-methylsorbitylamide

Figure 19 Succinic acid derivatives. (From Refs. 64–67.)

context [68]. The possibility of using these derivatives as synthetic barrier lipids was recognized at a later stage [69,70].

The third structure in Figure 21 is based on the N-behenylamide of gluconic acid, which is obtained from the amine with D-gluconolactone and in the next step receives its second long alkyl chain by acylation with stearoyl chloride [71].

The final compound is the stearic acid ester of dodecylglucoside and, as such, has been known for a long time. Because the alkyl glycosides, referred to as APG (alkyl polyglycosides), have been commercially available for some time, this variant should also be mentioned here. A similar type is derived from 10-hydroxydecylglucoside and is then acylated at the two primary OH groups to give the corresponding dilaurate [72]. The syntheses of the two first-mentioned substances are shown in Figure 22.

In both cases, the synthesis starts with the condensation of glucose and dodecylamine to give N-alkylglucosylamine. In the first case, the glucosylamine is hydrogenated to N-sorbitylamine and then acylated. In the second case, the

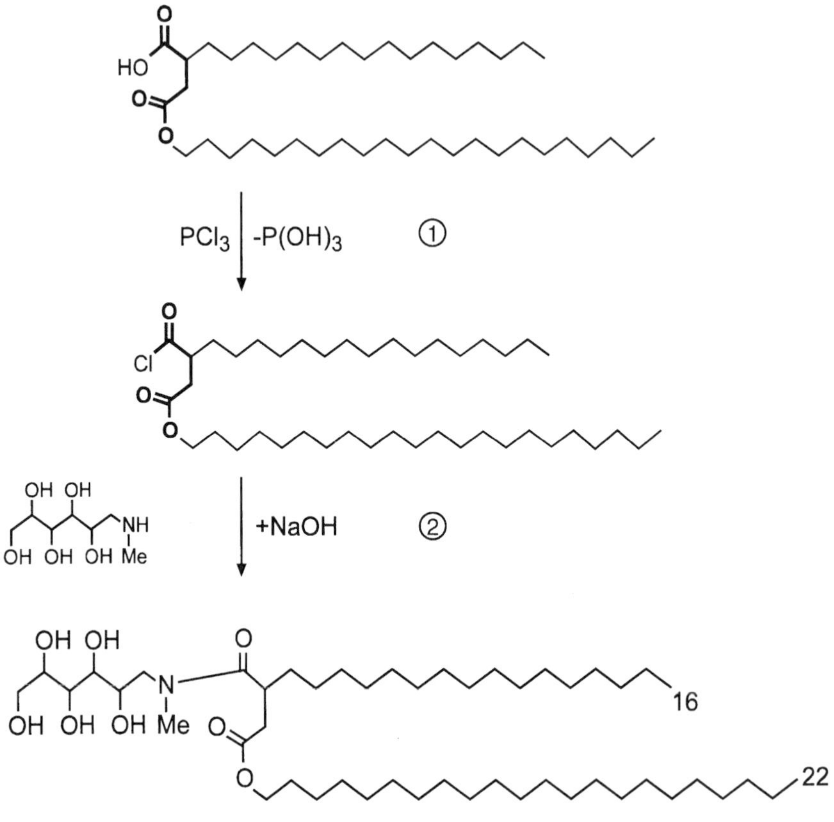

Figure 20 Synthesis of C16-Alkylsuccinic acid monobehenylester N-methylsorbitylamide. (From Ref. 64.)

glucosylamine is directly acylated to stearic acid N-dodecylglucosylamide. Although the second method involves one less step, the hydrogenation yields a hydrolysis-resistant product with an additional OH group. Both syntheses make use of favorably priced natural substances.

The compound in the final structure diagram is an especially interesting prospect for commercial production in view of its simple synthesis [73] (see Fig. 23).

The $C_{32/36}$-Guerbet acid, which is obtainable from the corresponding Guerbet alcohol through alkali fusion, was converted into the acid chloride. Aminolysis of the chloride in an aqueous environment with N-methylsorbitylamine resulted in the formation of $C_{32/36}$-Guerbet acid N-methylsorbitylamide. This synthesis was optimized for production on a commercial scale.

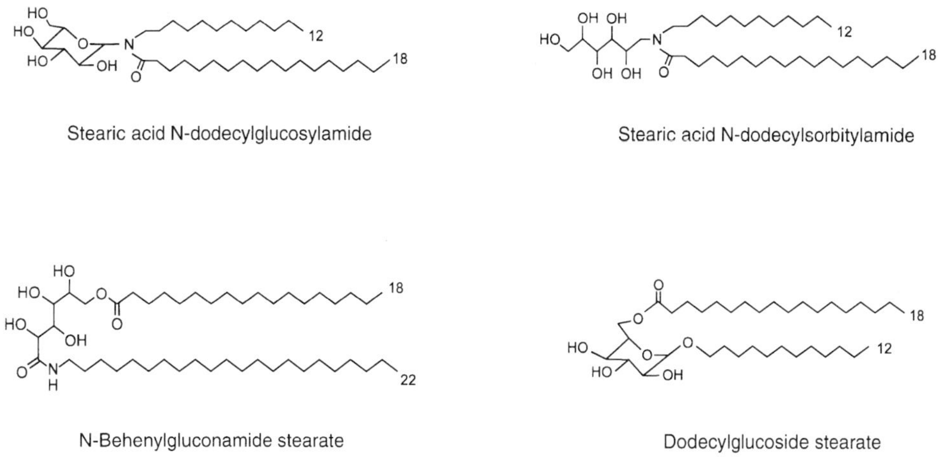

Chemistry of Natural and Synthetic Skin Barrier Lipids 27

Stearic acid N-dodecylglucosylamide

Stearic acid N-dodecylsorbitylamide

N-Behenylgluconamide stearate

Dodecylglucoside stearate

Figure 21 Sugar and sugar acid derivatives.

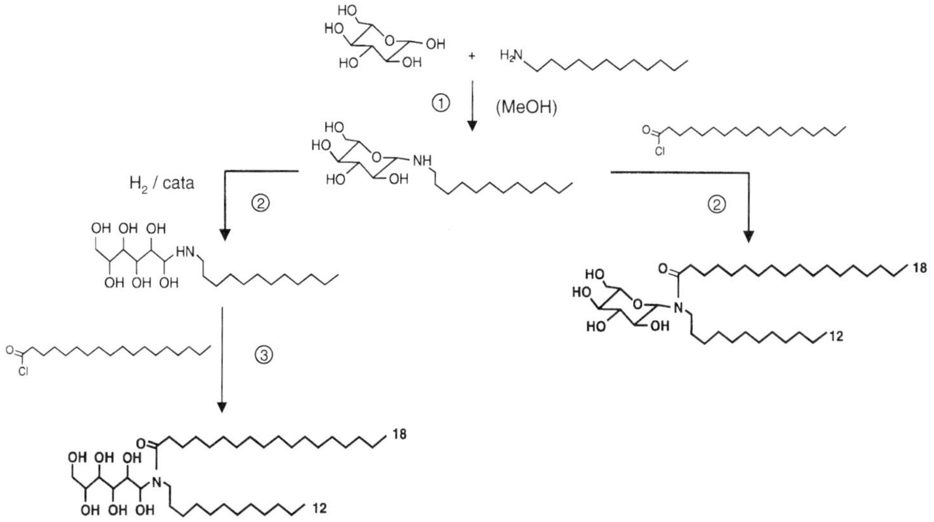

Figure 22 Syntheses on the basis of N-alkylglucosylamines and -sorbitylamines. (From Refs. 69,70.)

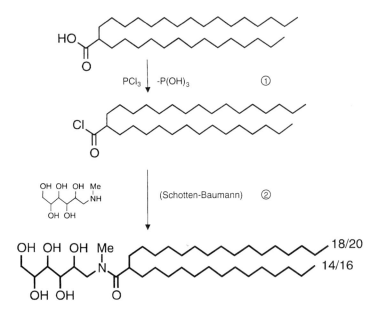

Figure 23 Synthesis on the basis of $C_{32/36}$-Guerbet-alkanoic acid. (From Ref. 73.)

VIII. PRECURSOR MOLECULES (SO-CALLED PRO-DRUGS)

As mentioned at the beginning, the glycosylceramides become deglycosylated as the differentiation of the keratinocytes proceeds. This gave rise to the idea that the glycosylceramides themselves could be used as so-called pro-drugs because they are metabolized to ceramides in the stratum corneum by glycosidases. Laboratoires Serobiologique was able to demonstrate this for a galactosylceramide, ceramide LS 3773 [74].

In another example, from Unilever, phosphoric acid esters and sulfuric acid esters of pseudoceramides are claimed as pro-drugs, whose active substance is said to be liberated by hydrolytic enzymes of the horny layer [75]. Structures previously described by Kao [57,58] served as the pseudoceramide basis. These substances can be formulated much better in cosmetic preparations than the underivatized parent compounds because they are more hydrophilic and similar to emulsifiers.

The sulfonation product of glucamide SBL (Ke 3193) shown in Table 7 [76] also belongs to this group. This sulfate readily forms liposomes in lipid/water systems. Some structures of the above-mentioned precursor molecules are given in Table 7.

Chemistry of Natural and Synthetic Skin Barrier Lipids

Table 7 Synthetic Ceramide Precursors

Ceramides 3773 (*Lab. Serobiologiques S.A.*) [74]	[structure: glycosylated ceramide with CH$_3$ (18) and CH$_3$ (18–20) chains]
Phosphoryl-SLE (*Unilever*) [75]	[structure: phosphate diester with Na$^+$ counterions, CH$_3$ (16) chains]
Sulfo-SLE (*Unilever*) [75]	[structure: sulfate ester with Na$^+$, CH$_3$ (16) chains]
Sulfoglucamid-SBL (*Henkel*) [76]	[structure: sulfated glucamide with Me group, CH$_3$ (16) and CH$_3$ (22) chains]

The development of further pseudoceramide precursors seems to be an interesting task for the future.

IX. PROPERTIES OF SOME PSEUDOCERAMIDES

Table 8 shows a comparison of the typical properties of a few selected pseudoceramides and the natural type-3 ceramide. The compared properties are the melting point and the effect on the lamellar structure of the model lipid mixture after Friberg [14]. The melting points are all lower than the melting point of the natural ceramide. For formulation purposes, an optimal melting point is in the range between 40° and 55°C. The high melting point of the natural ceramide is unusual, although it should be remembered that it does not occur alone but in mixtures with other lipids, when it melts at a significantly lower temperature. The final three substances lie in a favorable melting range. The effect of the pseudoceramides on the lamellar structure is in all cases similar to that of the natural ceramide—i.e., the thickness of the lipid bilayer of the nonaqueous system is increased and the increase in the total interlayer distance as a function of water

Table 8 Properties of Some Pseudoceramides

Structure Name	Melting point [°C]	Influence on lamellar structure[a]	Source
Ceramide type 3 (native)	105	yes	standard Sigma
Ceramide H 03 (Sederma)	92	yes	pseudo-ceramide (Sederma)
Ke 3039	36/48	yes	pseudo-ceramide (Henkel)
Glucamid-SBL	55	yes	pseudo-ceramide (Henkel)
C$_{32/36}$-Guerbet-acid N-methylsorbitylamide	39	—	pseudo-ceramide (Henkel)

[a] added to the lipid model of Friberg [14]

concentration weakens in comparison with a ceramide-free lipid mixture. As the water content increases, these molecules migrate from the zone between the ends of the fatty alkyl chains into the parallel-aligned alkyl chain zone. Moreover, as the water content increases, the liquid crystalline character decreases, thus modifying the permeability of the barrier layer. In this respect the pseudoceramides H03 and glucamide SBL [64] exhibit the greatest similarity to the natural type-3 ceramide. Guerbet acid N-methylsorbitylamide [73] has a favorable melting point comparable to that of the Henkel pseudoceramide Ke 3039, but it was not studied any further.

X. SUMMARY

The special structure and the physical and physiological functions of the epidermal lipids with special focus on ceramides were described. Ceramide analogues were then introduced and a general structural principle for pseudoceramides based on fatty and oligohydroxy components was explained. A few examples were selected from the large number of pseudoceramides described in the patent literature. These examples can be synthesized by simple, technically feasible methods and are also described as being used in cosmetic applications. In vitro tests showed that some of them behave very similarly to natural ceramides. Dermatological studies demonstrated properties such as regeneration of the barrier function, reduced skin roughness, and improved skin elasticity.

REFERENCES

1. Elias PM, Friend DS. The permeability barrier in mammalian epidermis. J Cell Biol 1975; 65:180–191.
2. Lampe MA, Williams ML, Elias PM. Human epidermal lipids: characterization and modulation during differentiation. J Lipid Res 1983; 24:131–140.
3. Wertz PW, Swartzendruber DC, Abraham W, Madison KC, Downing DT. Essential fatty acids and epidermal integrity. Arch Dermatol 1987; 123:1381a–1384a.
4. Downing DT, Wertz PW, Stewart ME. The role of sebum and epidermal lipids in the cosmetic properties of skin. Int J Cosmetic Sci 1986; 8:115–123.
5. Wertz PW, Swartzendruber DC, Kitko DJ, Medison KC, Downing DT. The role of the corneocyte lipid envelops in cohesion of the stratum corneum. J Invest Dermatol 1989; 93:169–172.
6. Stewart ME, Downing DT, Pochi PE, Strauss JS. The fatty acids of human sebaceous gland phosphatidylcholin. Biochim Biophys Acta 1978; 529:380–386.
7. Nicolaides N. Skin lipids: Their biochemical uniqueness. Science 1974; 186:19–26.
8. Guy R, Ridden C, Kealey T. The improved organ maintenance of the human sebaceous gland: modeling in vitro the effects of epidermal growth factor, androgens, estrogens, 13-cis-retinoic acid, and Phenol red. J Invest Dermatol 1996; 106:454–460.
9. Osborne DW, Friberg SE. Role of stratum corneum lipids as moisture retaining agents. J Dispers Sci Technol 1987; 8:173–179.
10. Fulmer AW, Kramer GJ. Stratum corneum abnormalities in surfactant-induced dry scaly skin. J Invest Dermatol 1986; 86:598–602.
11. Bouwstra JA, Thewalt J, Gooris GS, Kitson N. A model membrane approach to epidermal permeability barrier: an X-ray diffraction study. Biochemistry 1997; 36:1770.
12. Wertz PW, Miethke MC, Long SA, Strauss JS, Downing DT. The composition of the ceramides from human stratum corneum and from comedones. J Invest Dermatol 1985; 84:410–412.
13. Yardley HJ. Epidermal lipids. Int J Cosmetic Sci 1987; 9:13–19.

14. Friberg SE, Osborne DW. Small angle X-ray diffraction patterns of stratum corneum and a model structure for its lipids. J Disp Sci Technol 1985; 6:485–495.
15. Friberg SE. Micells, microemulsions, liquid crystals, and the structure of stratum corneum lipids. J Soc Cosmet Chem 1990; 41:155–171.
16. Friberg SE, Kayali IH, Margosiak M. Stratum corneum structure and transport properties. Drugs Pharm Sci 1990; 42:29–45.
17. Friberg SE, Suhaimi H, Goldsmith LB, Rhein LL. Stratum corneum lipids in a model structure. J Disp Sci Technol 1988; 9:371–389.
18. Moore DJ, Rerek ME, Mendelsohn R. Role of ceramides 2 and 5 in the structure of stratum corneum lipid barrier. Int J Cosmet Sci 1999; 21:353–368.
19. Forslin B, Engström S, Engblom J, Norlén L. A novel approach to the understanding of human skin barrier function. J Dermatol Sci 1997; 14:115–125.
20. Bouwstra JA, Gooris GS, Cheng K, Weerheim A, Bras W, Ponec M. Phase behavior of isolated skin lipids. J Lipid Res 1996; 37:999–1011.
21. Bouwstra JA, Cheng K, Gooris GS, Weerheim A, Ponec M. Role of ceramide 1 and 2 in stratum corneum lipid organization. Biochim Biophys Acta 1996; 1300:177–186.
22. Bouwstra JA, Gooris GS, Dubbelaar FER, Weerheim AM, Ijzerman AP, Ponec M. Role of ceramide 1 in the molecular organization of the stratum corneum lipids. J Lipid Res 1998; 39:186–196.
23. Bouwstra JA, Dubbelaar FER, Gooris GS, Weerheim AM, Ponec M. The role of ceramide composition in the lipid organization of the skin barrier. Biochim Biophys Acta 1999; 1419:127–136.
24. Lieckfeldt R, Villailain J, Gomez-Fernandez JC, Lee G. Diffusivity and structural polymorphism in some model stratum corneum lipid systems. Biochim Biophys Acta 1993; 1151:182–188.
25. Kitson N, Thewalt J, Lafleur M, Bloom M. A model membrane approach to epidermal permeability barrier. Biochemistry 1994; 33:6707–6715.
26. Friberg SE, Kayali I, Rhein LD, Simion FA, Cagan RH. The importance of lipids for water uptake in stratum corneum. Int J Cosmet Sci 1990; 12:5–12.
27. Motta S, Monti M, Mellesi L, Ghidoni R, Caputo R. Abnormality of water barrier function in psoriasis: role of ceramide fractions. Arch Dermatol 1994; 130:452–456.
28. Lavrijsen APM, Higounenc IM, Weerheim A, Oestmann E, Tuinenburg EE, Boddé HE, Ponec M. Validation of an in vivo extraction method for human stratum corneum ceramides. Arch Dermatol Res 1994; 286:495–503.
29. Holleran W, Mao-Qiang M, Gao WN, Menon GK, Elias PM, Feingold KR. Sphingolipids are required for mammalian epidermal barrier function. J Clin Invest 1991; 88:1338–1345.
30. Lintner K, Mondon P, Girard F, Gibaud C. The effect of synthetic ceramide-2 on transepidermal water loss after stripping or sodium lauryl sulfate treatment: an in vivo study. Int J Cosmet Sci 1997; 19:15–25.
31. Tokimitsu I, Asshi M, Suzuki T. European patent 277 641. 1988. Kao Corporation. CA 1989; 110:101532.
32. Imokawa G, Akasaki S, Kawamata A, Yano S, Takaishi N. Water-retaining function in the stratum corneum and its recovery properties by synthetic pseudoceramides. J Soc Cosmet 1989; 40:273–285.

33. Lambers JWJ, Koger HS. WO 95/29151. 1995. Gist-Brocades BV. CA 1996; 124: 176810.
34. Marsh NL, Elias PM, Holleran WM. Glycosylceramides stimulate murine epidermal proliferation. J Clin Invest 1995; 95:2903–2909.
35. Bektas M, Dullin Y, Wieder T, Kolter T, Sandhoff K, Brossmer R, Ihrig P, Orfanos CE, Geilen CC. Induction of apoptosis by synthetic ceramide analogs in the human keratinocyte cell line HaCat. Exp Dermatol 1998; 7:342–349.
36. Kaipia A, Chun S-Y, Eisenhauer K, Hsueh AJW. Tumor necrosis factor-α and its second messenger, ceramide, stimulate apoptosis in cultures of ovarian follicles. Endocrinology 1996; 137:4864–4870.
37. Ghahremani M, Foghi A, Dorrington JH. Activation of Fas ligand/receptor system kills ovarian cancer cell lines by an apoptic mechanism. Gynecol Oncol 1998; 70: 275–281.
38. Inokuchi JI, Kuroda Y, Kosaka S, Fujiwara M. L-Threo-1-phenyl-2-decanoylamino-1-propanol stimulates ganglioside biosynthesis, neurite outgrowth and synapse formation in cultured cortical neurons, and ameliorates memory deficits in ischemic rats. Acta Biochim Pol 1998; 45:479–492.
39. Sweeny EA, Sakakura C, Shirahama T, Masamune A, Ohta H, Hakomori S, Igarashi Y. Sphingosine and its methylated derivative N,N-dimethylsphingosine (DMS) induce apoptosis in a variety of human cancer cell lines. Int J Cancer 1996; 66:358–366.
40. Gupta AK, Fisher GJ, Elder JT, Nickoloff BJ, Voorhees JJ. Sphingosine inhibits phorbol ester-induced inflammation, ornithine decarboxylase activity and activation of protein kinase C in mouse skin. J Invest Dermatol 1988; 91:486.
41. Bibel DJ, Aly R, Shinefield HR. Antimicrobial activity of sphingosines. J Invest Dermatol 1992; 98:269–273.
42. Mathias S, Dressler S, Kenneth A, Kolesnick N. Characterization of a ceramide-activated protein kinase: stimulation by tumor necrosis factor α. Proc Natl Acad Sci USA. 1991; 82:10009–10013.
43. Schmidt RR, Escher B. Neuere Erkenntnisse über Glycosphingolipide. Fat Sci Technol 1992; 94:534–541.
44. Geilen CC, Wieder T, Orfanos CE. Ceramide signalling: regulatory role in cell proliferation and apoptosis in human epidermis. Arch Dermatol Res 1997; 289:559–566.
45. Rawlings AV, Davies A, Carlomusto M, Pillai S, Zhang K, Kosturko R, Verdejo P, Feinberg C, Nguyen L, Chandar P. Effect of lactic acid isomers on keratinocyte ceramide synthesis, stratum corneum lipid levels and stratum corneum barrier function. Arch Dermatol Res 1996; 288:383–390.
46. Davies A, Verdejo P, Feinberg C, Rawlings AV. Increased stratum corneum ceramide levels and improved barrier function following topical treatment with tetraacetylphytosphingosine. J Invest Dermatol 1996; 106:918.
47. Conti A, Rogers J, Verdejo P, Harding CR, Rawlings AV. Seasonal influence of stratum corneum ceramide 1 fatty acids and the influence of topical essential fatty acids. Int J Cosmet Sci 1996; 18:1–12.
48. Zhang K, Kosturko R, Rawlings AV. Effect of thiols on epidermal ceramide biosynthesis. J Invest Dermatol 1996; 104:687.

49. Harding CR, Watkinson A, Rawlings AV, Scott IR. Review article: Dry skin, moisturization and corneodesmolysis. Int J Cosmet Sci 2000; 22:21–52.
50. Schmidt RR, Baer T, Sandhoff K. German patent 4025330. 1992. BASF. CA 1992; 116:236094.
51. Crawford DNK, Rawlings AV, Scott IR. WO 93/22281. 1993. Unilever NV. CA 1994; 121:57220.
52. Philippe M, Garson JC, Gilard P, Hocquaux M, Hussler G, Leroy F, Mahieu C, Semeria D, Vanlerberghe G. Synthesis of N-2-oleoylamino-octadecane-1,3-diol: a new ceramide highly active for the treatment of skin and hair. Int J Cosmet Sci 1995; 17:133–146.
53. Wakita H, Tokura Y, Yaki H, Nishimura K, Furukawa F, Takigawa M. Keratinocyte differentiation is induced by cell-permeant ceramides and its proliferation is promoted by sphingosine. Arch Dermatol Res 1994; 286:350–354.
54. Möller H, Knörr W, Weuthen M, Guckenbiehl B, Wachter R. Neue Pseudoceramide durch Verknüpfung von speziellen Fettstoffen mit Kohlenhydrat-Derivaten. Fett/Lipid 1997; 99:120–129.
55. Watkins S, Motion K, Rossiter K, Critchley P. Questamide H—A designed ceramide analogue. SÖFW-J 1995; 121:228–238.
56. Motion KR, Janousek A, Watkins S. WO 94/07844. 1994, Quest International BV. CA 1994; 121:141205.
57. Theis H, Göring S. Use of sphingolipids in skin care products. SÖFW-J 1995; 121:343–348.
58. Kawamata A, Yano S, Hattori M, Akazaki S, Imokawa G, Takishi N. European patent 227994. 1989. Kao Corporation. CA 1987; 107:153944.
59. Critchley P, Rawlings AV, Scott IR. European patent 495624. 1992. Unilever. CA 1992; 117:137448.
60. Cho SH, Frew LJ, Chandar P, Medison SA. WO 95/16665. 1995. Unilever. CA 1995; 123:314397.
61. Pillai S, Cho SH, Rawlings AV. U.S. patent 5476661. 1995. Elisabeth Arden Co. CA 1996; 124:97266.
62. Eierdanz H, Busch P, Tesmann H, Knörr W, Wachter R. German patent 4238032. 1992. Henkel KGaA. CA 1994; 121:179102.
63. Möller H, Engels T, Wachter R, Busch P. WO 95/34531. 1995. Henkel KGaA. CA 1995; 123:340959.
64. Möller H, Wachter R, Busch P. WO 95/23807. 1995. Henkel KGaA. CA 1995; 123:256169.
65. Möller H, Wachter R, Busch P. German patent 4403258. 1995. Henkel KGaA. CA 1995; 123:122740.
66. Möller H, Wachter R, Busch P. German patent 4430851. 1995. Henkel KGaA. CA 1996; 124:37369.
67. Möller H, Wachter R, Busch P. WO 95/21851. 1995. Henkel KGaA. CA 1995; 123:122741.
68. Lockhoff O. Glycolipide als Immunomodulatoren—Synthese und Eigenschaften. Angew Chem 1991; 103:1639–1649.
69. Möller H, Biermann M. German patent 4326958. 1995. Henkel KGaA. CA 1995; 122:196562.

70. Möller H, Biermann M. German patent 4326959. 1995. Henkel KGaA. CA 1995; 122:196561.
71. Möller H, Wachter R. German patent 19523479. 1996. Henkel. KGaA. CA 1996; 125:329282.
72. Weuthen M, Riegels P, Wachter R. WO 95/19366. 1995. Henkel KGaA. CA 1995; 123:199297.
73. Möller H, Wachter R, Busch P. WO 96/01807. 1996. Henkel KGaA. CA 1996; 124: 261613.
74. Pauli M, Pauli G. Glycoceramides—epidermal physiological role. Interest in care cosmetics. Objectivation of their mode of action/functional efficacy on man. SÖFW-J 1995; 121:557–580.
75. Critchley P, Kirsch SE, Rawlings AV, Scott IR. WO 92/06982. 1992. Unilever NV. CA 1992; 117:97093.
76. Möller H, Föster T, Class M. German patent 19748399. 1999. Henkel KGaA. CA 1999; 130:339721.

2
Structure of Stratum Corneum Lipid Layers and Interactions with Lipid Liposomes

Joke Bouwstra
Leiden University, Leiden, The Netherlands

I. LIPID ORGANIZATION IN STRATUM CORNEUM

A. Introduction

The natural function of the skin is the protection of the body against the loss of endogenous substances such as water and against undesired influences from the environment caused by exogenous substances. The main barrier for diffusion through the skin is the outermost layer of the skin, the stratum corneum (SC) [1]. The SC consists of keratin-filled dead cells, the corneocytes, which are entirely surrounded by crystalline lamellar lipid regions. Furthermore, the cell boundary, the cornified envelope, is a very densely cross-linked protein structure, which prevents easy absorption of drugs into the cells. This has been illustrated by confocal studies, in which it was demonstrated that the penetration pathway of dyes is mainly through the intercellular regions in the SC [2,3]. For these reasons the lipid regions are considered to be of great importance for the skin barrier function [4,5]. In the past 20 years, many studies have been carried out to increase knowledge of the barrier function of the skin. These studies have mainly been focused on the lipid composition and organization in the SC and the changes involved in this lipid organization as a consequence of topical application of drugs in various formulations and patches. This chapter will mainly focus on the structure in the SC and the changes in this structure as a consequence of water, surfactants, and liposomes.

B. Lipid Organization in Stratum Corneum

The composition of the SC lipids strongly differs from that of cell membranes of living cells. The major lipid classes [6–8] in the SC are ceramides (CER), cholesterol (CHOL), and free fatty acids (FFA). The length of the acyl chains in the CER varies between 16 and 33 carbons; the most abundant acyl chain lengths of FFA are C22 and C24. These chain lengths are much longer than those of the phospholipids present in plasma membranes. Furthermore, the CER head groups are very small and contain several functional groups that can form lateral hydrogen bonds with adjacent ceramide molecules and with other lipids as well. The increased chain length and the small size of the head group results in a very densely packed structure (see below) compared to that present in membranes of living cells. This is closely related to differences in functionality. In cell membranes, transport across the membrane is required for a proper functioning of the cells; in SC, however, the lipid lamellae are designed to keep many substances outside the body and act therefore as permeability barriers. There are at least eight subclasses of ceramides (HCER) present in *human* stratum corneum [9]. These HCER, often referred to as HCER 1 to 8, differ from each other by the head-group architecture (mainly sphingosine or phytosphingosine base linked to either a fatty acid, an α-hydroxy fatty acid or an ω-hydroxy fatty acid) and the hydrocarbon chain length. HCER1 and HCER4 both have a very exceptional molecular structure: a linoleic acid is linked to an ω-hydroxy fatty acid with a chain length of approximately 30–32 C-atoms. In this respect, the HCER are different from ceramides isolated from pig SC (pigCER), in which only pigCER1 has this exceptional molecular structure [10] and in which pigCER5 has an unusual short chain length (main population around C16–C18).

Many efforts have been undertaken to characterize the lipid lamellar regions. At the end of the 1950s, and early 1960s, the first studies using x-ray diffraction were carried out with human SC [11,12]. The diffraction patterns of these excellent measurements were interpreted as being from lipids organized in tubes surrounding keratin filaments. In the 1970s, however, freeze fracture electron microscopic studies revealed that the lipids are organized in lamellae [13,14] located in the intercellular regions in the SC. It took another 10 years before RuO_4 has been introduced as a post-fixating agent to preserve the saturated lipids in the SC during the embedding procedure. The subsequent electron microscopic studies revealed an unusual lamellar arrangement of a repeating pattern with electron translucent bands in a broad-narrow-broad sequence [15–20]. These lipid lamellae are oriented along the surfaces of the cells. In addition differential scanning calorimetry, Fourier transformed infrared studies, and x-ray diffraction revealed that the lipids are arranged in a crystalline sublattice and that a small proportion of lipids formed a liquid phase [21–23]. Furthermore, the lipid bilayers were oriented along the surfaces of the cells. In 1988, a study was reported in

Structure of SC Lipid Layers and Interactions with Lipid Liposomes   39

which x-ray diffraction was used to determine the lipid organization in murine SC [24]. These studies revealed a lamellar organization of the lipids in the intercellular regions with an unusual long periodicity of the structure being approximately 13 nm. Furthermore, the lipids in the lamellae are mainly organized in a crystalline sublattice, which is in agreement with the FTIR studies. In subsequent studies, the lipid organization in pig SC and in human SC has also been assessed [25–28].

In Figure 1A the small angle x-ray diffraction curves of human SC measured at room temperature are plotted as a function of Q (for a brief explanation of the technique, see appendix A). Q is a measure for the scattering angle. The diffraction curve is characterized by a strong and a weak diffraction peak, both peaks having a shoulder on the right-hand side. The positions of these diffraction peaks are directly related to the periodicity of the phase by the equation: $Q = 2\pi n/d$, in which n is the order of the diffraction peak and d the periodicity of the lamellar phase. In the diffraction pattern of human SC, the number of peaks are limited, very broad, and partly overlap, so interpretation of this diffraction pattern is quite complicated. However, from this curve it is obvious that an increase in the water content from 20% to 60% w/w does not lead to a change in the peak positions, and therefore not to a change in periodicity of the lamellar phases. From this observation it was concluded that addition of water to the SC does not lead to a swelling of the lipid lamellae (for more detailed information with respect to water, see below). To analyze the diffraction curves in more detail, additional information was required. For this purpose, the SC lipids were crystallized from 120°C to room temperature, after which the x-ray pattern was monitored. As shown in Figure 1B, the diffraction curves revealed a series of peaks that were located at the same inter-peak distance. Such a diffraction profile is characteristic of a lamellar phase. From the positions of the peaks, the periodicity of the lamellar phase (d) was calculated. These calculations revealed that after recrystallization, the lipids in SC were organized in a lamellar phase with a periodicity of 13.4 nm. Comparing the peak positions of the diffraction pattern prior to and after recrystallization revealed that in untreated SC two lamellar phases are present, one with a periodicity of approximately 6.4 nm, and the other with a periodicity of approximately 13.4 nm, respectively [10]. Similarly, two lamellar phases with periodicities of 6 and 13.2 nm were present in pig SC [26]. Because the 13 nm phase is always present in all the species studied so far, and is very characteristic for the SC lipid phase behavior, this phase is probably very important for the skin barrier function [29].

In addition to the wide angle x-ray diffraction measurements, studies were carried in which the lateral packing of the lipids was examined. The information on the lateral packing of the lipids in the lamellae can be gained by wide angle x-ray diffraction [26]. The spacings of the reflections indicative for the lateral lipid packing are in the sub-nanometer range and are therefore located at much

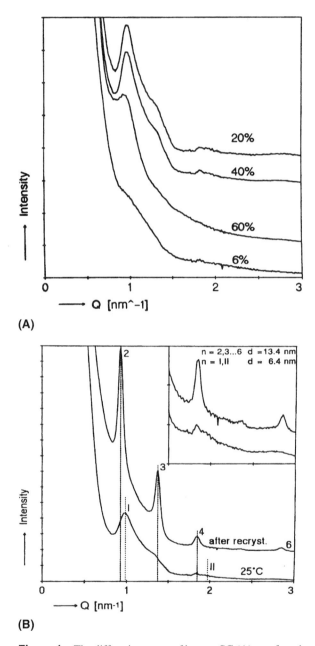

Figure 1 The diffraction curve of human SC (A) as a function of hydration level varying between 6% and 60%, (B) prior to and after recrystallization. The diffraction curve is plotted as a function of Q defined as $Q = 2\pi \sin\theta/\lambda$. The positions of the peaks are directly related to the periodicity of the lamellar phases by the relation: $Q_n = 2n\pi/d$; d is the periodicity and n the order of the reflection; Q_n is the position of the nth order reflection.

higher scattering angle in the diffraction pattern (see appendix) than those attributed to the lamellar phases. With respect to the SC, three sublattices are of importance: namely, the liquid, the hexagonal, and the orthorhombic lateral packing. These sublattices are characterized by different sets of reflections. The orthorhombic phase is characterized by spacings at 0.37 and 0.41 nm. The strongest reflection of the hexagonal phase is located at 0.41 nm. Finally, the pattern of the liquid phase is characterized by a very broad reflection at 0.46 nm reflecting the variation in interchain distance in the liquid phase. An example of the wide-angle x-ray diffraction pattern of human SC is provided in Figure 2.

Several features can be distinguished in this diffraction pattern. Two strong rings are observed at 0.378 and 0.417 nm, respectively. These reflections indicate that the lipids are organized in an orthorhombic lattice. However, the strongest reflection of the hexagonal packing is located at 0.417 nm reflection, so as a consequence it cannot be concluded from these measurements whether or not a

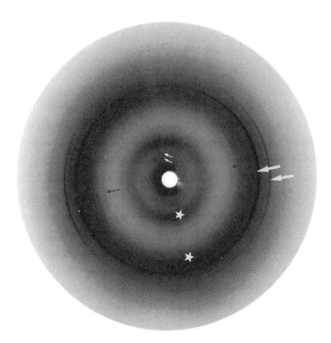

Figure 2 The wide-angle diffraction pattern of human stratum corneum. The reflections are indicated in the figure. Large white arrows: the 0.417 and 0.378 nm spacing attributed to the orthorhombic lateral packing. Black arrows: cholesterol reflections; stars: broad keratin reflections (at 0.92 and 0.46 nm). Small white arrows: higher order reflections of the lamellar phases.

hexagonal lateral packing is present as well (see below). The reflections attributed to the orthorhombic sublattice obscure the reflection of the hexagonal phase. In addition, several other features are present in the diffraction pattern. Two broad rings are present at approximately 0.46 and 0.96 nm. These reflections are also present in extracted SC and are assigned to soft keratin, which is present mainly in the corneocytes. Whether a population of lipids forms a liquid phase cannot be concluded from these studies, because the broad reflection of keratin at 0.46 nm obscures the broad reflection of lipids in a liquid phase. This ring is expected to be located at 0.46 nm as well. In addition, two reflections are present at 0.36 nm and 0.72 nm. These reflections are present only approximately 20–30% of the studies samples and are assigned to proteins. Finally, a series of rings are present in the diffraction pattern that can be attributed to phase-separated crystalline cholesterol. The CHOL reflections are present in the diffraction pattern of approximately 50% of the human SC samples.

C. Lipid Organization in Model Membranes

In order to understand the SC lipid phase behavior in normal and diseased skin in more detail, knowledge of the role the various lipid classes and subclasses play in the SC lipid organization is required. Because it is impossible to extract selectively lipids from the SC, these studies have to be carried out with model membrane systems, in which the lipid composition varies systematically. Several studies have been carried out in which the model membranes were prepared from palmitic acid, cholesterol, and bovine-brain ceramides. The choice for palmitic acid was based on a study by Lampe et al. [30], in which it was stated that the most abundant fatty acids present in SC have a chain of C16 or C18:1. However, in more recent studies it has become clear that this was an erroneous observation and that the main fatty acid chain lengths are C22 and C24 [31,32]. In several studies, the lateral packing of model membrane mixtures has been studied. NMR studies carried out with equimolar mixtures of CHOL and bovine-brain ceramide 3 in the absence and presence of palmitic acid revealed that the main population of lipids formed a crystalline phase, but that also a small proportion of lipids formed a more mobile phase [33,34]. Upon an increase in the temperature, the crystalline phase transformed to a liquid phase at around 60°C. This is in contrast to phospholipid-containing mixtures, in which in the presence of a substantial amount of cholesterol a liquid-ordered phase has been observed [35–38]. The presence or absence of this liquid phase may play an important role in the interactions between exogenous substances and the skin because exogenous substances might more easily mix with a fluid phase than with a crystalline phase [28,39]. In an additional study, it has been shown that sphingomyelin-containing mixtures form a liquid lateral packing, whereas ceramides having the same chain length distribution in the presence of cholesterol form a crystalline phase [40]. From this observation it was concluded that the presence of a liquid phase depends strongly on the size of the headgroup. The ceramide-containing mixtures were studied at either pH 5 or 7.4. Slight differences in phase behavior were observed at these two pH values. Furthermore, in these studies it was shown that more than 80% of the lipids was in a crystalline state, while a small portion formed a more mobile phase [41]. A liquid phase was formed at around 50°C. As far as the lateral packing is concerned, these bovine-brain ceramide mixtures mimic SC lipid phase behavior quite closely. However, as shown in an additional study using the x-ray diffraction technique, no lamellar phase was observed with a periodicity of approximately 12 to 13 nm [42,43]. It seems that in this respect the bovine-brain ceramide mixtures differ in phase behavior from the lipid organization in intact SC. In addition, in several studies Fourier transformed infrared (FTIR) spectroscopy measurements were carried out using mixtures prepared from either bovine-brain ceramide type III, synthetic ceramide 2, or synthetic ceramide 5 [44,45]. The ceramides were mixed with cholesterol and perdeuterated palmitic acid. The FTIR spectroscopy measurements were carried out as a function of temperature and revealed the presence of orthorhombic phases. At physiological temperature, the CD_2 scissoring mode of palmitic acid and the CH_2 scissoring mode of ceramides are each split into two components, indicating that these molecules are located in different lattices. Increasing the temperature led to a disordered lattice for palmitic acid at around 40°C and for bovine-brain ceramide type III at around 60°C, confirming that the components were not properly mixed in one orthorhombic lattice. A similar behavior was found when mixing fatty acids with either synthetic ceramide 3 or synthetic ceramide 5 [44]. Similarly, in another study using FT-Raman spectroscopy, phase separation was observed between ceramides and saturated fatty acid in the absence of CHOL. Again, a single chain length was used [46]. Recently, we observed that addition of C22 fatty acid to a mixture of isolated CER and CHOL does not mix properly in the solid state, whereas addition of FFA with a chain length distribution between 16 and 22 to a CHOL:CER mixture (CER isolated from SC) did not result in a phase separation (Bouwstra et al., unpublished results). Therefore, it seems essential when mimicking lipid SC phase behavior that FFA mixtures with a chain length distribution similar to that present in SC are used instead of a single chain length of the fatty acids. In several studies, liposomes were used as a model system for SC lipid lamellae. Although permeation across a bilayer, stability, and fusion can be studied accurately when using liposomes [47–49], the lamellar lipid arrangement is expected to be different from that observed in SC. Therefore, in our opinion, results obtained with liposomes cannot directly be extrapolated to the situation in stratum corneum. For example, skin lipid liposomes do not form a stable suspension at around pH 5; the pH at the skin surface is approximately 5–6.

Several studies have been carried out in which mixtures prepared from isolated pig ceramides (pigCER) were examined. The first studies, using liposomes, set out to unravel the mechanisms that play a role in the formation of lamellar phases [50–52].

More recently, mixtures were prepared from either CHOL and pigCER or from CHOL, pigCER, and FFA [53–58]. A chain length distribution of the FFA was used according to Wertz and Downing [7]. The composition of these mixtures varied systematically. Small- and wide-angle x-ray diffraction studies were performed. The small-angle x-ray diffraction curve of an equimolar mixture prepared from CHOL and pigCER at a pH of 5 is provided in Figure 3 [57]. In the x-ray pattern, the presence of a large number of sharp peaks was noticed. The peaks indicated by I and II have been assigned to a lamellar phase with a periodicity of 5.2 nm, and peaks indicated by 1, 2, 3, 5, and 7 to a lamellar phase with a periodicity of 12.2 nm. Furthermore, two additional peaks at 3.35 nm and 1.68

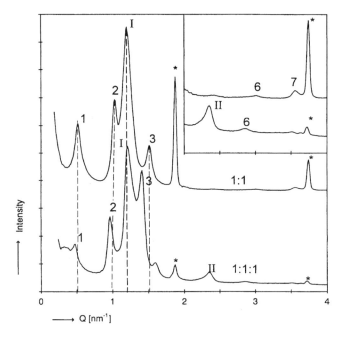

Figure 3 The small-angle x-ray diffraction curves of the 1:1 CHOL:CER and 1:1:1 CHOL:CER:FFA lipid mixtures. The arabic numbers indicate the diffraction orders of the long periodicity phase (repeat distance of 12.2 and 12.8 nm for the equimolar CHOL:CER and CHOL:CER:FFA mixtures, respectively). The roman numbers indicate the diffraction orders of the short periodicity phase (repeat distance between 5.2 and 5.5 nm).

nm have been detected, which can be assigned to hydrated crystalline CHOL that phase separated. Reducing the CHOL/pigCER molar ratio to 0.4 did not change the phase behavior, except that a smaller amount of CHOL phase-separated. Only a reduction to a molar ratio of 0.2 or an increase to a molar ratio of 2 weakened the 12.2 nm phase and increased the proportion of lipids forming the 5.2 nm lamellar phase. At a molar ratio of 0.2, the periodicity of the 5.2 nm phase increased slightly to 5.6 nm (see Fig. 4). From these observations, it was concluded that over a wide range the phase behavior is remarkably insensitive to changes in the CHOL/pigCER molar ratio. Furthermore, it was concluded that mixtures with CHOL and pigCER mimic lipid phase behavior in intact SC quite closely, not only with respect to the lateral packing (see below) but also with respect to the formation of two lamellar phases with periodicities of 5–6 and 12–13 nm. Addition of long-chain FFA mixture (dominant chain lengths in the mixture C22 and C24), the third class of lipids that is prominently present in SC, did not change the lipid phase behavior dramatically (see Fig. 3). At an equimolar CHOL:pigCER:FFA molar ratio, the periodicity of the 12.2 nm phase increased to approximately 12.8–13.0 nm; the 5.2 nm lameller phase increased its periodicity to 5.4 nm.

The wide-angle x-ray diffraction patterns of CHOL:pigCER mixtures revealed a strong reflection at approximately 0.409 nm, which indicates a hexagonal lateral packing. Varying the CHOL:pigCER molar ratio between 0.2 to 2 did not change the lateral packing, except for an increase in the intensities of the reflections attributed to hydrated crystalline CHOL. Addition of FFA did not induce changes in the lamellar phase behavior (see above); however, the lateral packing of the lipids changed drastically. FFA induced a phase transition from a hexagonal to orthorhombic lateral packing. This means that FFA increase the packing density of the lipids. This is a very important observation for the skin barrier function, because it is expected that the permeability through a crystalline phase is much lower than through a hexagonal phase. The transition from hexagonal to orthorhombic phase was observed only after incorporation of long-chain fatty acids (C22/C24 chain length) and not with short-chain fatty acids (C16/C18 chain length).

In diseased skin, a deviation in CER composition often has been found [59–62], so insight in the role of CER subclasses play in SC lipid phase behavior is also of great importance. To examine in more detail the role of the individual ceramides, mixtures prepared from CHOL and pigCER with varying pigCER composition were examined. For this purpose, mixtures prepared with pigCER (1–5), pigCER(2–6), or pigCER(1–2) have been used. Because in native SC, CHOL and CER are present in an approximately equimolar ratio, we first examined the phase behavior of the equimolar mixtures. These studies revealed that the lipids were organized in two lamellar phases with periodicities of approximately 12–13 nm and 5–6 nm, similarly as observed in intact SC (Fig. 4). The

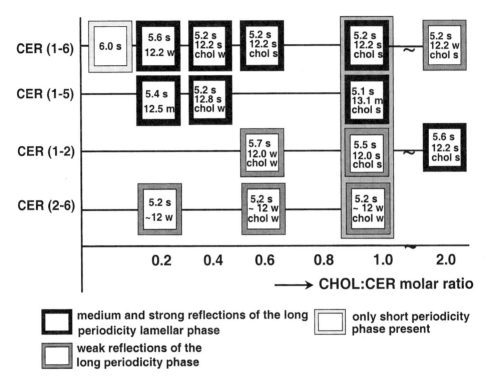

Figure 4 A schematic overview of the lamellar phases of several CHOL:pigCER mixtures as function of the molar ratio. Note the similarity in the lamellar phases of the various CHOL:CER mixtures at an equimolar ratio. The indication weak (w), medium (m) and strong (s) of the 12–13 nm phase and CHOL denote the presence of these phases compared to the 5-5.5 nm phase.

exception was found with an equimolar CHOL/CER mixture in which the pigCER1 was absent. In this mixture, the 12 nm phase was only weakly present [57], indicating that in equimolar CHOL/CER mixtures the pigCER1 plays a crucial role in the formation of the 12-13 nm lamellar phase. Similar observations have been made when FFA was incorporated into CHOL/pigCER mixtures at an equimolar ratio [63]. This is also in line with the absence of the 12-13 nm phase in synthetic ceramide mixtures, because none of the synthetic ceramides used in the studies until now mimic the structure of pigCER1. Again, only pigCER1 has been found to play a crucial role in the formation of the 12-13 nm phase. Reduction of the CHOL content made the lipid phase behavior in CHOL/pigCER mixtures more sensitive to the pigCER composition. For example, in mixtures with a CHOL:pigCER(1-2) molar ratio of 0.6, the 12.2 nm phase is hardly present [63]. This means that only a dramatic change in lipid phase behavior can be expected in intact SC when simultaneously the CHOL:pigCER molar ratio strongly deviates from the equimolar ratio and the pigCER composition deviates from the composition in normal pig SC. However, whether these findings can be extrapolated to the in vivo situation in humans is not yet known, because the CER composition in human SC is slightly different from that in pig SC. For example, in human SC, two CER are present with a linoleic acid moiety linked to the ω-hydroxy fatty acid. This might indicate that the role of HCER1 in the lipid phase behavior in human SC is less prominent than the role of pigCER1 in lipid organization in pig SC.

II. WATER AFFECTS LIPID ORGANIZATION IN STRATUM CORNEUM

One of the first reports on the interactions between the lipid bilayers and water has been published by van Duzee using thermal analysis [21]. They detected four endothermic transitions that were attributed either to lipid-phase transitions or to protein denaturation. The temperature at which these transitions occur depends on the water content in the stratum corneum. Such water-dependent thermal transitions have also been found in phospholipid membranes. However, in phospholipid membranes, the transition temperatures are much more strongly reduced by water than those observed in the SC. In subsequent studies, the effect of water on the SC lipid organization was examined by other techniques. The x-ray diffraction technique revealed that upon increasing the water content, almost no swelling of the lamellae took place [25–28], indicating that if water is present between the head groups, it will be present only in very small quantities. Using infrared spectroscopy [64], it was found that water did not affect lipid alkyl chain order at room temperature; with electron spin resonance [65], it was found that an increase in the water content increased the mobility of the hydrocarbon chains. However, whether this increase in mobility is due to an indirect effect of the swelling of the corneocytes or to a very small amount of water located between the head groups is not clear yet. In a separate publication, it was suggested that the increased chain mobility might be limited to certain domains in the intercellular regions [22]. This would be in agreement with the observation using freeze fracture electron microscopy. Using this technique, the lipid lamellae can be visualized as smooth areas at the plane of fracture. However, after extensive treatment with a phosphate buffer saline solution, next to these smooth regions, areas with a rough surface were present in the intercellular domains, indicating a change in the lipid structure (see Fig. 5).

Figure 5 After treatment with a phosphate buffer saline solution water pools as well as rough structures were observed in the intercellular lipid domains in the stratum corneum. C = corneocyte; RS = rough structure; W = waterpool; IL = intercellular lipids and D = desmosome. Bar represents 100 nm.

Furthermore, this article [66] reports the presence of water domains not only in the corneocytes but also in the intercellular regions. The water domains were often present between the bound lipids of the cornified envelope and the lipid bilayer regions (unpublished observations). The presence of the water domains indicates a phase separation between the lipid lamellae and the water. This is not unlikely to occur, because the lipids present in the intercellular domains are quite lipophilic, especially at those pH values at which the FFA are not dissociated. Separate domains of water are also able to change the penetration pathway of the penetrant and increase its permeability. It is very unlikely that the water domains are continuous channels across the stratum corneum: this would require large lipid-water interfacial regions, which would be energetically unfavorable.

III. VESICLES AFFECT DRUG TRANSPORT ACROSS SKIN

A. Introduction

The first publications on interactions between liposomes and skin appeared in 1980 and 1982 [67,68]. In these publications it was reported that liposomes applied to skin of white rabbits in vivo favored the disposition of drugs in the epidermis and dermis, whereas the amount of drug found in the various organs was reduced. Although in these studies it was strongly suggested that vesicles penetrated the skin, this suggestion was received with a lot of skepticism. After the first papers by Mezei and Gulasekharam, a large number of studies were carried out, a description of which will be provided below.

B. Vesicles Affect Skin Penetration in Vitro

After the introduction of liposomes as drug delivery systems for the transdermal route, Ganesan and Ho [69,70] published two studies. In these studies, liposomes were prepared from DPPC (dipalmitoyldiphosphatidylcholine) that forms gel-state bilayers. The authors proposed that the drug could be transported (1) as free solute, (2) as free solute but also associated with the liposomes, or (3) by a direct transfer of the drug from liposomal bilayer to SC without partitioning into the water phase. They assumed that the vesicles were neither absorbed intact nor fused with the SC surface. The permeation studies were performed in vitro with mouse skin using glucose, progesterone, and hydrocortisone. The amount of (model) drug in the receiver compartment was sampled as function of time. No radiolabeled DPPC was detected in the receiver of the diffusion cell and glucose entrapped in liposomes resulted in smaller skin permeability than glucose applied in a normal saline solution. When focusing on the lipophilic drugs the investigators found almost no release from the liposomes compared to the saline solution. When applied to the skin, the lipophilic drug penetrated through the skin, suggesting a direct transfer from vesicles to SC. Very convincingly, the results of these studies did not confirm intact liposome penetration as suggested by Mezei and Gulasekharam [67,68].

In a study by Knepp et al. [71,72], the vesicles were suspended in an agarose gel. They reported that progesterone release from an agarose gel alone was very fast compared to the release from liposomes embedded in the agarose gel and that vesicles reduced the progesterone transport rate across hairless mouse skin compared to application in an agarose gel alone. As far as the lipid composition is concerned, progesterone was applied in egg-phosphatidylcholine liposomes, DOPC (dioleoylphosphatidylcholine) liposomes, DMPC (dimirystoyl-phosphatidyl choline) liposomes, or in DPPC (dipalmitoylphosphatidyl choline) liposomes. The gel-state DPPC resulted in a slower skin penetration of progesterone than the egg-phosphatidylcholine liposomes. Furthermore, intercalation of *cis* unsaturated fatty acid (oleic acid) in phospholipid bilayers increased the drug transport rate across the skin by an order of magnitude (see Fig. 6). In two more recent articles [73,74], it has been reported that a gel immobilizes the liposomes and therefore might affect the interactions between the liposomes and the skin. However, the trend observed in the studies of Knepp et al. [28,29]—in which incorporation of drugs in liquid-state liposomes results in a higher skin permeation rate than when incorporated in gel-state liposomes—was also observed for gel-free formulations (see below).

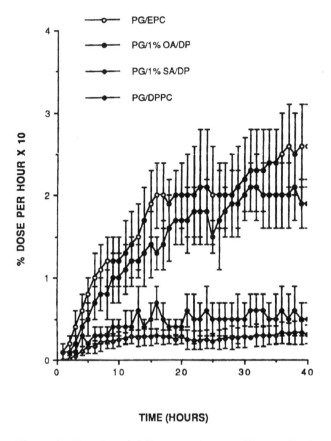

Figure 6 Transdermal delivery rate (mean ± SD; n = 6) of progesterone (PG) across hairless mouse skin in vitro encapsulated in EPC and DPPC liposomes containing 1% stearic (SA) acid. (From Ref. 72.)

Several other studies were carried out to evaluate whether liposome composition affects skin penetration of drugs. In the early 1990s, Dowton et al. [75] compared the effect of liposomal composition on the disposition of encapsulated cyclosporin A in mouse skin. They applied the vesicles nonocclusively in vitro. Two nonionic surfactant vesicles were compared with bovine-brain ceramide liposomes and liposomes prepared from saturated and unsaturated phosphatidylcholine. The various liposomes were saturated with cyclosporin A. In this way, an equal thermodynamic activity of cyclosporin A was achieved in all formulations. After 24 hours of application, the distribution of cyclosporin A was determined. They found that application of the drug in nonionic surfactant vesicles

Structure of SC Lipid Layers and Interactions with Lipid Liposomes 51

prepared from glyceryl dilaurate/cholesterol/polyoxyethylene-10 stearyl ether, the amount of cyclosporin A in deeper skin strata and receiver compartment was highest compared to the other formulations. These studies confirmed the results of the studies of Knepp et al. [71,72], although in the studies by Dowton et al. [75] the vesicles were applied nonocclusively without a gel. The findings of Knepp et al. [71,72] were also confirmed by studies of Hofland et al. [76]. They found that estradiol incorporated in gel-state nonionic surfactant resulted in a low drug transport rate through human SC compared to estradiol in liquid-state bilayers. These results were obtained with saturated estradiol formulations. In contrast to the studies of Knepp et al., the studies of Hofland et al. [72] revealed that a drug applied in liquid-state vesicles resulted in higher penetration rates than when applied in a phosphate buffer saline solution. This might be caused by a difference in the study design. Klepp et al. [71,72] incorporated the vesicles in agarose and used an *equal drug concentration* in the formulations. Hofland et al. suspended the vesicles in a buffer solution and used an equal *estradiol thermodynamic activity*. In the latter case an equal driving force from formulation to stratum corneum has been achieved. This difference in study design has a great impact when comparing drug permeation in vesicles compared to the control, such as a gel or buffer solution. In a subsequent study, van Hal et al. [77] reported that a decrease in cholesterol content in liquid state bilayers, which increases the fluidity of the vesicles bilayers, resulted in an increase in estradiol transport across SC. Meuwissen et al. [78] examined the diffusion depth of the bilayer label fluorescein-dipalmitoylphosphatidylethanolamine (fluorescein-DPPE) in the skin when applied in gel-state liposomes and in liquid-state liposomes using confocal laser scanning microscopy. They reported that fluorescein-DPPE applied in liquid-state liposomes penetrated deeper in the skin than the label applied in gel-state liposomes. Recently, Fresta and Puglisi [79] reported that corticosteroid dermal delivery with skin-lipid liposomes was more effective than delivery with phospholipid vesicles. This concerned the higher drug concentrations in deeper layers of the skin as well as the therapeutic effectiveness. From all the above-mentioned studies, it seems very clear that the *thermodynamic state of the bilayers* of the vesicles plays a crucial role in the effect of vesicles on drug transport rate across skin in vitro.

In a few studies, *occlusive* application was compared to *nonocclusive* application. These studies revealed that occlusive application of vesicle suspensions was less effective than nonocclusive application [80,81] (see Fig. 7). The results were somewhat unexpected: it has been reported that water is an effective permeation enhancer. However, in case of nonocclusive application, the increased drug transport can be caused by (1) a more profound interaction between the liposomal constituents and the skin and/or (2) the presence of a hydration gradient in the skin. According to Cevc et al. [80], the water gradient is an important driving force for drug diffusion. They claim intact vesicle penetration through the skin,

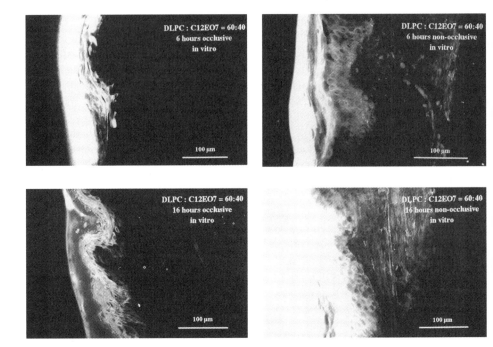

Figure 7 Occlusive versus nonocclusive application of vesicles prepared from dilaurylphosphatidylcholine (DLPC) and heptaoxyethylene alkylethers (C12E07). A cross section of human skin visualized using confocal laser scanning microscopy after 6 or 16 hours application of fluorescein-phosphatidyl ethanol amine in vesicles.

as long as flexible liposomes, Transfersomes®, have been used (see below). In a few studies, pretreatment with vesicles was compared to application of the drug associated with the vesicles. In all studies, it was found that drug association with vesicles was more effective than pretreatment with vesicles. *When using an equal thermodynamic driving force, these findings strongly suggest that the vesicles do not act just as penetration enhancers [76] but that when choosing the proper vesicle composition, an additional effect can be expected.*

The size and lamellarity of liposomes may affect the skin penetration rate. In a study by du Plessis et al. [82], the effect of vesicle size on drug disposition was evaluated. They found no favored disposition of the drug in lower skin strata and the receiver compartment when the drug was applied in small vesicles compared to large vesicles. The authors concluded that intact penetration of liposomes does not occur. Furthermore, it was found that vehicle lamellarity had little effect on the disposition of cholesterol and cholesterol sulfate in the various skin strata. In a more recent study, the role of the size of vesicles on transport was studied by ESR. They found that smaller vesicles disintegrate faster in the skin than large liposomes [83].

It seems that the physical parameters such as vesicle size and lamellarity are less important than the application method and the thermodynamic state of the bilayers. These findings are in favor of *the absence of intact vesicle* penetration across the skin. There are a number of studies [84–88] in which liposomes have been compared to lotion, creams, or penetration enhancers. Although information can be obtained about the relative effectiveness of liposomes compared with other formulations, no information can be obtained about mechanisms because in these studies (1) the thermodynamic activity of the drug in the various formulations might differ quite extensively and (2) the different components in the liposomes, lotions, or creams might have different interactions with the skin. Although these studies are of interest because they provide insight into the possible applications of liposomes as drug formulation compared to other ointments or creams that might already be on the market, these studies will not be discussed in this chapter.

Touitou et al. [89] compared penetration enhancers with liposomes; they found that liposomes mainly deposit the drug in the skin and therefore can act as an excellent reservoir, whereas the penetration enhancers increased the transcutol transport through the skin. A few years ago, the terms proliposomes and proniosomes were introduced. In the case of proliposomes, lecithin was mixed with sorbitol particles: in the case of proniosomes, surfactants were mixed with levonorgestrel. In both systems, vesicles were formed upon hydration [90,91].

C. Vesicles Affect Skin Penetration in Vivo

In 1986 Komatsu et al. [92] applied butylparaben to the skin in a liposome formulation prepared from egg phosphatidylcholine, cholesterol, and dicetylphosphate. The liposomes were applied occlusively to the dorsal skin of guinea pig in vivo. The fate of the liposomes was determined using autoradiography. The radioactive DPPC (dipalmitoylphosphatidylcholine) was found mainly on the application site, even after 48 hours; in contrast, the applied ^{14}C butylparaben radioactivity was found in the small intestine, feces, gallbladder, and urinary bladder. In an additional in vitro study [93], the mechanisms involved in ^{14}C butylparaben penetration was examined very systematically using a flow-through diffusion cell. The amount of lipids as well as the amount of ^{14}C butylparaben was varied. The investigators found that (1) the percentage ^{14}C butylparaben transported across the skin decreased as a function of increasing amount of lipids, and (2) the transport rate of radioactive phospholipids (DPPC) across guinea pig skin was much lower than that of ^{14}C butylparaben, indicating that the lipids remain in the SC and that co-penetration of butylparaben and phospholipids to deeper layers in the skin did not occur, which confirmed the results of their in vivo experiments [92].

Again it was concluded that most likely the liposomes do not penetrate as intact entities across the skin. From the finding that the ^{14}C butylparaben flux decreased at increasing lipid content, it was concluded that only ^{14}C butylparaben dissolved in the water phase contributed to the percutaneous penetration. However, another mechanism may also play a role. An increase in the lipid content reduced the thermodynamic activity of butylparaben in the bilayers, which might lead to a decrease in the driving force from the lipid phase to either the water phase or the SC. This may result in a reduced penetration through the skin as well.

In another more recent study [94], double labeling was carried out. ^{3}H phosphatidylcholine and ^{14}C tretinoin were intercalated in soybean lecithin. It was found that the ratio of the labels in SC was approximately constant, and that the ^{3}H/^{14}C ratio in epidermis was lower and decreased steeply until a skin depth of approximately 200 µm was reached. The authors concluded that co-penetration of a drug-liposome bilayer is possible in the SC, but that based on the reduced ^{3}H/^{14}C ratio in deeper skin strata, a separate diffusion of the drug and liposomal bilayers in these layers occurs.

In a series of studies carried out in the early 1990s, it was clearly demonstrated that the transport of protein across the skin was facilitated by NAT 106 liposomes [95] and that these liposomes also increased the penetration of ^{35}S-heparine and ^{99m}Tc [96]. In a related study, it was shown that these liposome suspensions also resulted in a decrease in the corneal blood supply as measured by laser Doppler flowmetry [97], indicating that *empty* liposomes induce changes in the lower regions of the skin. When examining the composition of the NAT 106 formulation, in which single and double chain phospholipids are present, it is possible that these vesicles consist of elastic bilayers and therefore approaches the properties of Transfersomes (see below).

The group of Cevc published a series of papers on Transfersomes, vesicles prepared from lipids and an edge activator that might be a single-chain lipid or surfactant. The edge activator makes the vesicles very elastic. It is stated that as a result of the hydration force in the skin Transfersomes can be squeezed through SC lipid lamellar regions [80]. Investigators compared occlusive with nonocclusive application and found nonocclusive application more effective than occlusive application (see above). This is in line with their theory on lipid transport under the water gradient based driving force. In order to detect the mass flow across the skin and the disposition in the animal, they used tritiated DPPC as a radioactive compound and reported that Transfersomes were much more effective than the "conventional" rigid liposomes when applied nonocclusively. After 8 hours of Transfersome application in mice, significant amounts were found in blood (approximately 8% of the *applied* dose) and liver (20% of the *recovered* dose). The group claimed that 50–90% of the dermally deposited lipids can be transported beyond the level of the SC. Although there is no doubt that their Transfersomes have clear advantages with respect to increasing the transport of active material *across the skin*, it was claimed in these studies that the elastic Transfersomes penetrated *intact* through SC and viable skin into the blood circulation. This statement, in particular, resulted in much skepticism in the research field. In fact, this statement was and is still not proven. The skepticism is also caused by the knowledge that such large associates are extremely difficult to transport across the skin as intact entities, especially because the SC consists of a very tight structure that is designed to act as a barrier (see previous paragraph). In more recent studies of this group, several drugs were associated with the vesicles. In the first published study on drugs associated with Transfersomes [98], the Transfersomes suspension contained relatively high amounts of lidocaine or tetracaine. The Transfersomes were tested on rats and on humans and in both studies the Transfersomes appeared to be more efficient than the conventional liposomes or solutions. The differences found using corticosteroids were somewhat less encouraging [99]. In another report, insulin was associated with Transfersomes [100,101]. Blood glucose levels could be lowered after about 3 hours of application. A clear difference in glucose level in blood was observed between application of insulin associated with micelles, conventional liposomes, or Transfersomes in the sense that Transfersomes resulted in a significantly lower glucose level. This was observed in studies of mice, but in more recent studies of humans, a decrease in glucose level also was observed [102]. Not only relatively large molecules such as insulin are claimed to be transported but even big molecules such as FITC-BSA [103] or I-BSA can be transported across the skin when associated with Transfersomes. The biodistribution of the derived radioactivity after an 8-hour application to mice skin of I-BSA associated with Transfersomes is very similar to that of Transfersome-associated I-BSA injected subcutaneously. These results are very encouraging and certainly show that Transfersomes indeed have advantages over vesicles prepared from only double-chained phospholipids and cholesterol. It would be of extreme interest to study the mechanisms involved in these vesicles in more detail.

A very extensive and well-designed study was described by Ogiso et al. [104], who compared gel systems in which D-limonene or laurocapram was present as a penetration enhancer and studied the transport of beta-histine in the presence or absence of liposomes in the gels. The liposomes were prepared from either egg phosphatidylcholine or hydrogenated soybean phosphatidylcholine. The bioavailability of the drug in the presence of the egg PC liposomes was higher than in the absence of the egg phosphatidylcholine liposomes, confirming earlier reports. It seems that the fluidity of the membranes is an important factor for the penetration enhancement, which confirmed the results of earlier studies [71,72,76–78]. Furthermore, lipid analysis revealed that the amount of ceramides decreased dramatically in the SC after application of a d-limonene–containing formulation, indicating extraction of lipids from the skin, and the phospholipid content in the SC increased after liposomes treatment. This suggests that the liposomes replace the ceramides in the SC. The investigators assumed that in their formulations the replacement of ceramides by phospholipids plays an important role in the mechanisms involved in the increased penetration of beta-histine.

D. Visualization and Related Studies

Several studies were performed in order to get a detailed insight into the interactions between vesicles and skin. Most studies focused on penetration characteristics of vesicles or vesicle constituents.

The fate of liposomes after application to the skin was studied by γγ angular correlation spectroscopy with Indium encapsulated in liquid-state vesicles. The measurements revealed that the vesicles disintegrate on the surface of the skin and that only a negligible amount of marker penetrated into the deeper layers of the SC. These results are in agreement with the results of studies carried out by Lasch et al. [105]: they found, using fluorescence spectroscopy, that no intact vesicle penetration occurred. In a more recent study, vesicle-skin interactions were examined using confocal laser scanning microscopy (CLSM) [106]. A large number of liposome compositions were examined. These studies revealed that the liposome constituents were present only in the outermost layers of the SC and that the constituents acted only as penetration enhancers.

In additional studies [107], the interactions between liposomes and skin were examined by freeze fracture electron microscopy and Fourier transformed infrared spectroscopy (FTIR). For this purpose, human SC was treated with liposomes prepared from either NAT 106, NAT 50, or NAT 89. The liposomes prepared from NAT 50 fused on the surface of the skin, whereas vesicles prepared from NAT 89 penetrated between the upper SC cell layers. In addition, quite frequently distinct regions with a rough surface were detected, indicating a mixing between the SC lipids and the liposomes or separate domains of phospholipids. Application of vesicles prepared from NAT 106 revealed large changes in the lipid organization in the SC (see Fig. 8). Almost no intact lipid bilayers were observed between the corneocytes. The intercellular regions were characterized by "flattened isolated regions."

It seems that liquid-state liposomes prepared from different lipid mixtures result in different interactions with the SC. Interestingly, the NAT 106 contained the largest fraction PC. The strong interaction with the SC might be explained by the presence of a single-chain lipid, lysophosphatidyl choline, in this formulation, which may act as an edge activator and consequently may increase the elasticity of the liposomal bilayers. In a related study [108], in vivo skin hydration was monitored after occlusive application of either NAT 106 liposomes prepared in D_2O or after application of pure D_2O. The skin hydration and penetration were monitored by FTIR. The liposomes were superior compared to pure D_2O in driving D_2O into the skin, and the phospholipid components could be detected in

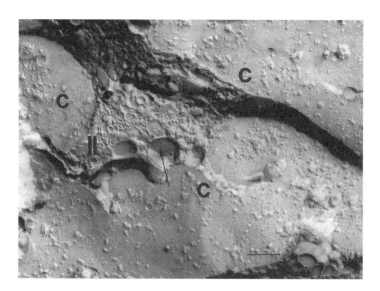

Figure 8 Nat 106 liposomes had a very strong effect on the microstructure of the stratum corneum. The corneocytes (C) were swollen considerably and the smooth ultrastructure of the intercellular lipid lamellae (LL) showed flattened spherical structures, see arrow. The linear arranged keratin filaments along the cell boundary of the corneocytes are absent. The scale bar indicates 0.1 μm.

deeper layers in the SC. The presence of phospholipids deep in the SC confirmed the results observed with freeze fracture electronmicroscopy, in which strong interactions between phospholipid layers and SC were observed. In a study by Korting et al. [109], liposomes prepared from soya-lecithin were prepared and applied on reconstructed epidermis. After application, the skin was visualized on an ultrastructural level using OsO_4 staining and fixation. The investigators observed no intact liposomes in the lower layers of the SC, but disposition of lipids derived from liposomes between and within the corneocytes. These findings are in agreement with those of Hofland et al. [105]. Korting et al. did not agree with the interpretations of Foldvari et al. [110], who claimed that intact liposomes penetrate into the skin. Vrhonik et al. [111] found that dermal delivery increased when the substance is applied in vesicles with a bilayer domain structure. In another study, the interaction between vesicles prepared from cholesterol and surfactants was investigated [112]. When applied occlusively to human skin, the vesicles absorbed onto the SC surface and frequently changed the lipid organization between the SC superfacial cell layers. Occasionally, changes in the deeper layers of the SC, such as the presence of vesicular structures, were observed. The authors asserted that penetration of intact vesicles in the SC would be unlikely to occur and explained the presence of vesicles as the penetration of vesicle constituents that might be able to re-form vesicle in the SC. Various water pools were observed in the lipid regions between the corneocytes, indicating that phase separation between SC lipids and water occurs. This is in agreement with the studies of van Hal et al., who claimed that also in the absence of surfactants, water pools were observed. In a more recent study [113], it was found that the vesicles were only very occasionally present in the deeper layers of the skin and, therefore, are not expected to affect the diffusion characteristics of drugs across the skin. In two recent studies, CLSM was used to study the penetration of fluorescent labels across the skin in more detail. Schätzlein and Cevc [114,115] intercalated Rhodamine-DHPE (1,2-dihexadecanoyl-sn-glycero-3-phosphatidylethanolamine-N-lissamine rhodamine B sulfonyl, triethylammonium salt) in the bilayers of Transfersomes prepared from soya phosphatidylcholine and sodium cholate. The suspensions were applied nonocclusively on mice skin, and after 4 to 12 hours, the skin was examined ex vivo using CLSM. The investigators reported the presence of transport routes in the SC, the inter-cluster pathway between groups of cells, which should be a preferred route of transport between corneocytes that only partly overlap. However, the skin surface is not flat but contains many wrinkles. In recent experiments from our own group, we found also "clefts" with a bright fluorescent appearance. However, in our studies these clefts may correspond to wrinkles in the skin [116]. Intra-cluster pathways were also observed by Schätzlein and by Cevc [107,108], because regions are observed with high fluorescence intensity across the lipid bilayer regions. The authors interpret these regions as being virtual pores between the corneocytes. In all interpretations, one should realize that CLSM is a technique that *cannot* be used to visualize the transport of vesicles as intact entities—it can only be used to visualize the transport of the label, which is not necessarily still associated with the vesicle. In addition, it was reported that by using in vitro time-resolved infrared ATR (attenuated total reflection) spectroscopy, Transfersomes resulted in a faster skin penetration than liposomes. However, whether intact vesicles or liposomal fragments penetrate is not clear from these experiments either, because in this case only the lipids are detected and not the structure that is formed by the lipids [117]. Recently, estradiol penetration across human skin after application in Transfersomes was studied in vitro [118]. In this excellent study, the ratio between the lipids and surfactants was varied systematically in order to vary the elasticity of the bilayers. A clear correlation was found between the molar ratio of the bilayer and surfactant-forming lipids and the penetration rate of estradiol across the skin, suggesting that indeed the Transfersomes increased the transport across the skin more than the traditional liposomes or micelles.

Very recently the fate of vesicles prepared from single and double chain surfactants has been studied. These vesicles have *elastic* bilayers but are entirely

Structure of SC Lipid Layers and Interactions with Lipid Liposomes 59

prepared from surfactants. The vesicles that were most intensively studied were prepared from a sucrose ester surfactant (double chain surfactant), polyoxyethylene laurate ester and cholesterol sulfate. These vesicles were studied after application onto human skin in vitro nonocclusively [119]. Studies were carried out using RuO_4 fixation in combination with thin sectioning and transmission electron microscopy (TEM). Furthermore, the penetration pathway of a label intercalated in the bilayers was visualized by two-photon excitation fluorescent microscopy. The TEM results showed three types of interactions after nonocclusive application of the elastic vesicles on human skin in vitro.

1. The presence of spherical lipid structures containing or surrounded by electron dense spots, indicative of the presence of vesicle material, both on the surface of the skin and in between the upper 3–4 cell layers.
2. Oligolamellar vesicles were observed between the 2–3 upper corneocytes.
3. Large areas containing lipids, surfactants, and electron-dense spots were observed deeper down into the SC.

Furthermore, after treatment with vesicles containing PEG-8-L and a saturated C12-chain surfactant, small stacks of bilayers were found in the intercellular spaces of the SC. Treatment with conventional vesicles affected the most apical corneocytes only to some extent, whereas no lamellar stacks, oligolamellar vesicles, or vesicular material were observed deeper down in the SC.

In addition, none of the vesicle formulations affected the viable epidermis or dermis. Although not that frequently observed in vitro, in one donor treated with elastic vesicles a very peculiar phenomenon was observed. Namely, large areas of lamellar stacks were visualized that differed in appearance from the SC lipid bilayers and were present throughout the entire stratum corneum. In some regions, these lamellar stacks disorganized the intercellular skin lipid bilayers (Fig. 9,a,b) by pushing them apart. The bilayers in these stacks were occasionally oriented perpendicular to the bilayers of the SC (Fig. 9c) or were accumulated and oriented randomly (Fig. 9d).

This phenomenon was observed very frequently in SC of hairless mouse skin [120] when applied in vivo (nonocclusively). In various regions, isolated areas of stacks had a different appearance than the bilayers observed in SC (see Fig. 10).

It was speculated that under the influence of the hydration force, elastic vesicles partition into the stratum corneum, after which (due to the elasticity and the reduced water content in the stratum corneum) the vesicles might easily flatten by releasing water from the interior of the vesicles. When these flattened vesicles fuse together, the bilayer stacks are formed. Because these stacks contain mainly liquid-state bilayers, the stacks remain phase-separated from the crystalline lipid

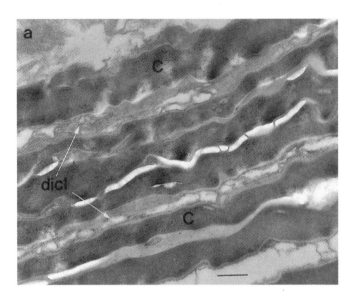

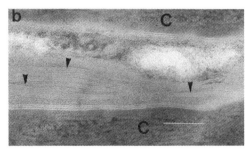

Figure 9 TEM micrographs of human skin incubated nonocclusively with PEG-8-L: L-595 vesicles (70:30) for 16 hours. (a) Overview of stratum corneum with disorganized intercellular lipids (dicl). Scale bar represents 200 nm. (b) A detailed micrograph of disorganized intercellular lipid bilayers. Bilayers appear to be detached from each other in units of 3 bilayers (arrowheads). Scale bar represents 100 nm. (c) Frequently, stacks of bilayers which resemble stacked flattened vesicles were observed with their bilayers oriented perpendicular to skin lipid bilayers and cell membranes (arrows); arrowhead = unit of 3 skin lipid bilayers. Scale bar represents 100 nm. (d) Detailed micrographs of intercellular spaces filled with round and oval stacks of bilayers (arrows) oriented at random. Scale bar represents 100 nm.

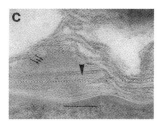

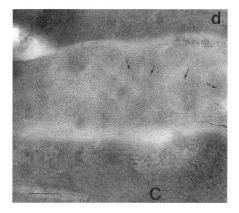

Figure 9 Continued

lamellae from the stratum corneum. A schematic overview of how such a lamellar stack may be assembled is illustrated in Figure 11. Possibly, the vesicles flatten in the direction in which the water will be removed most easily from the vesicles, which is most probably water diffusion along the head groups. This speculation might justify the perpendicular orientation of the lamellar stacks compared to the skin bilayers. Once the skin lipid bilayers are disrupted, the lamellar stacks are oriented randomly.

In addition to these extraordinary features, two-photon excitation microscopy revealed that the fluorescent label intercalated in the bilayers of conventional vesicles resulted in a homogeneous distribution of the label in the intercellular lipid domains. However, when the same label was intercalated in the bilayers of the elastic vesicles, an inhomogeneous label distribution was observed (see Fig. 12). It seems that thread-like channels are formed that serve as penetration pathways for the dyes. These results certainly demonstrate that elastic vesicles exert another interaction with the SC than the conventional vesicles. It would be

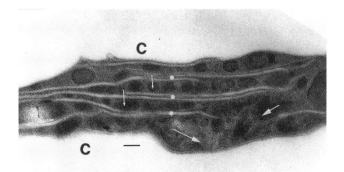

Figure 10 Transmission electron micrographs of hairless mouse skin treated with liquid-state PEG-8-L:L-595:CS (70:30:5) vesicles, non-occlusively. Detailed micrographs of skin treated for 6 hours. Lamellar stacks are present in the intercellular spaces. The lamellar stacks are often oriented perpendicular to the lipid bilayers (arrows) and seem to 'push' the lipid bilayers apart. These lamellar stacks were observed after 1, 3 and 6 hours of application. Arrows = lamellar stack; * = lipid bilayer; C = corneocyte. Bar represents 100 nm.

of great interest to gain more information on the fate of these vesicles in vivo in humans.

E. Conclusions

Considering all the studies performed, we can generally conclude that drug transport can be adjusted on demand by association of the drug with vesicles. One of the central parameters is the thermodynamic state of the bilayers, which affects quite dramatically drug permeation across the SC and, in addition, the interactions with the SC. It seems that the size and the lamellarity of vesicles affect drug

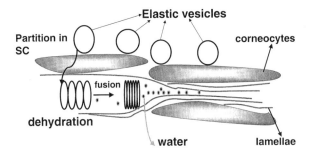

Figure 11 Schematic presentation of the possible working mechanism of elastic vesicles in SC.

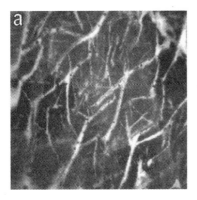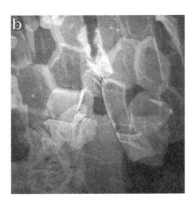

Figure 12 TPE images of skin treated with PEG-8-L:L-595:CS (70:30:5) liquid-state vesicles. Threadlike channels were formed in the entire stratum corneum after application with elastic vesicles non-occlusively (a). In contrast, application of rigid vesicles non-occlusively (b) resulted in a homogenous distribution of the fluorescent dye in the lipid regions.

transport only slightly. In general, it can be concluded that using "traditional" liposomes, intact vesicle transport across SC does not occur. This is concluded from visualization and permeation studies as well as from biophysical measurements. Drug transport increases when single-chain surfactants are intercalated in the vesicles. These single-chain surfactants can act as penetration enhancers, such as in gel-state vesicles, or might result in a decrease in the interfacial tension and make the vesicles more deformable. Compared to traditional liquid-state vesicles, these deformable vesicles, often referred to as Transfersomes, have been shown to increase drug transport across the skin even further. However, questions arise about the mechanism involved in this increased drug transport. A first step has been taken in studies of van den Bergh [119,120]. However, it would be of high interest to determine the fate of the liposomes in human skin in vivo. Only then can more clear indications be made about the working mechanisms of these vesicles.

VI. APPENDIX

Using the x-ray diffraction technique, the scattered intensities are measured as a function of $\theta/2$, the scattering angle (see Fig. 13). The intensity of the scattered x-rays as a function of θ is directly related to the electron density differences in the sample. If the electron density differences have a repeating pattern, the diffraction pattern is characterized by a series of peaks (intensity maxima of scattered x-rays) (see Figs. 14 and 15).

Examine the structure:

⇒ **X-ray Diffraction**

Wide Angle (WAXD)

Small Angle (SAXD)

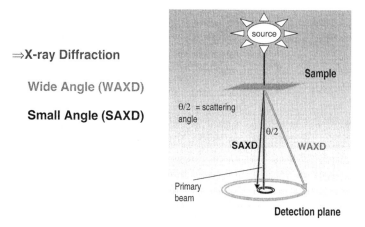

Figure 13 A schematic presentation of the x-ray diffraction technique.

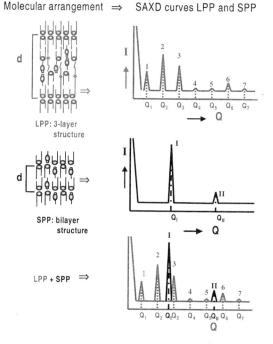

Figure 14 The relationship between a lamellar phase periodicity (d) and its diffraction pattern. In the diffraction curve the intensity is plotted as function of Q. Q is a measure for the scattering angle. n is the order of the diffraction peak denoted by 1, 2, 3 etc. (long periodicity) or I and II (short periodicity). An increase in d results in a smaller inter-peak distance in the diffraction pattern.

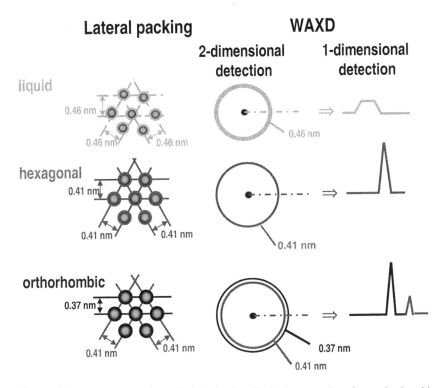

Figure 15 The position of the alkyl chains in a liquid, hexagonal, and an orthorhombic lattice (direction perpendicular to the chains) and the related diffraction pattern.

REFERENCES

1. Wertz PW, Downing DT. Stratum corneum: biological and biochemical considerations. In: Hadgraft J, Guy RH, eds. Transdermal Drug Delivery. New York, Basel: Marcel Dekker, 1989:1–17.
2. Simonetti O, Hoogstraate AJ, Bialik W, Kempenaar JA, Schrijvers AHGJ, Bodde HE, Ponec M. Visualization of diffusion pathways across the stratum corneum of native and in-vitro-reconstructed epidermis by confocal laser scanning microscopy. Arch Dermatol Res 1995; 287:465–473.
3. Meuwissen MEMJ, Janssen J, Cullander C, Junginger HE, Bouwstra JA. A cross-section device to improve visualization of fluorescent probe penetration into the skin by confocal laser scanning microscopy. Pharm Res 1998; 15:352–356.

4. Boddé H, Kruithof M, Brussee J, Koerten H, Visualisation of normal and enhanced HgCl2 transport through human skin in vitro. Int J Pharm 1989; 253: 13–24.
5. Potts RO, Guy RH. Predicting skin permeability. Pharm Res 1992; 9:663–669.
6. Ponec M, Weerheim A, Kempenaar J, Mommaas AM, Nugteren DH. Lipid composition of cultured human keratinocytes in relation to their differentiation. J Lipid Res 1988; 29:949–996.
7. Wertz PW, Downing DT. Physiology, biochemistry and molecular biology in the skin, 2d ed. In: Goldsmith LA, ed. Oxford/New York: Oxford University, 1991; 205–236.
8. Schurer NY, Elias PM. The biochemistry and function of stratum corneum. In: Elias PM, ed. Advances in Lipid Research. Academic Press, 1991; 24:27–56.
9. Robson KJ, Stewart ME, Michelsen S, Lazo ND, Downing DT. 6-Hydroxy-4-sphingenine in human epidermal ceramides. J Lipid Res 1994; 35:2060–2068.
10. Wertz PW, Downing DT. Ceramides of pig stratum epidermis, structure determination. J Lipid Res 1983; 24:753–758.
11. Swanbeck G, Thyresson N. An x-ray diffraction study of scales from different dermatosis. Acta Derm Venereol 1961; 41:289–296.
12. Swanbeck G, Thyresson N. A study of the state of aggregation of the lipids in normal and psoriatic horny layer. Acta Derm Venereol 1962; 42:445–557.
13. Breathnach AS, Goodman T, Stolinsky C, Gross M. Freeze fracture replication of cells of stratum corneum of human cells. J Anat 1973; 114:65–81.
14. Breathnach AS. Aspects of epidermal structure. J Invest Dermatol 1975; 65:2–12.
15. Madison KC, Swarzendruber DC, Wertz PW, Downing DT. Presence of intact intercellular lipid lamellae in the upper layers of the stratum corneum. J Invest Dermatol 1987; 88:714–718.
16. Hou SY, Mitra AK, White SH, Menon GK, Ghadially R, Elias P. Membranes structure in normal and essential fatty acid-deficient stratum corneum; characterization of ruthenium tetroxide staining and X-ray diffraction. J Invest Dermatol 1991; 96: 215–223.
17. van den Bergh BAI, Swartzendruber DC, Bos-van der Geest A, et al. Development of an optimal protocol for the ultrastructural examination of skin by transmission electron microscopy. J Microsc 1997; 187:125–133.
18. Downing DT. Lipid and protein structures in the permeability barrier of mammalian epidermis. J Lipid Res 1992; 33:301–314.
19. Fartasch M, Bassakas ID, Diepgen TL. Disturbed extrusion mechanism of lamellar bodies in dry non-eczematous skin atopics. Br J Dermatol 1992; 127:221–227.
20. Swartzendruber DC. Studies of epidermal lipids using electron microscopy. Semin Dermatol 1992; 11:157–161.
21. van Duzee BF. Thermal analysis of human stratum corneum. J Invest Dermatol 1975; 65:404–408.
22. Gay CL, Guy RH, Golden GM, Mak VM, Francoeur ML. Characterization of low-

temperature (i.e., <65 degrees C) lipid transitions in human stratum corneum. J Invest Dermatol 1994; 103:233–239.
23. Ongpipattanakul B, Burnette R, Potts RO. Evidence that oleic acid exists as a separate phase within stratum corneum. Pharm Res 1991; 8:350–354.
24. White SH, Mirejovsky D, King GI. Structure of lamellar lipid domains and corneocytes envelopes of murine stratum corneum: an X-ray diffraction study. Biochemistry 1988; 27:3725–3732.
25. Bouwstra JA, Gooris GS, van der Spek JA, Bras W. The structure of human stratum corneum as determined by small angle X-ray scattering. J Invest Dermatol 1991; 96:1006–1014.
26. Bouwstra JA, Gooris GS, Bras W, Downing DT. Lipid organization in pig stratum corneum. J Lipid Res 1995; 36:685–695.
27. Bouwstra JA, Gooris GS, Salomons- de Vries MA, van der Spek JA, Bras W. Structure of human stratum corneum as a function of temperature and hydration: a wide-angle x-ray diffraction study. Int J Pharm 1992; 84:205–216.
28. Bouwstra JA, Gooris GS, van der Spek JA, Lavrijsen S, Bras W. The lipid and protein structure of mouse stratum corneum: A wide and small angle diffraction study. Biochim Biophys Acta 1994; 1212:183–192.
29. Bouwstra JA, Gooris GS, Dubbelaar FER, Weerheim A, Ponec M. pH, cholesterol sulfate and fatty acids affect stratum corneum lipid organisation. J Invest Dermatol Symposium proceedings, 1998; 3:69.
30. Lampe MA, Williams ML, Elias P. Human epidermis lipids: characterization and modulation during differentiation. J Lipid Res 1983; 24:131–140.
31. Wertz PW, Downing DT. In: Goldsmith LA, ed. Physiology, Biochemistry, and Molecular Biology of the Skin. 2d ed. Oxford University Press, 1991; 205–236.
32. ten Grotenhuis E, Demel RA, Ponec M, de Boer DR, van Miltenburg JC, Bouwstra JA. Phase behaviour of stratum corneum lipids in Langmuir-Blodget monolayers. Biophys J 1996; 71:1389–1399.
33. Fenske DB, Thewalt JL, Bloom M, Kitson N. Models of stratum corneum intercellular membranes: 3H NMR of microscopically oriented multilayers. Biophys J 1994; 67:1562–1573.
34. Kitson N, Thewalt J, Lafleur M, Bloom M. A model membrane approach to the epidermal permeability barrier. Biochemistry 1994, 33:6707–6715.
35. Engelman DM, Rothman J. The planar organization of lecithin-cholesterol bilayers. Nature 1972; 247:3694–3697.
36. Demel RA, de Kruyff B. The function of sterols in membranes. Biochim Biophys Acta 1976; 457:109–132.
37. Liu F, Sugar IP, Chong L-G. Cholesterol and ergosterol superlattices in three-component liquid crystalline lipid bilayers as revealed by dehydroergosterol fluorescence. Biophys J 1997; 72:2243–2254.
38. Nielsen M, Miao L, Ipsen JH, Zuckermann MJ, Mouritsen OG. Off-lattice model for the phase behaviour of lipid-cholesterol bilayers. Physical Reviews 1999; E 59: 5790–5803.
39. Engblom J, Engström S, Fontell K. The effect of the skin penetration enhancer

Azone® on fatty acid-sodium soap-water mixtures. J Control Rel 1995; 33:299–305.
40. Thewalt J, Kitson N, Araujo C, MacKay A, Bloom M. Models of stratum corneum intercellular membranes: the sphingolipid headgroup is a determinant of phase behavior in mixed lipid dispersions. Biochem Biophys Commun 1992; 188:1247–1252.
41. Kitson N, Thewalt J, Lafleur M, Bloom M. A model membrane approach to the epidermal permeability barrier. Biochemistry 1994; 33:6707–6715.
42. Bouwstra JA, Thewalt J, Gooris GS, Kitson N. A model membrane approach to the epidermal permeability barrier: an x-ray diffraction study. Biochemistry 1997; 36:7717–7725.
43. Schückler F, Bouwstra JA, Gooris GS, Lee G. An x-ray diffraction study of some model stratum corneum lipids containing azone and dodecyl-L-pyroglutamate. J Contr Rel 1993; 23:27–36.
44. Moore DJ, Rerek ME. Insights into the molecular organisation of lipids in the skin barrier from infrared spectroscopy studies of stratum corneum lipid models. Acta Derm Venereol, Supp 2000; 208:16–22.
45. Moore DJ, Rerek ME, Mendelsohn R. Lipid domains and FTIR spectroscopy studies of the conformational order and phase behaviour of ceramides. J Phys Chem 1997; 101:8933–8940.
46. Neubert R, Rettig W, Warterig S, Wegener M, Wienholdt A. Structure of stratum corneum lipids characterized by FT-Raman spectroscopy and DSC. II. Mixtures of ceramides and saturated fatty acids. Chem Phys Lipids 1997; 89:3–14.
47. Hatfield RM, Fung LW-M. A new model system for lipid interactions in stratum corneum vesicles: effect of lipid composition, calcium and pH. Biochemistry 1999; 38:784–791.
48. Hatfield RMI, Fung LW-M. Molecular properties of a stratum corneum model lipid system: large unilamellar vesicles. Biophys J 1995; 68:196–207.
49. de la Maza A, Lopez O, Coderch L, Parra JL. Permeability changes in liposomes modeling the stratum corneum lipid composition due to C12-alkyl betaine/sodium dedecyl sulfate mixtures. Int J Pharm 1998; 171:63–74.
50. Abraham W, Wertz PW, Landmann L, Downing DT. Stratum corneum lipid liposomes: calcium-induced transformation into lamellar sheets. J Invest Dermatol 1987; 88:212–214.
51. Kuempel D, Swarzendruber DC, Squier CA, Wertz PW. In vitro reconstruction of stratum corneum lipid lamellae. Biochim Biophys Acta 1998; 1372:135–140.
52. Abraham W, Downing DT. Lamellar structures formed by stratum corneum lipids in vitro: A deuterium nuclear magnetic resonance study. Pharm Res 1992; 9:1415–1421.
53. Bouwstra JA, Gooris GS, Cheng K, Weerheim A, Bras W, Ponec M. Phase behavior of isolated skin lipids. J Lip Res 1996; 37:999–1011.
54. Bouwstra JA, Gooris GS, Dubbelaar FER, Weerheim A, Ponec M. pH, cholesterol sulfate and fatty acids affect stratum corneum lipid organisation. J Invest Dermatol Proceedings 1998; 3:69–74.
55. Bouwstra JA, Gooris GS, Dubbelaar FER, Weerheim A, Ponec M. Cholesterol

sulfate and calcium affect stratum corneum lipid organisation over a wide temperature range. J Lip Res 1999; 40:2303–2312.
56. Bouwstra JA, Gooris GS, Dubbelaar FER, Weerheim AM, Ijzerman AP, Ponec M. Role of ceramide 1 in the molecular organization of the stratum corneum lipids. J Lipid Res 1998; 39:186–196.
57. Bouwstra JA, Cheng K, Gooris GS, Weerheim A, Ponec M. The role of ceramides 1 and 2 in the stratum corneum lipid organisation. Biochim Biophys Acta 1996; 1300:177–186.
58. Bouwstra JA, Gooris GS, Dubbelaar FER, Ponec M. Phase behaviour of skin barrier model membranes at pH 7.4, 2000; 46:979–992.
59. Motta S, Monti M, Sesana S, Mellesi L, Caputo R, Carelli S, Ghidoni R. Ceramide composition of the psoriatic scale. Biochim Biophys Acta 1993; 1182:147.
60. Lavrijsen APM, Bouwstra JA, Gooris GS, Boddé HE, Ponec M. Reduced skin barrier function parallels abnormal stratum corneum lipid organisation in patients with lamellar ichthyosis. J Invest Dermatol 1995; 105:619–624.
61. Elias PM, Williams ML, Maloney ME, Bonifas JA, Brown BE, Grayson S, Epstein EH. Stratum corneum lipids in disorders of cornification. J Clin Invest 1984; 74:1414–1421.
62. Melton JL, Wertz PW, Swartzendruber DC, Downing DT. Effects of essential fatty acid deficiency on epidermal O-acylsphingolipids and transepidermal water loss in young pigs. Biochim Biophys Acta 1987; 921:191–197.
63. Bouwstra JA, Dubbelaar FER, Gooris GS, Weerheim AM, Ponec M. The role of ceramide composition in the lipid organisation of the skin barrier. Biochim Biophys Acta 1999; 1419:127–136.
64. Mak VM, Potts RO, Guy RH. Does hydration affect intercellular lipid organisation in the stratum corneum? Pharm Res 1991; 8:1064–1065.
65. Alonso A, Meirelles N, Tabak M. Effect of hydration upon the fluidity of intercellular membranes of stratum corneum: an EPR study. Biochim Biophys Acta 1995; 1237:6–15.
66. van Hal DA, Jeremiasse EE, Junginger HE, Spies F, Bouwstra JA. Structure of fully hydrated human stratum corneum: a freeze fracture electron microscopy study. J Invest Dermatol 1996; 106:89–95.
67. Mezei M, Gulasekharam V. Liposomes—a selective drug delivery system for the topical route of administration. I. Lotion dosage forms. Life Science 1980; 26:1473–1477.
68. Mezei M, Gulasekharam V. Liposomes—a selective drug delivery system for the topical route of administration. II. Gel dosage form. J Pharm Pharmacol 1982; 34:473–474.
69. Ho NFH, Ganesan MG, Flynn GL. Mechanism of topical delivery of liposomally entrapped drugs. J Contr Rel 1985; 2:61–65.
70. Ganesan MG, Weiner ND, Flynn GL, Ho NFH. Influence of liposomal drug entrapment on percutaneous absorption. Int J Pharm 1984; 20:139–154.
71. Knepp VM, Hinz RS, Szoka FC, Guy RH. Controlled drug release from a novel liposomal delivery system. I. Investigations of transdermal potential. J Contr Rel 1988; 5:211–221.

72. Knepp VM, Szoka FC, Guy RH. Controlled drug release from a novel liposome delivery system. II. Transdermal delivery characteristics. J Contr Rel 1990; 12:25–30.
73. Lasch J, Bouwstra JA. Interactions of external lipids (lipid vesicles) with the skin. J Liposome Res 1995; 5:543–569.
74. Foldvari M, Jarvis B, Ogueijofor CJN. Topical dosage form of liposomal tetraciane: effect of additives on the in vitro release and in vivo efficacy. J Contr Rel 1993; 27:193–205.
75. Dowton SM, Hu Z, Ramanchandran C, Wallach DFH, Weiner N. Influence of liposomal composition on topical delivery of encapsulated cyclosporin A. I. An in vitro study using hairless mouse skin. STP Pharma Sciences 1993; 3:404–407.
76. Hofland HEJ, van der Geest R, Boddé HE, Junginger HE, Bouwstra JA. Estradiol permeation from nonionic surfactant vesicles through human stratum corneum in vitro. Pharm Res 1994; 11:659–664.
77. van Hal D, van Rensen A, de Vringer T, Junginger H, Bouwstra J. Diffusion of estradiol from non-ionic surfactant vesicles through human stratum corneum in vitro. S.T.P. Pharma Sci 1996; 6:72–78.
78. Meuwissen MM, Mougin L, Junginger HE, Bouwstra JA. Transport of model drugs through skin in vitro and in vivo by means of vesicles. Proceed Intern Symp Control Rel Bioact Mater 1996; 23:303–304.
79. Fresta M, Puglisi G. Corticosteroid dermal delivery with skin-lipid liposomes. J Contr Rel 1997; 44:141–151.
80. Cevc G, Blume G. Lipid vesicles penetrate into intact skin owing to the transdermal osmotic gradients and hydration force. Biochim Biophys Acta 1992; 1104:226–232.
81. van Kuyk-Meuwissen MEMJ, Junginger HE, Bouwstra JA. Interactions between liposomes and human skin in vitro, a confocal laser scanning microscopy study. Biochim Biophys Acta 1998; 1371:13–39.
82. Du Plessis J, Ramanchandran C, Weiner N, Muller DG. The influence of particle size of liposomes on the disposition of drugs into the skin. Int J Pharm 1994; 103:277–282.
83. Sentjurc M, Vrhovnik K, Kristl J. Liposomes as a topical delivery system: the role of size on transport studied by the EPR imaging method. J Contr Rel 1999; 59:87–97.
84. Lasch J, Wohlrab W. Liposome-bound cortisol: a new approach to cutaneous therapy. Biomed Biochim Acta 1986; 45:1295–1299.
85. Wohlrab W, Lasch J. Penetration kinetics of liposomal hydrocortisone in human skin. Dermatologica 1987; 174:18–22.
86. Michel C, Purmann T, Mentrup E, Seiller E, Kreuter J. Effect of liposomes on percutaneous penetration of lipophilic materials. Int J Pharm 1992; 84:93–105.
87. Singhy R, Vyass SP. Topical liposomal system for localized and controlled drug delivery. J Dermatol Sci 1996; 13:107–111.
88. Kim M-K, Chung S-J, Lee M-H, Cho A-R, Shim C-K. Targeted and sustained delivery of hydrocortisone to normal and stratum corneum-removed skin without enhanced skin absorption using a liposome gel. J Contr Rel 1997; 46:243–251.

89. Touitou E, Levi-Schaffer F, Dayan N, Alhaique F, Riccieri F. Modulation of caffeine skin delivery by carrier design: liposomes versus permeation enhancers. Int J Pharm 1994;131–136.
90. Vora B, Khopada AJ, Jain NK. Proniosome based transdermal delivery of levenorgestrol for effective contraception. J Contr Rel 1998; 54:149–165.
91. Hwang B-Y, Jung B-H, Chung S-J, Lee H-H, Shim C-K. In vitro skin permeation of nicotine from proliposomes. J Contr Rel 1997; 49:177–184.
92. Komatsu H, Higazi K, Okamoto H, Miyakwa K, Hashida M, Sezaki H. Preservative activity and in vivo percutaneous penetration of butylparaben entrapped in liposomes. Chem Pharm Bull 1986; 34:3415–3422.
93. Komatsu H, Okamoto H, Miyagawa K, Hashida M, Sezaki H. Percutaneous absorption of butylparaben from liposomes in vitro. Chem Pharm Bull 1986; 34:3423–3430.
94. Masini V, Bonte F, Meybeck A, Wepierre J. Cutaneous bioavailability in hairless rats of tretinoin in liposomes or gel. J Pharm Sci 1993; 82:17–21.
95. Artmann C, Roding J, Ghyczy M, Pratzel HG. Liposomes from soya phospholipids as percutaneous drug carriers. Arzneim Forsch Drug Res 40(II) 1990; 12:1363–1365.
96. Artmann C, Roding J, Ghyczy M, Pratzel HG. Liposomes from soya phospholipids as percutaneous drug carriers. Arzneim Forsch Drug Res 40(II) 1990; 12:1365–1368.
97. Gehring W, Ghyzhy M, Gloor M, Heitzler Ch, Roding J. Significance of empty liposomes alone and as drug carriers in dermatology. Arzneim Forsch Drug Res 40(II); 1990; 12:1368–1371.
98. Plana ME, Gonzalez P, Rodriquez L, Sanchez S, Cevc G. Noninvasive percutaneous induction of topical analgesia by a new type of drug carrier, and prolongation of local pain insensitivity by anesthetic liposomes. Anesth Analg 1992; 75:615–621.
99. Cevc G, Blume G, Schatzlein A. Transfersomes-mediated transepidermal delivery improves the regio-specificity and biological activity of corticosteroids in vivo. J Cont Rel 1997; 45:211–226.
100. Cevc G. Dermal insulin. In: Berger M, Gries A, eds. Frontiers in Insulin Pharmacology. Stutgart: Georg Thieme; 1993;61–74.
101. Cevc G, Gebauer D, Stieber J, Schatzlein A, Blume G. Ultraflexible vesicles, Transfersomes, have an extremely low pore penetration resistance and transport therapeutic amounts of insulin across the intact mammalian skin. Biochim Biophys Acta 1998; 1368:201–215.
102. Cevc G. Transfersomes, liposomes and other lipid suspensions on the skin: permeation enhancement, vesicle penetration, and transdermal drug delivery. Crit Rev Ther Drug Carrier Sys 1996; 13:257–388.
103. Paul A, Cevc G. Non-invasive administration of protein antigens: transdermal immunization with the bovine serum albumin in transfersomes. Vaccine Res 1995; 4:145.
104. Ogiso T, Niinaka N, Iwaki M. Mechanism for enhancement effect of lipid disperse system on percutaneous absorption. J Pharm Sci 1996; 85:57–64.

105. Lasch J, Laub R, Wohlrab W. How deep do intact liposomes penetrate into human skin? J Control Release 1991; 18:55–58.
106. Kirjaivainen M, Urtti A, Jääskelänen I, Suhonen TM, Paronen P, Valjakka-Koskela R, Kiesvaara J, Mönkkönen J. Interactions of liposomes with human skin in vitro—the influence of lipid composition and structure. Biochim Biophys Acta 1996; 1304: 179–189.
107. Hofland HEJ, Bouwstra JA, Bodde HE, Spies F, Junginger HE. Interactions between liposomes and human stratum corneum in vitro: freeze fracture electron microscopical visualization and small-angle x-ray scattering studies. Br J Derm 1995; 132:853–856.
108. Bodde HE, Pechtold LARM, Subnel MTA, de Haan FHN. Monitoring in vivo skin hydration by liposomes using infrared spectroscopy in conjunction with tape stripping. In: Braun Falco O, Korting HC, eds. Liposome Dermatics. Berlin: Springer Verlag; 1992; 137–149.
109. Korting HC, Stolz W, Schmid MH, Maierhofen G. Interactions of liposomes with human epidermis reconstructed in vitro. Br J Dermatol 1995; 132:571–579.
110. Foldvari M, Gesztes A, Mezei M. Dermal drug delivery by liposome encapsulation: clinical and electron microscopic studies. J Microencaps 1990 7:479–489.
111. Vrhovnik K, Kristl J, Senjurc M, Smid-Korbar J. Influence of liposome bilayer fluidity on the transport of encapsulated substance into the skin as evaluated by EPR. Pharm Res 1998; 15:525–530.
112. Hofland HEJ, Bouwstra JA, Spies F, Bodde HE, Nagelkerke J, Cullander C, Junginger HE. Interactions between non-ionic surfactant vesicles and human stratum corneum in vitro. J Lip Res 1995; 5:241–263.
113. van Hal DA. Nonionic surfactant vesicles for dermal and transdermal drug delivery. Thesis. Leiden University; 1994:149–176.
114. Schätzlein A, Cevc G. Skin penetration by phospholipid vesicles, transfersomes, as visualized by means of confocal laser scanning microscopy. In: Cevc G, Paltauf F, eds. Phospholipids: Characterization, Metabolism, and Novel Biological Applications. AOCS Press: Champaign; 1993:189–207.
115. Cevc G. Transfersomes, liposomes and other lipid suspensions on the skin: permeation enhancement, vesicle penetration, and transdermal drug delivery. Crit Rev Ther Drug Carrier Sys 1996; 13:257–388.
116. Meuwissen MMEJ, Junginger HE, Bouwstra JA. Application of vesicles in human skin in vitro: a confocal laser scanning microscopy study. Biochim Biophys Acta 1990; 1371:31–39.
117. Reinl HM, Hartinger A, Dettmar P, Bayerl TM. Time-resolved infrared ATR measurements of liposome transport kinetics in human keratinocyte cultures and skin reveals a dependence on liposome size and phase state. J Invest Dermatol 1995; 105:291–295.
118. El Maghraby GMM, Williams AC, Barry BW. Oestradiol delivery from ultradeformable liposomes: refinement of surfactant concentration. Int J Pharm 2000; 196: 63–74.
119. van den Bergh BAI, Vroom J, Gerritsen H, Junginger H, Bouwstra JA. Interactions of elastic and rigid vesicles with human skin in vitro: electron microscopy

and two-photon excitation microscopy. Biochim Biophys Acta 1999; 1461:155–173.
120. van den Bergh BAI, Bouwstra JA, Junginger HE, Wertz PW. Elasticity of vesicles affects hairless mouse skin structure and permeability. J Contr Rel 1999; 62:367–379.

3
Computer Modeling of Skin Barrier Lipid Layers by Molecular Dynamics

Monika Höltje
Institute of Pharmaceutical Chemistry, Heinrich-Heine University, Düsseldorf, Germany

I. INTRODUCTION

A molecular modeling investigation of stratum corneum skin lipids is described in this chapter. For all the readers who are not familiar with the details of computational chemistry, I want to give a brief introduction to the techniques we have used.

What is molecular modeling? Molecular modeling is a general term that covers a wide range of computer graphics and computer programs used to build, display, manipulate, simulate, and analyze "realistic" molecule structures and to calculate physicochemical properties of these structures. The main purpose of molecular modeling is to provide three-dimensional model pictures that fit the known experimental facts and can be used to rationalize the behavior of the molecules or to help in the design of further experiments.

What is molecular dynamics? Molecular dynamics (MD) is a modeling technique that can be used to simulate the thermal motion of a structure as a function of time, using the forces acting on the atoms to drive the motion. The masses of the atoms are known, so Newton's second law of motion may be used to compute the accelerations and thus the velocities of the atoms. The accelerations and velocities can be used to calculate new positions for the atoms over a short time (normally 1–2 femtoseconds), thus moving each atom to a new position in space. This process can be done many thousands of times, generating a series of conformations of the structure. The velocities of the atoms are related directly to the temperature at which the simulation is performed. A simulation run at 310 K provides information on the structural fluctuations that occur around the starting conformation at body temperature, perhaps to illustrate which parts of a molecule are most flexible or to examine the number of conformational transitions [more detailed introductions to the principles and applications of molecular modeling and molecular dynamics can be found in Refs. 1 through 4].

MD simulations can provide a detailed atomic-level understanding of the structural and dynamic aspects of the investigated molecular assemblies. They have been used very successfully during the past years to study phospholipid membrane systems in molecular detail [for reviews on this field, see Refs. 5–7]. We wanted to benefit from these studies, trying to transfer the approved methods to stratum corneum (SC) lipid model systems. The goal of our present study was to receive a molecular picture of SC lipid structures that may assist the interpretation of experimental results.

In order to understand the complex nature of the SC lipid lamellae as well as the relationship between lipid composition and functionality, there have been many experimental investigations using small- and wide-angle x-ray diffraction, electron microscopy, infrared spectroscopy (IR), and nuclear magnetic resonance techniques (NMR). The results of these techniques, however, although used very successfully to provide detailed atomic insights into protein structures, are often difficult to interpret in the case of lipid layers. This is because of the heterogeneity and semifluid character of lipid layers under physiological conditions. Usually, lipid bilayers have two distinct phases in the physiological temperature range. The lower-temperature phase is a highly ordered gel state in which the hydrocarbon chains are extended and closely packed, with little mobility. The higher-temperature phase is the more disordered liquid crystalline phase. In addition, skin lipids can arrange in hexagonal phases, even at physiological temperatures. Thus, SC lipids are characterized by manifold chain-melting transitions at different temperatures, and complex polymorphism seems to be common [8]. The preference of a lipid for a given phase is governed by the interaction forces that are the combined results of volume, polarity, headgroup interactions, and van der Waals attractions between the hydrocarbon chains under a certain condition of concentration, temperature, pressure, water content, and pH. These interaction forces appear to be a determinant of the aggregation form of the lipids. Among the well-known macroscopic properties of a lipid system are analytical values that are difficult to obtain by experiment. MD simulations offer an approach to a detailed description of the packing parameters and other structural properties of lipid assemblies.

Skin lipids are complex systems, composed of very diverse molecular structures. MD simulations are complicated too; a certain number of issues need to

Modeling of Skin Barrier Lipid Layers by MD

be resolved before a reasonable simulation can be performed. Thus, one has to decide what is the appropriate choice of input (size of the system, number of molecules, starting conformation, solvent type) and what is the appropriate choice of technical parameters (time, force field, electrostatics, pressure, temperature, etc.). Moreover, MD has a number of limitations. The treatment of the interaction forces is a very simplified form compared to the "real" function. Another limitation is the time and size problem. Current computational power allows simulations of system sizes of roughly 5–10 nm on a time scale up to a few nanoseconds (ns). Processes such as phase transitions at body temperature or phase separations in lipid mixtures are out of reach in normal MD simulations.

To not get lost among all the various skin data and MD parameters, extensive testing of the parameter sets on simple model systems, which can be compared to experimental data, is necessary. We decided to build our skin lipid model step by step and have chosen simple models as starting points. These simplified skin barrier models have been used successfully in numerous experiments testing skin permeability, toxicity, and barrier-perturbing effects of drugs and cosmetic formulations [9].

Skin lipid compositions vary significantly, depending on sample origin and sample preparation. It appeared to us that an equimolar mixture of fatty acids, cholesterol, and ceramides might be a reasonable model. The fatty acids chosen are palmitic acid and stearic acid. This choice is based on similar skin lipid models widely utilized in experiments, which we wanted to use for the experimental validation of our simulation results [10]. A certain amount of the fatty acid chains in stratum corneum is certainly longer than in our model. Long-chain saturated fatty acids ought to form more or less rigid crystalline domains in SC at body temperature, thus being not so interesting from a "molecular dynamic" point of view. They may therefore be excluded from our models at this stage.

Our study consists of three parts. First, the simulation behavior of palmitic and stearic acid is tested; this is followed by a study on the influence of cholesterol on the fatty acids; and finally, the effect of ceramides is investigated. We started our calculations with a fully hydrated free fatty acid bilayer (model A). In the next step, this model was enlarged by adding an appropriate amount of cholesterol (model B). In a further step, the components of model B were mixed with ceramide 2 (model C).

Model A was chosen to prove the simulation setup and to investigate the dynamic behavior of free fatty acids. Subsequently, we compared the simulations of model B and model C to determine the influence of cholesterol and ceramide 2 on a number of bilayer properties: conformational and orientational order parameters, areas per headgroup, and atom densities.

Simulation analysis data were compared to experimental data taken from literature and to data of thermal analysis and FTIR experiments, which were performed by Förster et al. [11]. We would like to point out that the aim of our investigations is an overall qualitative description of the model systems—not a detailed quantitative evaluation. The reason for this is the lack of accurate experimental data on analogous lipid mixtures on which our simulation results could be fitted. We have focused mainly on hydrocarbon chain dynamics, with details of headgroup and water interactions being omitted in the present study.

II. METHODS

A. Initial Structures

Starting from the crystal coordinates of stearic acid [12], we used the space group operations and the lattice transformations of the stearic acid crystal to generate a monolayer of 141 fatty acid molecules. The molecules in the x-ray crystal structure are tilted and exhibit strong hydrogen bonds among the headgroups. We adjusted the coordinates so that all hydrocarbon chains were perpendicular to the layer plane and the crystal headgroup packing was disrupted in order to save computational time. Two monolayers were used to construct a bilayer consisting of 282 lipid molecules; 136 randomly chosen stearic acids (68 in each monolayer) were changed to palmitic acids by removing the last two carbon atoms in the chain. The fatty acid bilayer (model A) was transformed into model B by exchanging 112 (= about 50 weight %) of the free fatty acids randomly for cholesterol.

Model C was composed of 132 palmitic acid molecules, 85 cholesterol molecules, and 44 ceramide 2 molecules. In this model, only palmitic acid was used, because model A and model B simulations indicated nearly no difference in the overall behavior of stearic and palmitic acid.

In human skin, a pH gradient is found, reaching from pH 7 in the innermost SC layers to a pH of about 5 on the SC surface. Thus, moving outwards, across the SC, the degree of ionization will drop from 90% to less than 10% at pH 5 [13]. Because we are interested in the SC surface, we have used neutral fatty acids and no corresponding soaps in the present study.

The models were put in rectangular boxes, with the z plane defining the direction of the bilayer normal. The boxes were then enlarged on both tops to receive water caps. The water content of SC is reported to be about 10 weight % [14]. Simulations with a number of water molecules according to this value gave unstable runs, because the very small hydration level of about 2.5 water molecules per lipid led to insufficient solvation of the headgroups in the boundary region. Thus we increased the water content to 26 weight % in model A (1515 H_2O molecules) and model B (1770 H_2O molecules). Model C received 2163 H_2O molecules (29%) because the ceramides needed more water molecules per headgroup. The chemical formulas of the lipid compounds are shown in Figure 1.

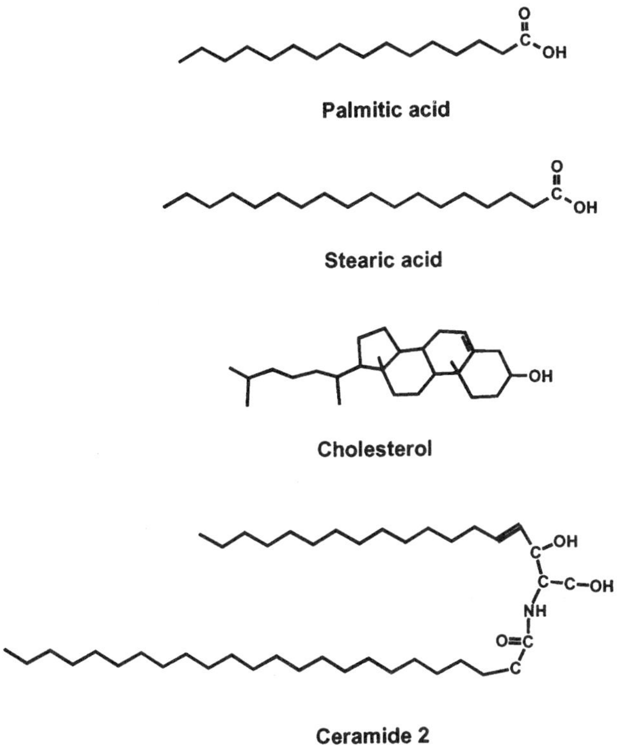

Figure 1 The chemical structures of the molecules used in the simulations.

Figure 2 is a schematic diagram of one bilayer surface of model B, illustrating the positions of the cholesterol molecules among the fatty acids. It should be pointed out that our choice of a random cholesterol distribution among the lipids is only one of several possibilities. Chong et al. [15] have suggested a maximum separation of steroid molecules in phosphatidylcholine bilayers. Smondyrev et al. [16] have compared regular cholesterol arrays and cholesterol-rich stripes in their MD studies on diphosphatidylcholine/cholesterol mixtures and found differences in equilibration time and stability of the bilayer models. However, we do not have any experimental data concerning an ideal distribution of cholesterol molecules in skin lipid models. So we took a random initial distribution because we did not want to introduce an artificial order that would obey particular rules, thus making the results more complex. For the construction of model C, ceramide 2 was taken as a representative of the ceramides family and again a random distribution was applied.

Figure 2 Schematic diagram of the surface of model B showing the placing of cholesterol among the fatty acids: # cholesterol, ■ palmitic acid, ★ stearic acid.

B. Simulation Conditions

The initial molecular geometry of the lipid layers together with the water caps was first optimized with a conjugate gradient method to reduce any unfavorable steric interactions among the molecules. All simulations were performed under NPT conditions, which means that the number of particles N, the pressure P, and the temperature T were kept fixed, allowing the volume to find its size according to system behavior.

Periodic boundary conditions were applied in all three dimensions. *Periodic boundary* is a term used to describe a method for dealing with the surface conditions in molecular dynamics simulations. Computer simulations can track only a very small number of particles (compared to experimentally used samples). As a result, most of the molecules are near the edge of the sample—that is, near its surface and the surrounding vacuum. The system size would have to be extremely large to ensure that the surface has only a small influence on the bulk properties, but such a system would be too large to simulate. To avoid artificial surface effects (e.g., to prevent molecules from migrating in the vacuum) one can use a maneuver, the periodic boundary conditions. In periodic boundary conditions, the simulation box is replicated throughout space to form an infinite lattice. By mirroring the contents of the central box, a periodic system is generated. When a molecule diffuses out of one side of the box, it reenters on the other. Thus, a constant density (number of particles N) can be maintained. Further details on the force field and the MD parameters have been published elsewhere [11].

It is well known that membrane simulations starting with lipids in the well-ordered and closely packed crystal lattice can require very long equilibration times and resemblance to the primary ordering can still be seen in the final structures even with simulations up to several hundred of picoseconds (ps) at physiological temperatures, a finding that agrees well with our own simulation experience. Thus, we have used a simulated annealing procedure to disrupt the crystal packing of model A during a high temperature phase followed by a cooling down to skin temperature, thus allowing the system to adjust its volume and shape and to find its adequate size.

First, a 200 ps MD run was performed at 350 K, a temperature that is above the melting temperature of palmitic acid (63°C = 336 K) and stearic acid (69.6°C = 343 K). Then, the temperature was gradually lowered to 303 K. After this temperature was reached, simulations were carried out at constant temperature for several nanoseconds.

The initial structure of model B was constructed from an ensemble taken from the simulation of model A at 350 K. Model C was generated from the starting structure of model B by inserting ceramide 2 molecules into the lipid layers. Molecule building was performed with the aid of the program SYBYL6.4 [Tripos Inc., St. Louis, USA]. All simulations were done with GROMACS [17] on a SGI indigo2 or a SGI Origin 2000 computer [Silicon Graphics, Mountain View, CA, USA].

III. RESULTS AND DISCUSSION OF THE MOLECULAR SIMULATION

A. Model A

We started our model A simulations from lipids in a solid crystalline phase with the alkyl chains in the all-*trans* conformation and with a close alignment to each other. When the system was heated to a temperature of 350 K, the highly ordered crystal arrangement started to melt. The occurrence of gauche conformations created kinks in the alkyl chains, which reduced the interaction between the chains. The lipid packing became considerably looser because chain-chain interactions got lost, due to the presence of gauche isomerizations. This evolution can be seen very clearly from Figure 3a, which shows the development of the average alkyl chain *trans* fraction during the heating/cooling cycle. Figure 3b shows the average surface area occupied by one fatty acid within the simulation box as a function of time (calculated by dividing the xy surface area of the simulation box by the number of lipid molecules in one layer). The number of *trans* bonds decreased from 100% to 80% during heating to the high temperature of 350 K. Connected with the conformational motions, the area per lipid increased from 20 Å^2 to 24 Å^2.

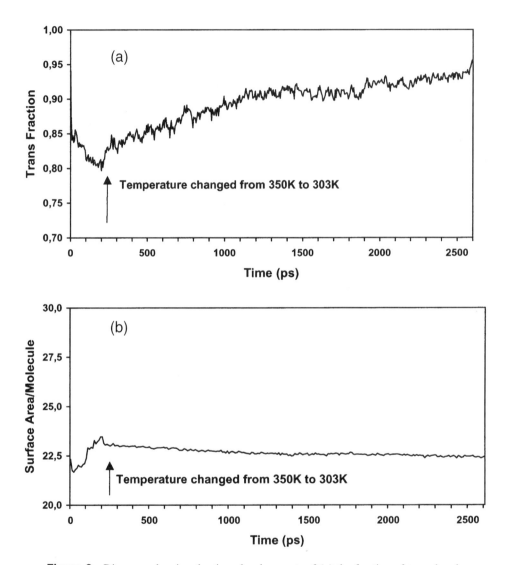

Figure 3 Diagrams showing the time developments of (*a*) the fraction of *trans* bonds in the hydrocarbon chains of the fatty acids and (*b*) the surface area per molecule during the melting and recrystallization simulation of model A.

When the system was cooled down to skin temperature (303 K), the lipids started to rearrange themselves in an ordered crystalline ensemble (Fig. 3a, b). The area per lipid moved continuously to smaller values and the *trans* fraction increased again. This process had not finished after 2600 ps, but because the tendency could be seen clearly, we stopped the simulation at the point where the *trans* fraction achieved 0.95 again and the area per lipid obtained a value of 22 Å^2. The structural changes of model A during the simulation procedure can be seen nicely from Figure 4, which shows a cross-section through the bilayer from three different simulation steps (0 ps, 200 ps, and 2.6 ns) and an assembly of stearic acids obtained from x-ray crystallographic coordinates [12].

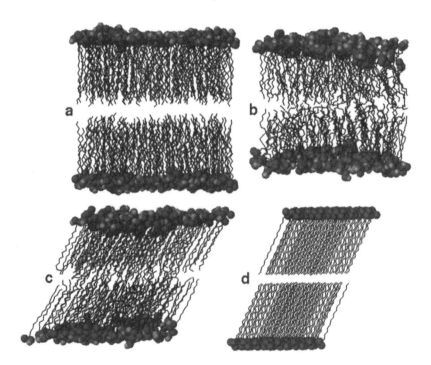

Figure 4 Snapshots of the fatty acid bilayer after melting and recrystallization. (*a*) The initial conformation of model A, composed of 136 palmitic acid molecules and 146 stearic acid molecules. (*b*) Resulting molecular assembly at 350 K (liquid crystalline state). (*c*) Final structure of model A at 303 K (gel state). (*d*) Molecular diagram of a unit cell of stearic acid as derived from crystal analyses [12]. Note the similar collective tilts of the fatty acid chains observed in the simulation and in the crystal structure. Hydrogen atoms and water molecules are omitted for clarity. Polar heads are plotted as balls.

Starting from fully extended solid crystalline hydrocarbon chains, which were arranged perpendicular (Fig. 4a), the bilayer reached an unordered, liquid crystalline–like phase during the high-temperature time steps (Fig. 4b). Cooling down resulted in ordered chains again, which exhibit now a strong tilt in the bilayer plane (Fig. 4c), a formation that looks similar to the $L_{\beta'}$ gel phase in which the polar headgroups are disordered but the mostly all-*trans* chains are packed regularly in a tilted manner [18]. This tilt is also observed in solid crystals, as can be seen from Figure 4d.

The behavior of the fatty acids in model A is in good agreement with experimental data. Neubert et al. [19] have carried out FT-Raman analyses on the melting of stearic acid. They found that in a double-layered subcell the highly ordered all-*trans* structure of the alkyl chains was present up to about 5°C below the melting point. Between 65°C and the melting point, gauche bonds appeared among *trans* sequences. On melting, the lamellae arrangement disappeared and the resulting structure was a random mixture of gauche and *trans* bonds. Bearing these results in mind, we interpret our simulation data as follows: within the 200 ps run at 350 K the lipids reached the premelting phase, the area per lipid and the number of gauche bonds increased by 20%. Cooling down the temperature after 200 ps to skin conditions stopped the melting. The lipids slowly went back to a high-order arrangement and reached a gel phase after 2.6 ns.

B. Model B

The starting configuration of model B was derived from premelting fatty acids in which the cholesterol molecules were inserted, so no high-temperature run was necessary. The total simulation was performed at 303 K for 2 ns. The resulting molecular structure of model B is presented in Figure 5. It is seen that the hydrocarbon chains stayed remarkably disordered. A large number of the chains have a substantial gradient away from the bilayer normal, but in contrast to the final structure of model A, no collective tilt can be found.

Figure 6a shows the corresponding *trans* fraction of the fatty acid alkyl chains. From the start, the *trans* fraction oscillates around 0.8 and converges to a value of 0.81 for the last 500 ps. No upward tendency can be seen, a result that is quite in contrast to model A. Thus, the data suggest that when the fatty acids are mixed with cholesterol in the premelted state, they reach a stable conformational state immediately during the simulation time. To characterize that state, we want to compare our *trans* fraction values with data from investigations on dipalmitoylphosphatidylcholine (DPPC), one of the most studied lipids. Experimental results for the liquid crystalline L_α phase of DPPC demonstrate *trans* fractions of 0.76–0.6, depending on the method used [20–22]. For the $L_{\beta'}$ gel phase, a *trans* fraction of above 0.9 has been derived from Raman measurements as well as from MD simulations [22,23]. The *trans* fraction in our model B simu-

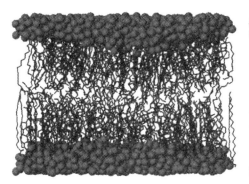 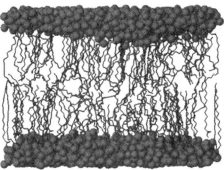

Figure 5 Model B at the end of the MD simulation. The full system, which consists of 87 stearic acids, 76 palmitic acids, 112 cholesterols, and 1770 water molecules, is shown on the left side. Cholesterol molecules have been removed in the right plot to demonstrate the conformations of the fatty acids. Water molecules are displayed as balls.

lation lies between the values of a gel and a liquid crystalline phase. The well-ordered gel phase arrangement of the pure fatty acids is shifted to less rigid conformations when cholesterol is added. Thus, cholesterol influences the fluidity of the fatty acid chains. This fluidizing effect of cholesterol, which can be recognized from our data, is experimentally well known [13,24]. But we should note that the chains are not fully liquid crystalline, because the resulting *trans* fraction of 0.8 is too high to indicate a complete L_α phase. Experimental investigations show similar results. Several authors find in SC lipids at skin temperature orthorhombic phases with alkyl chains in the predominantly all-*trans* conformation when no cholesterol is present and hexagonal phases as well as gauche alkyl conformations when cholesterol is added [8,13,25]. Förster and coworkers received from FTIR and DSC analysis data that fit well to the simulation results. The addition of cholesterol to an appropriate fatty acid mixture led to a slight increase in conformational disorder, seen by a small positive shift of the CH_2-stretching frequency from 2848.0 cm^{-1} to 2848.3 cm^{-1} in the FTIR spectrum and to a reduction in the melting enthalpy, which could be seen from thermal analysis [11].

Now we turn to the surface area per molecule in model B. The time evolution of this parameter converged very rapidly, as can be seen from Figure 6b. During the first simulation steps, the surface area was decreasing because the placement of the cholesterol molecules among the fatty acids led to a somewhat loose starting arrangement. After approximately 50 ps, the surface area had dropped to a value of about 29 Å^2 per molecule and stayed stable for the remaining run. We have analyzed the volume requirements of the cholesterol mole-

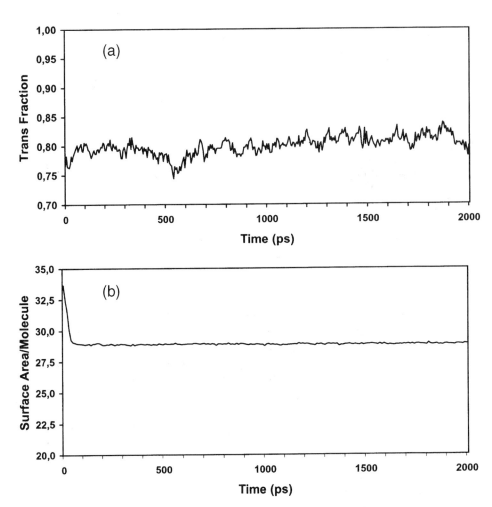

Figure 6 (*a*) Time evolution of the fraction of *trans* bonds in the hydrocarbon chains and (*b*) time evolution of the surface area per molecule during the simulation of model B at 303 K.

cules in our simulations. The average surface area for a cholesterol molecule has been determined experimentally by different authors and ranges from 26 to 39 Å^2 [15,26–29]. Recently performed studies on MD simulations of mixtures of phosphatidylcholine with cholesterol obtained values of 32.4 Å^2, 38 Å^2, and 39 Å^2 [16,23,30,31].

In model A we have obtained an area per fatty acid molecule of approximately 24 Å^2 when the *trans* fraction has reached 80%. Assuming that the area per fatty acid has not changed in model B, because the *trans* fraction is likewise 80%, we get an estimate for the area per cholesterol molecule by subtracting from the total area per monolayer the total average area occupied by the fatty acid molecules (141 × 24) and dividing the difference by the number of cholesterol molecules per monolayer. In this manner, we calculate an average surface area of 35 Å^2 for one cholesterol molecule in model B. This value is 4 Å^2 smaller than the value obtained by Hyslop et al. [28] in a pure cholesterol monolayer and 2.5 Å^2 larger than the values used by Tu et al. [23] and by Smondyrev et al. [16] in their simulations on DPPC/cholesterol mixtures.

Molecular dynamics calculations on a pure cholesterol bilayer, which we have performed, resulted in an area per cholesterol of 36 Å^2. We interpret the divergence in the overall data as follows: In pure cholesterol ensembles, the rigid molecules exhibit highly hindered motions, which do not allow very close contacts. When single cholesterol molecules are surrounded by flexible and pliant hydrocarbon chains, a much closer arrangement is possible, reducing the surface area to smaller values (e.g., down to 32 Å^2, depending on the concentration). Thus, a value of approximately 35 Å^2 for our model B cholesterols seems to be reasonable, because not all molecules are isolated and cholesterol-cholesterol contacts are found in the assembly (see Fig. 2).

C. Deuterium Order Parameters

Order parameters characterize the orientational order of the hydrocarbon chains and can be measured using the NMR technique. In molecular dynamics simulations, the deuterium order parameter (S_{CD}) can be calculated for each carbon in a chain, using the following expression [32]:

$$S_{CD} = (2/3 \cos^2 \Theta - 1/2), \tag{1}$$

where Θ is the angle between the *i*th molecular axis and the bilayer normal (z axis).

Calculated S_{CD} values can vary between −0.5 (full order along the bilayer normal) and zero (in the case of full disorientation). Figure 7 shows a comparison of the calculated S_{CD} values of the palmitic acids in models A and B. The curves are plotted as a function of the carbon position along the chain (the carbon atom next to the carbonyl group is denoted as number 1; the terminal methyl group is not calculated).

Let us first examine the order parameter profile of model A at the end of the 303 K run. The high order within the alkyl chains, which occurred when the temperature was reduced to skin values, can be clearly seen from the profile. All carbon atoms exhibit very similar order parameters. In spite of the strong conformational order within the chains, the average S_{CD} value is relatively small

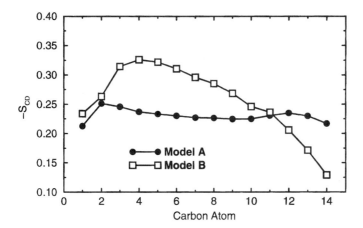

Figure 7 Calculated deuterium order parameters S_{CD} of the palmitic acid chains in model A and model B at 303 K, averaged over the last 100 ps. Values are plotted as a function of carbon atom position along the chain. Carbon number 1 corresponds to the first atom of the hydrocarbon tail; the terminal carbon has not been calculated.

(−0.25). This reflects the fact that the hydrocarbon chains are now tilted and their overall orientation is no longer parallel to the bilayer normal.

Experimental studies available for comparable fatty acids offer average S_{CD} values of soaps, ranging from −0.19 to −0.14 at room temperature [33]. Soaps, due to their strong headgroup repulsion, need larger areas per lipid than neutral fatty acids. This strongly influences the conformational behavior of the hydrocarbons (as can be seen from the low values). Therefore, we cannot compare our calculated order parameters of model A directly with experimental order profiles.

The second profile in Figure 7 demonstrates the effect of cholesterol on the order of the fatty acid alkyl chains. As can be seen, the influence is not uniform along the chain. There is little effect on the order parameters of the first two carbon atoms; these values are only slightly higher in model B than in model A. Subsequently, the profile shows a strong rise with a small plateau region extending over carbon atoms 3–6, after which the values drop significantly toward the end of the chain. It is apparent from the data that cholesterol induces a pronounced decrease in conformational order at the tail end of the lipids, thus producing an order profile with a general feature as observed for lipid bilayers in the liquid crystalline phase: a plateau region near the carbonyls, followed by a decrease in order to near zero at the terminal methyl groups. The average S_{CD} values do not differ so much (−0.25 for model A and −0.26 for model B), whereas the appearances of the profiles are strongly different. The experimentally known effect of cholesterol to disorder lipid chains in gel phase bilayers can be clearly

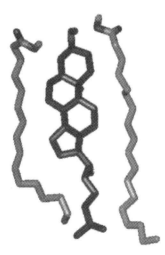

Figure 8 Example of a typical cholesterol–fatty acid ensemble. Cholesterol is drawn in black; the hydrocarbon chains of palmitic acid (*left*) and stearic acid (*right*) are plotted in gray.

seen from the order parameters. Adding 50 weight % cholesterol to the fatty acids generates a phase that resembles the order of liquid crystalline lipids, although no transition to a liquid crystalline phase has occurred, as already mentioned.

For a closer look at the influence of cholesterol on the alkyl chain order, we present a representative set of molecules extracted from our model B simulations (Fig. 8). It is clear from the picture why cholesterol has that profound effect on the conformation of the alkyl chains at the tail end. For lipids neighboring cholesterol, the upper half of the chains is in contact with the rigid steroid ring, which reduces the conformational freedom, whereas the lower part is mainly in contact with the skinnier and more flexible cholesterol tail, which allows kinks in that part of the chains.

D. Model C

The introduction of ceramide into our model raised questions concerning its integration into the other lipids. The major reason for this is the difference in chain length between ceramide 2 and the fatty acids and/or cholesterol and the difference in chain length within the ceramide itself. To date, the detailed organization of the lipid classes that make up the matrix of the SC has not been elucidated. A model for the molecular arrangement of ceramides among fatty acids and cholesterol has been proposed recently by Bouwstra et al. [34]. From this model, we adopted the idea of an interdigitation of two opposite ceramides (see Fig. 11). Thus, 22 interdigitated ceramide couples were randomly distributed in the bilayer. A dynamics run of 3.8 ns was executed under conditions identical to those of the model B simulation. A snapshot from the end of the simulation of model C is shown in Figure 9.

On a first view, the bilayer formation of model C seems not so different from model B. However, one can see from the picture that the tails of the ceramide chains occupy space in the bilayer center and that they are fairly disordered.

The time evolution of the *trans* fraction of the ceramides and the fatty acids is shown in Figure 10. In the case of the palmitic acids, the *trans* fraction rises from the initial value of about 0.8 to a value of about 0.89, whereas the value for the ceramides settled into oscillations around 0.80 during the simulation. The palmitic acid chains exhibit a higher conformational order compared with the chains of the ceramides.

A closer look at the conformational changes that happened to the ceramide molecules is given in Figure 11. We show two examples of ceramide pairs in

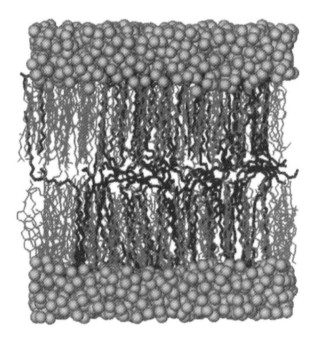

Figure 9 Model C at the end of the simulation at 303 K. The bilayer consists of 132 palmitic acids, 85 cholesterols, 44 ceramide-2 molecules, and 2163 water molecules. Water is displayed as balls; palmitic acids and cholesterol molecules are drawn in gray; ceramides are shown in black.

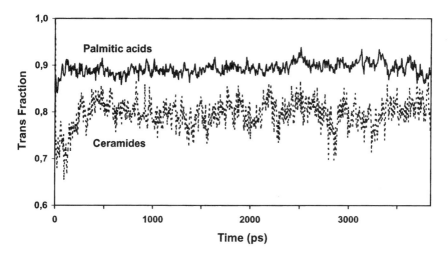

Figure 10 Time development of the fraction of *trans* bonds in the hydrocarbon chains of palmitic acid and ceramide-2 during the simulation of model C at 303 K.

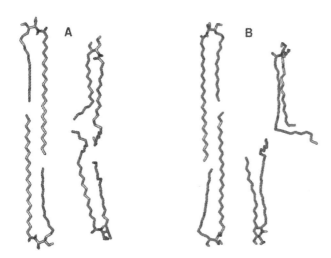

Figure 11 Snapshots of ceramide molecules from the simulation. *A and B, left part*: Interdigitated starting structures of two opposite ceramides. *A and B, right part*: Conformational arrangement at the end of the simulation. Note that the molecules in (*A*) stayed in an interdigitated position, whereas in (*B*) they have separated.

the interdigitated starting structure and their conformational arrangement at the end of the simulation. Some of the ceramides stayed in the interlocked position (example A), whereas in other couples the chains have turned away from each other (example B). Nearly all ceramides in the assembly present well-ordered upper chains and kinks in the end parts. The higher value for the palmitic acid *trans* fraction may be accounted for by neighboring ceramide chain parts in *trans* conformations.

The behavior of the ceramide molecules is somewhat in contrast to experimental data, which propose highly ordered all-*trans* ceramide conformations at body temperature [35,36]. We interpret the results as follows: In our model, the distribution of more or less isolated ceramide pairs among the other lipids forces the ceramides to resize to the shorter molecules in the neighborhood. If the ceramide chain length would be better compatible with that of the surrounding, one might speculate that the molecules would stay much more conformationally ordered. Such a view is compatible with the proposal of separated crystalline ceramide domains in the SC matrix [37]. Recently, Forslind has proposed the domain mosaic model of lipid organization: ceramides aggregate into the crystalline phase, providing large areas that constitute the watertight barrier [38]. However, at the border of these crystalline areas, lipids in a liquid crystalline phase can be found. These ''grain borders'' allow water to pass through the barrier.

The systems studied here are different from experimentally used SC mixtures in that only very small ''pieces'' with no particular domains could be investigated. However, it seems that the mixture in our model C offers a picture of a border situation, where long-chain ceramides meet short-chain molecules, with the consequence that the ceramides become partly disordered and adopt a more fluid phase.

Further support that our model C mimics well essential features of a ''real'' system comes from ^{2}H NMR data. Fenske et al. [39] performed a study on SC lipid models consisting of equimolar mixtures of bovine-brain ceramides, cholesterol, and palmitic acid. From the spectra, they obtained a S_{CD} profile for the palmitic acids, which we have compared with the calculated order parameters of our simulation. As can be seen from Figure 12, the comparison results in good agreement between the two profiles. The experimental data show somewhat lower S_{CD} values in the plateau region (carbon 3–8), which may be due to the higher temperature used in the measurement (the spectra were acquired at 50°C).

To establish the locations of the different molecule types in the bilayer, we show in Figure 13 the atom densities (along the z axis of the simulation box) of the oxygens of the ceramide OH-groups, together with those from the palmitic acids and the cholesterols. It is clear from the figure that the cholesterols and the palmitic acids are more likely to have their oxygens mainly below the ceramide OH-groups. Thus, they prefer to sit deeper in the bilayer and allow the more polar headgroups of the ceramides to reach for the water layer. Similar observa-

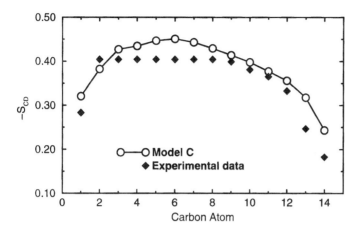

Figure 12 Calculated deuterium order parameters S_{CD} of the palmitic acid chains of the model C simulation in comparison to experimentally derived values. (From Ref. 39.)

tions have been reported for bilayers composed of phospholipids and cholesterol [24]. A detailed analysis of the interactions among the headgroups of the lipids and the water molecules at the interfacial region of the bilayer will be published elsewhere [40].

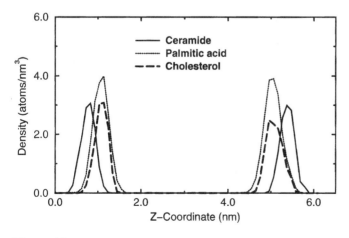

Figure 13 Oxygen atom density profiles for the OH-groups of model C lipids along the bilayer normal (Z-coordinate of the simulation box), averaged over the last 100 ps.

Finally, we compared our model C simulation to available lamellar spacing data recently published. Bouwstra and coworkers identified a short periodicity phase in mixtures of ceramides (all had the same chain length as ceramide 2) and cholesterol. They calculated from the electron density profiles a distance between the two headgroup regions of about 4.75 nm [34]. This value is in good agreement with our bilayer thickness: the average distance between the headgroups in model C is 4.65 nm.

IV. CONCLUSIONS

We have carried out molecular dynamics simulations of three different simple stratum corneum lipid models: a pure fatty acid bilayer (stearic acid, palmitic acid), a fatty acid/cholesterol mixture at a 1:1 ratio, and a palmitic acid/cholesterol/ceramide 2 (1:1:1) mixture. Biophysical aspects of the model structures have been investigated and compared with experimental results. At the atomic level the behavior of the different compounds showed the following picture:

1. Pure free fatty acid layers can adopt a well ordered $L_{\beta'}$ gel phase at skin temperature.
2. Cholesterol fluidizes the rigid gel phases by decreasing the conformational order of the adjacent hydrocarbon tails.
3. Ceramides consolidate the lipid assembly and seem to play an important role in the coherence of the bilayer because the tails of their long hydrocarbon chains occupy the inner space between both monolayers.

The computed properties, such as *trans* fractions, order parameters, or bilayer thickness, show reasonable agreement with experimental data. As can be seen from the simulations, computer modeling methods can offer useful information for validating or refining the interpretation of experimental data because they provide a very detailed picture of molecular interactions.

REFERENCES

1. Höltje HD, Folkerts G. Molecular Modeling—Basic Principles and Applications. New York: VCH Publisher, 1996.
2. Leach AR. Molecular Modelling—Principles and Applications. England: Longman, 1996.
3. Hinchliffe A. Modelling Molecular Structures. Chichester: Wiley, 1996.
4. van Gunsteren WF, Berendsen HJC. Computer simulation of molecular dynamics:

methodology, applications and perspectives in chemistry. Angew Chem Int Edit 1990; 29:992–1023.
5. Merz KM, Roux B. Biological Membranes: A Molecular Perspective from Computation and Experiment. Boston: Birkhäuser, 1996.
6. Tieleman DP, Marrink SJ, Berendsen HJC. A computer perspective of membranes: molecular dynamics studies of lipid bilayer systems. Biochim Biophys Acta 1997; 1331:235–270.
7. Tobias DJ, Tu K, Klein ML. Atomic-scale molecular dynamics simulations of lipids. Curr Opin Colloid Interface Sci 1997; 2:15–26.
8. Ongpipattanakul B, Francoeur ML, Potts RO. Polymorphism in stratum corneum lipids. Biochim Biophys Acta 1994; 1190:115–122.
9. Friberg SE, Kayali ICH, Margosiak M. Stratum corneum structure and transport properties. Drugs Pharm Sci 1990; 42:29–45.
10. Friberg SE, Suhaimi H, Goldsmith LB, Rhein LL. Stratum corneum lipids in a model structure. J Disp Sci Technol 1988; 9:371–389.
11. Höltje M, Förster T, Brandt B, Engels T, von Rybinski W, Höltje H-D. Molecular dynamics simulations of stratum corneum lipid models: fatty acids and cholesterol. Biochim Biophys Acta 2001; 1511:156–167
12. Goto M, Asada E. The crystal structure of the B-form of stearic acid. Bull Chem Soc Japan 1978; 51:2456–2459.
13. Lieckfeldt R, Villalaín J, Gómez-Fernández JC, Lee G. Apparent pK_a of the fatty acids within ordered mixtures of model human stratum corneum lipids. Pharm Res 1995; 12:1614–1617.
14. Warner RR, Myers MC, Taylor DA. Electron probe analysis of human skin: determination of the water concentration profile. J Invest Dermatol 1988; 90:218–224.
15. Chong PLG. Evidence for regular distribution of sterols in liquid crystalline phosphatidylcholine bilayers. Proc Natl Acad Sci USA 1994; 91:10069–10073.
16. Smondyrev AM, Berkowitz ML. Structure of dipalmitoylphosphatidylcholine/cholesterol bilayer at low and high cholesterol concentrations: molecular dynamics simulation. Biophys J 1999; 77:2075–2089.
17. van der Spoel D, van Buuren AR, Apol E, Meulenhoff PJ, Tieleman DP, Sijbers ALTM, van Drunen R, Berendsen HJC. Gromacs User Manual, Version 1.6. Nijenborgh 4, 9747 AG Groningen, The Netherlands, 1996.
18. Sun WJ, Suter RM, Knewtson MA, Worthington CR, Tristam-Nagle S, Zhang R, Nagle JF. Order and disorder in fully hydrated unoriented bilayers of gel phase dipalmitoylphosphatidylcholine. Phys Rev E 1994; 49:4665–4676.
19. Neubert R, Rettig W, Wartewig S, Wegener M, Wienhold A. Structure of stratum corneum lipids characterized by FT-Raman spectroscopy and DSC. 2. Mixtures of ceramides and saturated fatty acids. Chem Phys Lipids 1997; 89:3–14.
20. Mendelsohn R, Davies MA, Brauner JW, Schuster HF, Dluhy RA. Quantitative determination of conformational disorder in the acyl chains of phospholipid bilayers by infrared spectroscopy. Biochemistry 1989; 28:8934–8939.
21. Seelig J, Seelig A. Lipid conformations in model membranes and biological membranes. Q Rev Biophys 1980; 13:19–61.
22. Pink DA, Green TJ, Chapman D. Raman scattering in bilayers of saturated phosphatidylcholines. Experiment and theory. Biochemistry 1980; 19:349–356.

23. Tu K, Klein ML, Tobias DJ. Constant-pressure molecular dynamics investigation of cholesterol effects in a dipalmitoylphosphatidylcholine bilayer. Biophys J 1998; 75:2147–2156.
24. McMullen TPW, McElhaney RN. Physical studies of cholesterol-phospholipid interactions. Curr Opin Coll Int Sci 1996; 1:83–90.
25. Pilgram GSK, Engelsma-van Pelt AM, Oostergetel GT, Koerten HK, Bouwstra JA. Study on the lipid organization of stratum corneum lipid models by (cryo-)electron diffraction. J Lipid Res 1998; 39:1669–1676.
26. Craven BM, DeTitta GT. Cholesterol myristate: structures of the crystalline solid and mesophases. J Chem Soc Perkin Trans 1976; 2:814–822.
27. Pearson RH, Pascher I. The molecular structure of lecithin dihydrate. Nature 1979; 281:499–501.
28. Hyslop PA, Morel B, Sauerheber RD. Organization and interaction of cholesterol and phosphatidylcholine in model bilayer membranes. Biochemistry 1990; 29:1025–1038.
29. Almeida PFF, Vaz WLC, Thompson TE. Lateral diffusion in the liquid phases of dimyristoylphosphatidylcholine/cholesterol lipid bilayers: a free volume analysis. Biochemistry 1992; 31:6739–6747.
30. Robinson AJ, Richards WG, Thomas PJ, Hann MM. Behavior of cholesterol and its effect on head group and chain conformations in lipid bilayers: a molecular dynamics study. Biophys J 1995; 68:164–170.
31. Pasenkiewicz-Gierula M, Rog T, Kitamura K, Kusumi A. Cholesterol effects on the phosphatidylcholine bilayer polar region: a molecular simulation study. Biophys J 2000; 78:1376–1389.
32. Egberts E, Berendsen HJC. Molecular dynamics simulation of a smectic liquid crystal with atomic detail. J Chem Phys 1988; 89:3718–3732.
33. Boden N, Jones SA, Sixl F. On the use of deuterium nuclear magnetic resonance as a probe of chain packing in lipid bilayers. Biochemistry 1991; 30:2146–2155.
34. Bouwstra JA, Gooris GS, Dubbelaar FER, Weerheim AM, Ijzerman AP, Ponec M. Role of ceramide 1 in the molecular organization of the stratum corneum lipids. J Lipid Res 1998; 39:186–196.
35. Moore DJ, Rerek ME, Mendelsohn R. Role of ceramides 2 and 5 in the structure of the stratum corneum lipid barrier. Int J Cosmetic Lipids 1999; 21:353–368.
36. Wegener M, Neubert R, Rettig W, Wartewig S. Structure of stratum corneum lipids characterized by FT-Raman spectroscopy and DSC. I. Ceramides. Int J Pharm 1996; 128:203–213.
37. Mendelsohn R, Moore DJ. Vibrational spectroscopic studies of lipid domains in biomembranes and model systems. Chem Phys Lipids 1998; 96:141–157.
38. Forslind B. A domain mosaic model of the skin barrier. Acta Derm Vernereol 1994; 74:115–125.
39. Fenske DB, Thewalt JL, Bloom M, Kitson N. Models of stratum corneum intercellular membranes: ^{2}H NMR of macroscopically oriented multilayers. Biophys J 1994; 67:1562–1573.
40. Manuscript in preparation

4
Role of Lipids in Skin Barrier Function

Mitsuhiro Denda
Shiseido Life Science Research Center, Yokohama, Japan

I. INTRODUCTION

For terrestrial animals, the skin's most important role is to protect the water-rich internal organs from the dry environment. This cutaneous barrier function resides in the stratum corneum. The water impermeability of this thin (10–20 µm) layer is 1000 times higher than that of other membranes of living organisms [1]. This is the same level as that of a plastic membrane with the same thickness [2]. The stratum corneum is composed of two components: protein-rich nonviable cells and intercellular lipid domains (Fig. 1). The lipid molecules in the intercellular domain form a bilayer structure. Because of this specific ''brick and mortar'' structure, the stratum corneum shows high water impermeability [1,3].

When the barrier function is damaged by tape stripping or treatment with an organic solvent or detergent, a series of homeostatic processes in the barrier function is immediately accelerated, and the barrier recovers to its original level [3]. This homeostatic repair process is blocked by occlusion with a water-impermeable membrane such as plastic membrane or latex membrane [4]. The occlusion with a water-permeable membrane such as Gortex does not perturb the repair process [4]. The skin barrier function also has an ability to adapt to the environment. Under a low-humidity environment, the barrier function is enhanced [5]. The thickness of the stratum corneum increases in a dry environment. The content of the intercellular lipid in the stratum corneum increases and, consequently, the transepidermal water loss decreases—i.e., the water impermeability increases [5]. These results suggest that the skin barrier function senses the environmental change and reorganizes its function to adapt to the new environment

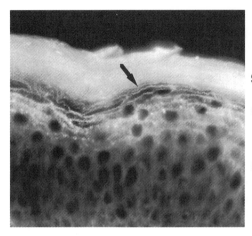

Figure 1 Human skin section stained with fluorescent probe Nile Red. Intense white membrane structure (neutral lipid: arrow) is seen in the uppermost layer of the epidermal granular layer.

(Table 1). Thus, skin barrier homeostasis is a self-referential and self-organizing system.

On the other hand, the skin barrier homeostatic system is under the influence of the central nervous system. Psychological stress delays barrier recovery, and sedative drugs can prevent the delay [6,7]. Glucocorticoid in the serum mediates the relationship between the stress and skin barrier homeostasis [7]. Odorants that have sedative effects also improve barrier homeostasis [8]. Various environmental factors that affect our emotions could influence the barrier function. Moreover, the barrier recovery rate shows a circadian rhythm [9]. The physiological stage of the whole body is related to the barrier homeostasis (Table 1).

Table 1 Environmental and Physiological Factors That Affect Skin Barrier Homeostasis

Environmental factors	Barrier recovery (ref.)	Physiological factors	Barrier recovery (ref.)
Low temperature	Delayed (95)	Aging	Delayed (69)
Low humidity	Accelerated (5)	Circadian rhythm	Altered (9)
High humidity	Delayed (5)	Increase of serum glucocorticoid	Delayed (7)
Psychological stress	Delayed (6,7)		

Role of Lipids in Skin Barrier Function

In a healthy state, the stratum corneum barrier function is able to recover from damage. After acute disruption of the stratum corneum barrier function by tape stripping or acetone treatment, epidermal DNA synthesis [10] or mRNA level of inflammatory cytokines [11] such as interleukin-1 (IL-1) or tumor necrosis factor (TNF), increase and then recover to the original level. However, even when the level of damage is relatively small, it may have a significant influence on the whole skin if it is repeated [12] or if it occurs in a dry environment [13]. Thus, improvement of the skin barrier is important not only for skin surface condition but also for whole skin pathology.

Lipids in the stratum corneum play a crucial role in the barrier function. The water impermeability and the water repellency of the lipids can protect our bodies from environmental dryness and the invasion of harmful environmental factors. Moreover, studies in the past few decades showed a biological activity of the lipid molecules in skin barrier homeostasis. This chapter mainly focuses on this water impermeable barrier function and the role of lipids in barrier homeostasis.

II. DEFINITION OF LIPIDS IN THE STRATUM CORNEUM

The lipids in the stratum corneum have two origins: one is the keratinocytes in the epidermis during its differentiation and another is the sebaceous gland. The epidermis is in a constant state of self-replacement. At the bottom layer, keratinocyte stem cells divide into daughter cells, which are displaced outward and which differentiate through successive overlying layers to enter the stratum corneum (Fig. 2). Then, the keratinocytes die, and their cellular organelles and cytoplasm disappear during the final process of differentiation. Intercellular lipids are primarily generated from exocytosis of lipid-containing granules called lamellar bodies, during the terminal differentiation. The secreted lipids spread over the intercellular domains and form a bilayer structure [14] (Fig. 3).

Sebaceous glands are usually associated with hair follicles and are holocrine in structure [15]. Sebum is formed when the lipid-rich cells die and disintegrate. Sebaceous glands open onto the skin surface. They are surrounded by connective tissue; on the peripheral basement membrane, there are germinative cells. The lipid-producing cells differentiate, form lipid globules in the cytosol, and at the final stage of differentiation, the nucleus disappears. Then the cellular structures are degraded, and the lipid is secreted through the sebaceous duct to the skin surface. In human skin, the sebaceous glands are concentrated in the face, forehead, and scalp and are absent on the palms and soles [15].

Thus, on the surface of human and other mammalian skin, there is a mixture of lipids that have different origins (see also Chapter 1). Sheu et al. [16] described in detail the structure of human skin surface lipids, an amorphous sheet of variable

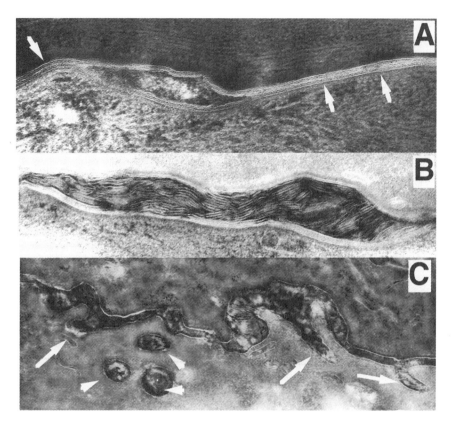

Figure 2 A: Electron micrograph of the intercellular lipid bilayer structure (arrows) in the human stratum corneum. B: Secreted lipids are stacked between the stratum corneum and the stratum granulosum. C: Lamellar bodies (arrowhead) and exyocytosis of lamellar bodies (arrow) in the human stratum granulosum.

thickness on the skin surface. The thickness was especially great on sebum-rich regions such as face areas. Even between the desquamating corneocytes in the uppermost several layers, a deranged lipid structure was found. On the other hand, the investigators reported intercellular lipid lamellar and lipid envelope in the desquamating cells. These findings suggested that the sebum from the sebaceous glands spread over the skin surface was mixed with lipids derived from epidermal lamellar bodies and formed a film on the surface of skin. The physicochemical studies of the lipid film might be important for further understanding of the protective function of the stratum corneum.

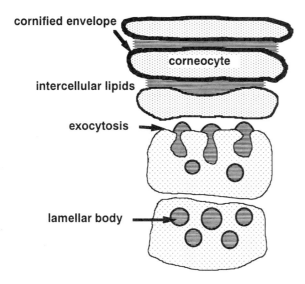

Figure 3 Process of the epidermal lipid synthesis and formation of the intercellular lipid bilayer structure. The intercellular lipids are primarily generated from exocytosis of lipid-containing granules called lamellar bodies during the terminal differentiation of the epidermal keratinocytes.

III. COMPOSITION OF LIPIDS IN THE STRATUM CORNEUM

The main components of the lipids in the stratum corneum originating from epidermal lamellar bodies are ceramides, cholesterol, and free fatty acid [14]. A small amount of cholesterol sulfate has also been reported as the intercellular lipids [14]. As described above, these lipids first appear in the lamellar body in the epidermal granular layer. The lamellar body lipids are phospholipids, sphingomyeline, cholesterol sulfate, glucosylceramide, and acylglucosylceramides. At the final stage of keratinocyte differentiation, the lamellar body lipids are secreted into the stratum granulosum–stratum corneum interface. During this process, the lipids are also processed to three types of lipids in the stratum corneum. When the barrier function is disrupted, the lipid synthesis, the lamellar body secretion, and the lipid processing are accelerated to repair the recovery. Inhibition of lipid synthesis [17,18], lipid processing [19,20], and migration of the lamellar body [21] induces barrier abnormalities. There are six to seven types of ceramides, and we previously demonstrated that the relative ratio of the species was altered in experimentally induced dry skin or with aging [22]. Bouwstra et al. [23] demonstrated that the difference in the ceramide species influences the lamellar phases of the cholesterol and creamed mixture in vitro. The reason why such a difference in ceramide species occurs in intact skin remains to be investigated.

Components of human sebum are squalene, wax esters, triglycerides, and free fatty acids [15,24]. The free fatty acids are formed from the triglycerides through the action of lipases. Cholesterol esters are also found in the skin surface lipids, but they might be processed from cholesterol derived from epidermal keratinocytes in the stratum corneum. The composition of sebum varies among mammals [24]. Sebaceous activity and fatty acid composition vary with age and gender because sebaceous gland activity is under the influence of sex hormones [15]. Tsuchiya et al. [25] demonstrated that the activity of the sebaceous glands was also influenced by psychological stress. Various environmental factors and systemic physiologic factors [15] might influence the sebum concentration on the skin surface.

IV. BIOCHEMISTRY OF STRATUM CORNEUM INTERCELLULAR LIPIDS

A. Lipid Synthesis

An important relationship between lipid synthesis and skin barrier function was first demonstrated by Grubauer et al. [26]. They presented an approximately threefold increase in the synthesis of cholesterol, fatty acid, and non-saponified lipids in the epidermis 1–4 hr after disruption of the skin barrier by acetone treatment. Importantly, those increases were perfectly blocked by occlusion with water-impermeable plastic membrane. Then, Proksch et al. [27] demonstrated a strong relationship between the rate-limiting enzyme in cholesterogenesis, HMG CoA (hydroxymethylglutaryl coenzyme A) reductase, and skin barrier function. The barrier disruption increased the enzyme activity within 15 min and the activity reached a maximum after 2.5 hr. Occlusion with a water-impermeable membrane reduced the increase. Application of an HMG CoA reductase inhibitor, lovastatin, inhibited the barrier recovery, and co-application of free cholesterol reversed the inhibition [18]. Disruption of the barrier function by acetone treatment or tape stripping increased the protein level and mRNA of HMG CoA [28]. The barrier disruption also increased the protein and mRNA levels of LDL (low-density lipoprotein) receptor [28]. The changes induced by barrier disruption were blocked by occlusion with a water-impermeable membrane [28]. Not only HMG CoA reductase but also other key enzymes for cholesterol synthesis are related to skin barrier function. Harris et al. [29] demonstrated that the barrier disruption increased the mRNA levels of HMG CoA synthase and squalene synthase in mouse epidermis and that the increase was prevented by occlusion with latex membrane. These results suggest that an acute barrier disruption induces lipid synthesis and lipid transport to repair the damage.

Fatty acid synthesizing enzymes are also associated with the barrier function. Ottey et al. [30] demonstrated that barrier disruption increased the activities of both acetyl CoA carboxylase and fatty acid synthase, and that the increase was prevented by occlusion with a water-impermeable membrane. The mRNA levels of these enzymes in the epidermis also increased after barrier disruption, and the increase was blocked by occlusion [29]. Yamaguchi et al. [31] suggested a role of fatty acid–binding protein (C-FABP) in the epidermis. This protein might transport intracellular fatty acids and induce fatty acid synthesis. They demonstrated that C-FABP expression was prevented by occlusion with a water-impermeable membrane after skin barrier disruption.

Ceramide synthesis is related to barrier homeostasis. Serine-palmitoyl transferase catalyzes formation of 3-ketosphinganin (i.e., the first step of sphingolipid synthesis). Topical application of β-chloro-L-alanin, an inhibitor of serine-palmitoyl transferase, inhibited the barrier recovery [17]. And mRNA of this enzyme in the epidermis increases barrier disruption, and the increase is prevented by occlusion with a water-impermeable membrane. Chunjor et al. [32] demonstrated the importance of glucosylceramide syntase activity on the barrier function. The activity of the enzyme was not altered by barrier disruption. However, the activity of this enzyme was higher in the outer epidermis and inhibition of the enzyme by topical application of d, 1-threo-1-phenyl-2-hexadecanoylamino-3-pyrrolidino-1-propanol delayed barrier repair. These results suggest that glucosylceramide synthase is also required for barrier homeostasis.

Man et al. [33] demonstrated that inhibition of the synthesis of both cholesterol and sphingolipid by fluvastatin (cholesterol synthesis inhibitor) and β-chloro-L-alanine did not cause further worsening of the skin barrier function 5–6 hr after barrier disruption. And 18–24 hr after treatment, the recovery rate of mice skin, in which both cholesterol and sphingolipid synthesis was inhibited, was better than in the animals in which only sphingolipid synthesis was inhibited. This suggests the importance of the relation of both sphingolipid (ceramide in the stratum corneum) and cholesterol.

Jensen et al. suggested an important role of sphingomyelinase [34]. Tumor necrosis factor (TNF) induces activation of sphingomyelinase via p55 receptor (TNF-R55). They demonstrated that TNF-R55–deficient mice showed delay of barrier repair compared with the wild type. The barrier disruption increased the sphingomyelinase activity in wild type and TNF-R75 (another type of TNF receptor)–deficient mice but not in TNF-R55–deficient mice. TNF-R55 signaling plays an important role in barrier homeostasis through the regulation of sphingomyelinase activity.

B. Lipid Processing

During terminal differentiation and stratum corneum formation, the lipids in the lamellar bodies are processed. This is also crucial for skin barrier homeostasis. Man et al. [20] reported that processing of phospholipids, precursors of free fatty acids, is necessary for barrier homeostasis. Topical application of phospholipase A2 after the barrier disruption delayed the recovery [20]. Co-application of palmitic acid with the inhibitor prevents the delay. Repeated application of these inhibitors induces barrier abnormalities. In this case, a significant decrease of free fatty acid content in the stratum corneum was observed. Again, co-application of palmitic acid normalized the barrier function.

Hydrolysis of glucosylceramide by β-glucocerebrosidase is also important for healthy barrier function [19]. This process produces free ceramide in the stratum corneum. Inhibition of β-glucocerebrosidase by topical application of an epoxide induced barrier abnormalities. In this case, the content of glycoceramide in the stratum corneum was 6 times higher than that in normal skin, whereas free ceramides in the stratum corneum did not show any significant difference. Electron microscopic studies showed an immature lipid bilayer structure in the stratum corneum of the mouse, which was treated by the inhibitor. Doering et al. [35] suggested an important role of sphingolipid activator protein, Pro-saposin, on skin barrier homeostasis. This protein stimulates enzymatic hydrolysis of sphingolipids such as glucosylceramide. They demonstrated that the pro-saposin knockout mouse showed an accumulation of glucosylceramide and abnormal structure of the intercellular lipid domain in the stratum corneum.

Abnormality of cholesterol sulfate processing also induces serious skin abnormalities [36]. In healthy epidermis, 5% of total lipids consists of cholesterol sulfate, which decreases in the stratum corneum to about 1%. Steroid sulfatase catalyzes the desulfation of cholesterol sulfate to cholesterol. In recessive X-linked ichthyosis, which is associated with a large amount of abnormal scales, the cholesterol sulfate in the stratum corneum was 10-fold higher than normal because of the absence of steroid sulfatase. Sato et al. [37] demonstrated that cholesterol sulfate inhibited both trypsin-type and chymotrypsin-type proteases, resulting in the reduced degradation of desmosomes, which play a crucial role in the adhesion of corneocytes and consequently the induction of abnormal scales. Moreover, Nemes et al. [38] reported another potential negative role of cholesterol sulfate in the stratum corneum. Involucrin cross-linking and involucrin esterification with ω-hydroxyceramides are crucial for cornified envelope formation. They demonstrated that both reactions were inhibited by cholesterol sulfate. The role of cholesterol sulfate in the normal epidermis should be studied.

Figure 4 shows a scheme of the synthesis and processing of the intercellular lipids.

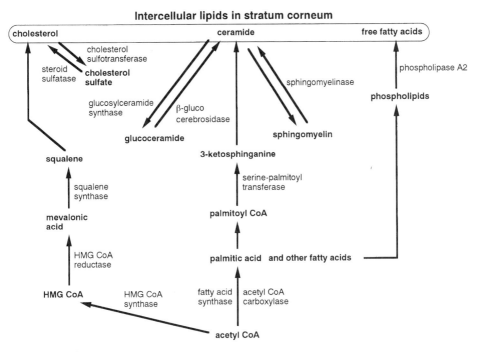

Figure 4 Diagram of synthesis and processing of the intercellular lipids in the stratum corneum.

V. ROLE OF LIPIDS IN THE STRATUM CORNEUM

The intercellular lipids in the stratum corneum play a crucial role in skin barrier function. The water impermeability is the result of the conformation of the lipid molecules and also the ordering of the corneocytes [3]. The cornified envelope, which is formed on the surface of the corneocytes, plays an important role in the structure of the barrier [39] (Fig. 5). Another role of the stratum corneum is the buffer function of water molecules in the corneocytes (Fig. 5). Decrease of free amino acids in the corneocytes is commonly observed in various kinds of dermatitis characterized by dry, scaly skin [40–43]. Decline of these functions leads to deterioration of skin pathology.

Previous studies have suggested that the lipid structure itself can absorb a huge amount of water [44]. However, this was later denied. Cornwell et al. [45] demonstrated the effect of hydration on the intercellular lipid structure of human stratum corneum using wide-angle x-ray diffraction. They monitored the packing

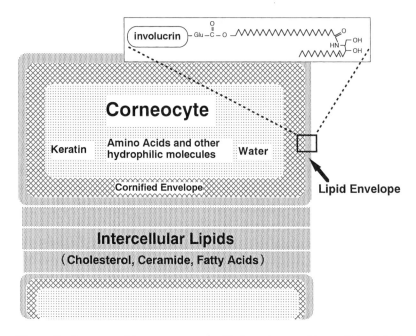

Figure 5 Structure of the corneocyte, cornified envelope, and intercellular lipid domain. The cross-linked protein envelope and covalently attached lipid envelope on the protein structure are formed around the corneocyte. Free amino acids and other water-soluble small molecules in the corneocyte play a crucial role in holding water molecules.

arrangement of the lipid bilayers on the stratum corneum—which were hydrated 0, 20–40, 40–60, 60–80, and 300%—and observed no effects of hydration on the lipid structure. The lipid bilayer structure contains some water molecules, but the water content is relatively small in comparison with the amount of water in the cornified cells. Moreover, in dry skin induced by detergent or tape stripping, the total amount of stratum corneum ceramide did not change, though the skin surface conductance, which is a measure of skin moisture, and barrier function decreased and the amino acid content decreased [42].

Tanaka et al. [43] reported that the amino acid content was reduced in the stratum corneum in atopic respiratory disease and the transepidermal water loss did not change. They suggested that the free amino acid content is a crucial factor in the dry, scaly features of not only experimentally induced dry skin but also atopic dermatitis. The ultrastructure of the intercellular lipids in the stratum corneum contributes to the barrier function of healthy skin, but the decline of the barrier function in dermatoses might be caused by various other factors. We previously evaluated the intercellular lipid alkyl chain conformation by attenuated total reflectance infrared spectroscopy on healthy skin and surfactant-induced scaly skin of human subjects [46]. In normal healthy skin, there was a correlation between the lipid conformation and the transepidermal water loss. However, no difference was observed in the surfactant-induced scaly skin. Disorders of the corneocytes or phase separation of the lipid domain might cause barrier dysfunction. Reichelt et al. [47] demonstrated a significant alteration of lipid composition in the stratum corneum of keratin 10–deficient mice. The mutual interaction of proteins and lipids in the epidermis should be investigated for further understanding the cause of abnormalities of the stratum corneum.

For the stabilization of the intercellular lipid bilayer structure, the hydrophobic envelope formed on the surface of the corneocytes plays an important role [39]. During the terminal differentiation of the keratinocyte, the protein envelope is formed on the keratinocyte. The cross-linked protein structure is mainly composed of involucrin, loricrin, and filaggrin. Then, ω-hydroxyceramide molecules covalently attach on the protein envelope. Any abnormality of the formation of this protein-lipid envelope on the surface of the corneocytes induces barrier abnormalities even when other lipid synthesis and processing systems are normal. Behne et al. [48] demonstrated that inhibition of ω-hydroxyceramide inhibited barrier repair after tape stripping. From their electron microscopic studies they suggested that it was due to the disruption of the cornified lipid envelope and lamellar membrane formation. Segre et al. [49] reported that a transcription factor, Klf4, is required for skin barrier formation. Klf4-deficient mice showed the absence of the barrier and an abnormal cornified envelope, whereas mutant mice showed a normal lipid profile. The process of the cornified envelope formation has been studied by Steinert and his coworkers, who demonstrated that glutamine-glutamate rich regions ordinarily come from involucrin as a major substrate for the attachment of ω-hydroxyceramide [50]. Then, they demonstrated by their series of in vitro studies that transglutaminase 1 played an important role for both involucrin cross-linking [51] and attachment of ω-hydroxyceramide to involucrin by ester bond formation [52]. However, Elias et al. [53] reported the existence of a cornified lipid envelope in the scales of lamellar ichthyosis patients with absence of transglutaminase 1. In the transglutaminase-deficient epidermis, other enzymes might play a role in lipid envelope formation.

The role of sebum on the skin remains unknown. Abrams et al. [54] denied the contribution of sebum to the water-impermeable barrier function of the stratum corneum. They demonstrated that extraction of sebum by acetone did not affect the barrier function, whereas extraction of ceramides, cholesterol, and fatty acids by a chloroform and methanol mixture lowered the barrier function significantly. Sebum might play a water-repellent role in human skin because it contains hydrophobic lipids such as squalene [24]. Of interest, other mammals such as mice, dogs, cats, and monkeys do not secrete squalene. Humans, and some waterside mammals such as the otter and beaver, secrete squalene. Hairless humans might need squalene to keep their bodies waterproof.

Although squalene in the sebum has been reported to be an oxygen-scavenging agent [55], Chiba et al. [56] demonstrated that squalene-monohydroxperoxide induced skin roughness and wrinkle formation on hairless mice. The biochemical role of squalene in human intact skin remains to be investigated. On the other hand, Thiele et al. [57] reported that the uppermost layer of sebum-rich facial stratum corneum showed significantly higher levels of the antioxidant α-tocopherol than the stratum corneum on the arm. They also demonstrated that sebaceous gland secretion was a major route of the α-tocopherol to the skin surface.

Another report by Metze et al. [58] suggested that the sebaceous glands secrete immunoglobulin A, which plays an important role in inactivation of invading viruses. Recently, Man et al. [59] demonstrated that the dry scaly skin of asebia mice, which have deficient sebaceous glands, could be improved by application of glycerol, which is normally metabolized from triglycerides. People in the cosmetic industry tend to focus on reduction of sebum secretion because they believe it could prevent acne and also prevent makeup from coming off. But such a reduction might cause a skin problem. The positive role of sebum and the sebaceous gland should be investigated.

VI. TOPICAL APPLICATION OF LIPIDS

Effects of topical application of SC lipids were evaluated by researchers at the University of California, San Francisco. They focused on the effects of lipids on the barrier repair process [60–62]. First, they demonstrated that the application of a single lipid among the main components of the intercellular lipids (i.e., ceramide, cholesterol, and free fatty acid) delayed barrier repair [60]. Application of a mixture with two of them also delayed the barrier recovery. Also, an equimolar mixture of ceramide, cholesterol, and free fatty acid did not delay barrier repair. Finally, they found the optimal molar ratio of the lipid mixture that could accelerate barrier recovery [61]. The molar ratios were ceramide:cholesterol:palmitic acid:linoleic acid = 1:1:1:3 or 1:1:3:1. Stearic acid, in place of palmitic acid, also showed the same effect. Acylceramide alone did not accelerate barrier repair, but a mixture of acylceramide and cholesterol accelerated barrier recovery. However, addition of a free fatty acid, such as linoleic acid, palmitic acid, or stearic acid, did not further accelerate barrier recovery. The investigators also compared the "physiologic lipids" (i.e., a mixture of the intercellular lipids) with petrolatum on the barrier repair process [21]. Petrolatum repaired the barrier more quickly than the physiologic lipids, but 2 or 4 hr later, the application of a physiologic lipid resulted in a better recovery than petrolatum. An histochemical study showed that the applied petrolatum stayed in the stratum corneum, whereas the physiologic lipids were incorporated into the nucleated layers of the epidermis. The applied physiologic lipids penetrated into the epidermis, followed by cellular uptake. Then they entered the lamellar bodies, were secreted into the stratum granulosum–stratum corneum interface, and reorganized the intercellular lipid structures.

The effect of the application of the physiologic lipids varied depending on the method of barrier disruption [62]. When the barrier was disrupted by tape stripping or treatment with an organic solvent, the topical application of physiologic lipids was effective. However, when the barrier was disrupted by treatment with a detergent such as sodium dodecyl sulfate or ammonium lauryl sulphosuccinate, the physiologic lipids did not accelerate barrier repair. When other detergents, such as N-laurosarcosine free acid or dodecylbenzensulfuric acid were used to disrupt barrier function, the lipid mixture could accelerate barrier repair. A detergent may cause not only barrier disruption but also more extensive damage in the nucleated layer of the epidermis.

Abnormalities of barrier function of recessive X-linked ichthyosis could be improved by topical application of cholesterol [36]. Patients displayed a 10-fold increase in cholesterol sulfate and a 50% decrease of free cholesterol in the stratum corneum. They also showed lower barrier function and delayed barrier repair. Studies have shown that abnormal cholesterol sulfate accumulation may cause these abnormalities [36,37]. In healthy young people, topical application of cholesterol alone delayed barrier repair as described above [60]. But in the case of recessive X-linked ichthyosis, cholesterol accelerated barrier repair and normalized the intercellular lipid ultrastructure [36].

Application of cholesterol is also effective for the abnormalities of aged skin. Details will be given in the following section.

VII. AGING AND LIPIDS

Previous studies demonstrated that sebum secretion tends to decrease with age. Saint Leger reported [63] an approximately 60% decrease in triglycerides on the lateral mid-calf of the subjects (45 female and 5 male, aged 20–76). They also showed an increase of skin dryness with aging. Nazzaro-Porro reported [64] that the accumulation of upper chest skin surface lipids over 24 hr reached highest values during ages 15–45 in both males and females and then started to decrease. The decrease was more obvious in female subjects. Wilhelm et al. [65] demonstrated the age-dependent surface sebum concentration on the skin at different anatomic sites. A significant decrease of sebum was observed on the ankle. Alteration of wax ester concentration on the skin surface with aging has been reported by several investigators. Jacobsen et al. [66] demonstrated the wax ester secretion rates on the forehead in male and female aged 15–97: rates were highest in those 15–35 years of age and then started to decrease with age. The rate of decrease in male subjects was 23% per decade and 32% in female subjects. Yamamoto et al. [67] also reported an alteration of sebaceous gland activity with aging by the ratio of wax ester/[cholesterol + cholesterol ester] and showed similar results. They also demonstrated a variation in the fatty acid composition of the wax ester in the subjects. C16 iso-branched saturated fatty acid, C14, C15, C16, C18:1 straight, and C16:1 iso-branched mono-unsaturated fatty acids showed a significant correlation with aging from infancy to the twenties. C15, C16:1 straight, and C16:1 iso-branched mono-unsaturated fatty acids also showed a significant correlation with aging after maturation.

Stratum corneum intercellular lipids also decrease with aging, but the decrease is relatively smaller than that of sebum. Hara et al. [68] demonstrated the stratum corneum lipid contents of subjects with senile xerosis and of young controls. The triglyceride concentration was 65% lower in the subjects with senile xerosis and ceramide concentration was 28% lower. The xerosis skin showed a significant decrease of water content in the stratum corneum, but the transepidermal water loss was not significantly different. An electron microscopic observation showed the decrease of keratohyaline granules in the granular layer of the skin of patients with senile xerosis. These results suggest that the decrease of water in the skin surface of the patients was the result of reduced skin surface lipids and amino acid content in the stratum corneum.

We demonstrated the alteration of ceramide composition with aging [22]. A significant difference in ceramide composition was observed only in female subjects. There was a significant increase in ceramide 1 and 2 with a corresponding decrease in ceramide 3 and 6 from prepubertal age to adulthood. Thereafter, the portion of ceramide 2 decreased with age, whereas ceramide 3 increased with age. The alteration of the ceramide composition with aging is opposite to the changes of ceramide composition in dry, scaly skin induced by tape stripping or detergent treatment. These results suggest that ceramide composition is influenced by epidermal proliferate activity and also by sex hormonal changes.

The transepidermal water loss in the elderly is the same or slightly lower than that in young individuals, suggesting that the barrier function in the elderly is not lower. However, Ghadially et al. [69] reported that the barrier function in elderly subjects was destroyed more easily than that in young individuals, and that the recovery rate of the barrier function after the disruption of the barrier in the elderly subjects was lower than that in the young subjects. They compared 15 subjects aged 20–30 years and 6 subjects older than 80. The basal transepidermal water loss of the volar arm of the elderly subjects was not lower than that in the young subjects. But barrier function was perturbed by 18 ± 2 tape strippings in the elderly subjects, whereas 31 ± 5 strippings were needed in the younger subjects. Moreover, barrier recovery 24 hours after barrier disruption was only 15% for the elderly subjects, whereas the young subjects showed 50–60% recovery at that time. The investigators observed the same tendency on aged hairless mice skin. An electron microscopic study showed a similar number of lamellar bodies in both young and aged epidermis, but a paucity of secreted lamellar body content in the aged epidermis both from humans and mice.

Although the *total lipid* content in the stratum corneum decreased 30% more in the aged mice than in the young mice, the composition of the main key lipids (i.e., ceramide, cholesterol, and free fatty acid) did not differ between young and aged mice. However, among the main *stratum corneum intercellular lipids* (i.e., ceramide, cholesterol, and free fatty acids), synthesis of cholesterol is reduced more than of the other two lipids [70]. Topical application of cholesterol also accelerated skin barrier repair after barrier disruption. The physiologic lipid mixture also improved barrier function and, moreover, a cholesterol:ceramide:palmitic acid:linoleic acid = 3:1:1:1 mixture with cholesterol as the dominant lipid further accelerated barrier recovery in chronologically aged human skin and hairless mouse skin. Electron microscopic observations showed that the acceleration of barrier repair in chronologically aged skin was associated with acceleration of replenishment of the interstices in the intercellular lipid domain with lamellar structures [70].

Haratake et al. [71] demonstrated that topical application of mevalonic acid, which is an intermediate of cholesterol biosynthesis, stimulates cholesterol synthesis in the epidermis of aged mice. They also reported improvement of barrier repair in aged mice by the treatment. The deterioration of the skin barrier function with aging could be improved by the regulation of cholesterol in the epidermis.

VIII. PERSPECTIVE

As described above, lipids play various roles in human skin. Regulation of the lipids on the skin surface can not only improve the superficial condition but also cure whole skin pathology. Topical application of suitable lipids is one method of lipid regulation. On the other hand, studies have suggested another way to improve the condition of the stratum corneum. Skin barrier repair can be accelerated not only by the topical application of the physiologic lipids but also by regulation of nonlipid factors such as enzymes and ions (Tables 2 and 3). As described above, repeated barrier disruption induces epidermal hyperplasia and inflammation. The acceleration of barrier repair will improve those skin conditions. Thus, biochemical and biophysical studies of epidermal barrier homeostasis are important for clinical dermatology.

Lipid metabolism is regulated by a series of enzymes in the epidermis [72], and each of them has its optimal pH [73] and other conditions such as ion balance [74]. For example, the pH value of the healthy stratum corneum is kept acid

Table 2 Substances That Affect Skin Barrier Recovery

Lipids	Barrier recovery (ref.)	Ions	Barrier recovery (ref.)
Single lipid	Delayed (60)	Calcium	Delayed (75)
Two-lipid mixture	Delayed (60)	Potassium	Delayed (75)
Equimolar mixture of three lipids	Normal (60)	Sodium	Normal (75)
		Magnesium	Accelerated (74)
Optimized mixture of three lipids	Accelerated (61)	Magnesium + calcium	Accelerated (74)
Nuclear hormone receptor activator		Acidic pH	Normal (73)
		Neutral or basic pH	Delayed (73)
PPARα activator	Accelerated (92)		

because the lipid-processing enzymes have a low optimal pH. Mauro et al. [73] demonstrated that topical application of basic buffer after the barrier disruption delayed the repair process because the basic condition perturbs lipid processing.

Other ions such as calcium and magnesium [72,75] also play important roles in the lipid metabolism of the epidermis. We demonstrated the heterogeneous distribution of calcium, magnesium, and potassium in human epidermis [76]. Both calcium and magnesium are localized in the granular layer, and potassium is localized in the spinous layer. Immediately after disruption of barrier function, this distribution disappeared. Calcium plays various roles in stratum corneum barrier formation [77]. For example, it induces terminal differentiation [78], formation of the cornified envelope, and also epidermal lipid synthesis [79]. Menon et al. demonstrated that alteration of the calcium gradient affects the exocytosis of the lamellar body at the interface between the stratum corneum and the epidermal granular layer [80]. Vicanova demonstrated [81] the improvement in barrier function of reconstructed human epidermis by the normalization of epidermal calcium distribution. The heterogeneous field that is formed by calcium and other ions with consumption of ATP might be crucial for terminal differentiation and barrier formation of the epidermis. Magnesium is required for the activity of Rab-geranylgeranyl transferase, which modifies Rab, a low-molecular-weight GTP-binding protein [82]. After the modification, Rab plays an important role on exocytosis and endocytosis [83]. For barrier formation, exocytosis of the lamellar body is an important process. Previous studies have indicated that Rab is modified by Rab-geranylgeranyl transferase during the terminal differentiation of the epidermis [84]. The topical application of calcium or potassium impaired barrier repair, whereas magnesium or a mixture of calcium and magnesium salts accelerated the repair process [74].

Table 3 Substances That Affect Skin Barrier Recovery

Protease inhibitors	Barrier recovery (ref.)	Histamine receptor agonist and antagonists	Barrier recovery (ref.)
Trypsin-like serine protease inhibitor	Accelerated (72)	H1 Receptor antagonist	Accelerated (94)
Other protease inhibitors	Normal (72)	H2 Receptor antagonist	Accelerated (94)
Plasminogen activator inhibitor	Accelerated (72)	H3 Receptor antagonist	Normal (94)
		H2 Receptor agonist	Delayed (94)
		H3 Receptor agonist	Normal (94)
		Histamine	Delayed (94)
		Histamine releaser	Delayed (94)

Recently, an important role of retinoid X receptors for the stratum corneum has been suggested by several workers. Topical application of retinoic acid decreased the skin barrier function. Imakado et al. [85] demonstrated that a dominant-negative retinoic acid receptor α targeting mutant in the epidermis of mice showed a severe decline of the barrier function. The transgenic mice showed thin and loosely packed stratum corneum and, of note, there was no lipid bilayer structure in the stratum corneum of the mutant mice.

Feingold and his coworkers have demonstrated an important role of nuclear hormone receptors on epidermal differentiation and stratum corneum barrier formation. First, Hanley et al. [86–88] demonstrated that activation of the nuclear hormone receptors PPARα (peroxysome-proliferator-activated receptor α) and FXR (farnesoid X–activated receptor) with oleic acid, linoleic acid, and clofibrate accelerated the development of the epidermis and barrier formation in the fetal skin organ culture system. Of note, activities of β-glucocerebrosidase and steroid sulfatase, which play an important role in barrier lipid processing, were increased by the activation of PPARα and FXR. Activation of PPARα by farnesol also stimulated the differentiation of epidermal keratinocyte [89–91]. The cornified envelope formation, involucrin and transglutaminase protein, and mRNA levels were also increased by the activation of PPARα [92]. Of interest, DNA synthesis was inhibited by the treatment [91]. Activation of PPARα in knockout mice that showed focal hyperkeratosis did not increase the epidermal differentiation [93]. These results suggest that the nuclear hormone receptor plays a crucial role in the development and differentiation of epidermis and stratum corneum barrier homeostasis. They also showed that topical application of PPARα activators accelerated barrier recovery after tape stripping or acetone treatment and prevented epidermal hyperplasia induced by repeated barrier disruption [92]. Regulation of the nuclear hormone receptor would open a new possibility for improvement of cutaneous barrier function.

Recently, Ashida et al. [94] reported the relationship between the histamine receptor and skin barrier function [94] (Table 3). The topical application of histamine H1 and H2 receptor antagonists accelerated the barrier repair. Histamine itself, H2 receptor agonist, and histamine releaser delayed the barrier repair. Neither a histamine H3 receptor antagonist nor agonist affected the barrier recovery rate. Topical application of H1 and H2 receptor antagonists prevented the epidermal hyperplasia induced by barrier disruption in a dry environment. The mechanism of the relationship between the histamine receptors and the barrier repair process has not been clarified. However, these findings provide another perspective for future skin care systems.

The regulation of epidermal lipid metabolism by regulation with other factors might be effective to improve skin pathology, because it can improve the endogenous homeostatic process. Occlusion or moisturization with artificial material could improve skin condition. However, these treatments potentially perturb the homeostasis of the skin. On the other hand, recovery of the original, endogenous skin function by acceleration of its homeostatic process results in natural healthy skin without side effects. A better understanding of skin homeostasis is required to develop an ideal skin care system.

IX. CONCLUSIONS

Stratum corneum lipids play a crucial role in skin barrier function. Synthesis and processing of lipids are associated with the barrier function. Although the epidermal barrier function is a self-referential and self-organizing system, it is also related to the physiological condition of the whole body. Abnormalities of the barrier function induced by various environmental or intrinsic factors may cause whole skin problems. Topical application of physiologic lipids or suitable regulation of the epidermal lipid metabolism could improve the skin barrier function and the condition of the whole skin.

REFERENCES

1. Potts RO, Francoeur ML. The influence of stratum corneum morphology on water permeability. J Invest Dermatol 1991; 96:495–499.
2. Tagami H. Stratum corneum as a barrier in the skin. The Japanese J Dermatol 1998; 108:713–727.
3. Elias PM, Menon GK. Structural and lipid biochemical correlates of the epidermal permeability barrier. In: Elias PM, ed. Advances in Lipid Research Vol. 24 Skin Lipids. San Diego: Academic Press; 1991:1–26.

4. Grubauer G, Elias PM, Feingold KR. Transepidermal water loss: the signal for recovery of barrier structure and function. J Lipid Res 1989; 30:323–333.
5. Denda M, Sato J, Masuda Y, Tsuchiya T, Koyama J, Kuramoto M, Elias PM, Feingold KR. Exposure to a dry environment enhances epidermal permeability barrier function. J Invest Dermatol 1998; 111:858–863.
6. Denda M, Tsuchiya T, Hosoi J, Koyama J. Immobilization-induced and crowded environmental-induced delay barrier recovery in murine skin. Br J Dermatol 1998; 138:780–785.
7. Denda M, Tsuchiya T, Elias PM, Feingold KR. Stress alters cutaneous permeability barrier homeostasis. Am J Physiol 2000; 278:R367–R372.
8. Denda M, Tsuchiya T, Shoji K, Tanida M. Odorant inhalation affects skin barrier homeostasis in mice and humans. Br J Dermatol 2000; 142:1007–1010.
9. Denda M, Tsuchiya T. Barrier recovery rate varies time-dependently in human skin. Br J Dermatol 2000; 142:881–884.
10. Proksch E, Feingold KR, Man MQ, Elias PM. Barrier function regulates epidermal DNA synthesis. J Clin Invest 1991; 87:1668–1673.
11. Wood LC, Jackson SM, Elias PM, Grunfeld C, Feingold KR. Cutaneous barrier perturbation stimulates cytokine production in the epidermis of mice. J Clin Invest 1992; 90:482–487.
12. Denda M, Wood LC, Emami S, Calhoun C, Brown BE, Elias PM, Feingold KR. The epidermal hyperplasia associated with repeated barrier disruption by acetone treatment or tape stripping cannot be attributed to increased water loss. Arch Dermatol Res 1996; 288:230–238.
13. Denda M, Sato J, Tsuchiya T, Elias PM, Feingold KR. Low humidity stimulates epidermal DNA synthesis and amplifies the hyperproliferative response to barrier disruption: implication of seasonal exacerbations of inflammatory dermatoses. J Invest Dermatol 1998; 111:873–878.
14. Elias PM, Feingold KR. Lipids and the epidermal water barrier: metabolism, regulation, and pathophysiology. Semin Dermatol 1992; 11:176–182.
15. Thody AJ, Shuster S. Control and function of sebaceous glands. Physiol Rev 1989; 69:383–416.
16. Sheu HM, Chao SC, Wong TW, Lee JYY, Tsai JC. Human skin surface lipid film: an ultrastructural study and interaction with corneocytes and intercellular lipid lamellae of the stratum corneum. Br J Dermatol 1999; 140:385–391.
17. Holleran WM, Man MQ, Gao WN, Menon GK, Elias PM, Feingold KR. Sphingolipids are required for mammalian epidermal barrier function. J Clin Invest 1991; 88:1338–1345.
18. Feingold KR, Man MQ, Proksch E, Menon GK, Brown BE, Elias PM. The lovastatin-treated rodent: a new model of barrier disruption and epidermal hyperplasia. J Invest Dermatol 1991; 96:201–209.
19. Holleran WM, Takagi Y, Menon GK, Legler G, Feingold KR, Elias PM. Processing of epidermal glucosylceramides is required for optimal mammalian cutaneous permeability barrier function. J Clin Invest 1993; 91:1656–1664.
20. Man MQ, Jain M, Feingold KR, Elias PM. Secretory phospholipase A2 activity is required for permeability barrier homeostasis. J Invest Dermatol 1996; 106:57–63.
21. Man MQ, Brown BE, Wu-Pong S, Feingold KR, Elias PM. Exogenous nonphysiologic vs physiologic lipids. Arch Dermatol 1995; 131:809–816.

22. Denda M, Koyama J, Hori J, Horii I, Takahashi M, Hara M, Tagami H. Age- and sex-dependent change in stratum corneum sphingolipids. Arch Dermatol Res 1993; 285:415–417.
23. Bouwstra JA, Dubbelaar FER, Gooris GS, Weerheim AM, Ponec M. The role of ceramide composition in the lipid organization of the skin barrier. Biochem Biophys Acta 1999; 1419:127–136.
24. Stewart ME, Downing DT. Chemistry and function of mammalian sebaceous lipids. In: Elias PM, ed. Advances in Lipid Research. Vol. 24, Skin Lipids. San Diego: Academic Press; 1991:263–302.
25. Tsuchiya T, Horii I. Immobilization-induced stress decreases lipogenesis in sebaceous glands as well as plasma testosterone levels in male Syrian hamsters. Psychoneuroendocrinology 1995; 20:221–230.
26. Grubauer G, Feingold KR, Elias PM. Relationship of epidermal lipogenesis to cutaneous barrier function. J Lipid Res 1987; 28:746–752.
27. Proksch E, Elias PM, Feingold KR. Regulation of 3-hydroxy-3-methylglutaryl-coenzyme A reductase activity in murine epidermis: modulation of enzyme content and activation state by barrier requirements. J Clin Invest 1990; 84:874–882.
28. Jackson SM, Wood LC, Lauer S, Taylor JM, Cooper AD, Elias PM, Feingold KR. Effect of cutaneous permeability barrier disruption on HMG CoA reductase, LDL receptor and apoprotein E mRNA levels in the epidermis of hairless mice. J Lipid Res 1992; 33:1307–1314.
29. Harris IR, Farrell AM, Grunfeld C, Holleran WM, Elias PM, Feingold KR. Permeability barrier disruption coordinately regulates mRNA levels for key enzymes of cholesterol, fatty acid, and ceramide synthesis in the epidermis. J Invest Dermatol 1997; 109:783–787.
30. Ottey K, Wood LC, Grunfeld C, Elias PM, Feingold KR. Cutaneous permeability barrier disruption increases fatty acid synthetic enzyme activity in the epidermis of hairless mice. J Invest Dermatol 1995; 104:401–404.
31. Yamaguchi H, Yamamoto A, Watanabe R, Uchiyama N, Fujii H, Ono T, Ito M. High transepidermal water loss induces fatty acid synthesis and cutaneous fatty acid-binding protein expression in rat skin. J Dermatol Sci 1998; 17:205–213.
32. Chunjor CSN, Feingold KR, Elias PM, Holleran WM. Glucosylceramide synthase activity in murine epidermis: quantitation, localization, regulation, and requirement for barrier homeostasis. J Lipid Res 1998; 39:277–285.
33. Man MQ, Feingold KR, Elias PM. Inhibition of cholesterol and sphingolipid synthesis causes paradoxical effects on permeability barrier homeostasis. J Invest Dermatol 1993; 101:185–190.
34. Jensen JM, Schutze S, Forl M, Kronke M, Proksch E. Roles of tumor necrosis factor receptor p55 and sphingomyelinase in repairing the cutaneous permeability barrier. J Clin Invest 1999; 104:1761–1770.
35. Doering T, Holleran WM, Potratz A, Vielhaber G, Elias PM, Suzuki K, Sandhoff K. Sphingolipid activator proteins are required for epidermal permeability barrier formation. J Biol Chem 1999; 274:11038–11045.
36. Zettersten E, Man MQ, Sato J, Denda M, Farrell A, Ghadially R, Williams ML, Feingold KR, Elias PM. Recessive x-linked ichthyosis: role of cholesterol-sulfate accumulation in the barrier abnormality. J Invest Dermatol 1998; 111:784–790.

37. Sato J, Denda M, Nakanishi J, Nomura J, Koyama J. Cholesterol sulfate inhibits proteases that are involved in desquamation of stratum corneum. J Invest Dermatol 1998; 111:189–193.
38. Nemes Z, Demeny M, Marekov LN, Fesus L, Steinert PM. Cholesterol 3-sulfate interferes with cornified envelope assembly by diverting transglutaminase 1 activity from the formation of cross-links and esters to the hydrolysis of glutamine. J Biol Chem 2000; 275:2636–2646.
39. Nemes Z, Steinert PM. Bricks and mortar of the epidermal barrier. Exp Mol Med 1999; 31:5–19.
40. Horii I, Obata M, Tagami H. Stratum corneum hydration and amino acid content in xerotic skin. Br J Dermatol 1989; 121:587–592.
41. Takahashi M, Ikezawa Z. Dry skin in atopic dermatitis and patients on hemodialysis. In: Loden M, Maibach HI, eds. Dry Skin and Moisturizers, Chemistry and Function. Boca Raton: CRC Press; 2000:135–146.
42. Denda M, Hori J, Koyama J, Yoshida S, Namba R, Takahashi M, Horii I, Yamamoto A. Stratum corneum sphingolipids and free amino acids in experimentally-induced scaly skin. Arch Dermatol Res 1992; 284:363–367.
43. Tanaka M, Okada M, Zhen YX, Inamura N, Kitano T, Shirai S, Sakamoto K, Inamura T, Tagami H. Decreased hydration state of the stratum corneum and reduced amino acid content of the skin surface in patients with seasonal allergic rhinitis. Br J Dermatol 1998; 139:618–621.
44. Imokawa G, Hattori M. A possible function of structural lipids in the water-holding properties of the stratum corneum. J Invest Dermatol 1985; 84:282–284.
45. Cornwell PA, Barry BW, Stoddart CP, Bouwstra JA. Wide-angle X-ray diffraction of human stratum corneum: effects of hydration and terpene enhancer treatment. J Pharm Pharmacol 1994; 46:938–950.
46. Denda M, Koyama J, Namba R, Horii I. Stratum corneum lipid morphology and transepidermal water loss in normal skin and surfactant-induced scaly skin. Arch Dermatol Res 1994; 286:41–46.
47. Reichelt J, Doering T, Schnetz E, Fatasch M, Sandhoff K, Magin TM. Normal ultrastructure, but altered stratum corneum lipid and protein composition in a mouse for epidermolytic hyperkeratosis. J Invest Dermatol 1999; 113:329–334.
48. Behne M, Uchida Y, Seki T, de Montellano PO, Elias PM, Holleran WM. Omega-hydroxyceramides are required for corneocyte lipid envelope (CLE) formation and normal epidermal permeability barrier function. J Invest Dermatol 2000; 114:185–192.
49. Segre JA, Bauer C, Fuchs E. Klf4 is a transcription factor required for establishing the barrier function of the skin. Nat Genet 1999; 22:356–360.
50. Marekov LN, Steinert PM. Ceramides are bound to structural proteins of the human foreskin epidermal cornified cell envelope. J Biol Chem 1998; 273:17763–17770.
51. Nemes Z, Marekov LN, Steinert PM. Involucrin cross-linking by transglutaminase 1. J Biol Chem 1999; 274:11013–11021.
52. Nemes Z, Marekov LN, Fesus L, Steinert PM. A novel function for transglutaminase 1: attachment of long-chain ω-hydroxyceramides to involucrin by ester bond formation. Proc Natl Acad Sci USA 1999; 96:8402–8407.
53. Elias PM, Uchida Y, Rice RH, Komuves L, Holleran WM. Formation of partial

cornified envelopes (CE) and replete corneocyte-lipid envelope (CLE) in patients with lamellar ichthyosis. J Invest Dermatol (abs.) 2000; 114:758.
54. Abrams K, Harvell JD, Shriner D, Wertz P, Maibach H, Maibach HI, Rehfeld SJ. Effects of organic solvents on in vitro human skin water barrier function. J Invest Dermatol 1993; 101:609–613.
55. Saint-Leger D, Bague A, Cohen E, Chivot M. A possible role for squalene in the pathogenesis of acne. I. In vitro study of squalene oxidation. Br J Dermatol 1986; 114:535–542.
56. Chiba K, Sone T, Kawakami K, Onoue M. Skin roughness and wrinkle formation induced by repeated application of squalene-monohydroxide to the hairless mouse. Exp Dermatol 1999; 8:471–479.
57. Thiele JJ, Weber SU, Packer L. Sebaceous gland secretion is a major physiologic route of vitamin E delivery to skin. J Invest Dermatol 1999; 113:1006–1010.
58. Metze D, Jurecka W, Gebhart W, Schmidt J, Mainitz M, Niebauer G. Immunohistochemical demonstration of immunoglobulin A in human sebaceous and sweat glands. J Invest Dermatol 1989; 92:13–17.
59. Man MQ, Wertz PW, Feingold KR, Elias PM. Barrier function and moisturization in asebia mice: function of cutaneous sebaceous glands. J Invest Dermatol (abstr.) 2000; 114:796.
60. Man MQ, Feingold KR, Elias PM. Exogenous lipids influence permeability barrier recovery in acetone-treated murine skin. Arch Dermatol 1993; 129:728–738.
61. Man MQ, Feingold KR, Thornfeld CR, Elias PM. Optimization of physiologic lipid mixtures for barrier repair. J Invest Dermatol 1996; 106:1096–1101.
62. Yang L, Man MQ, Taljebini M, Elias PM, Feingold KR. Topical stratum corneum lipids accelerate barrier repair tape stripping, solvent treatment and some but not all types of detergent treatment. Br J Dermatol 1995; 133:679–685.
63. Saint-Leger D, Francois AM, Leveque JL, Stoudemayer TJ, Grove GL, Kligman AM. Age-associated changes in stratum corneum lipids and their relation to dryness. Dermatologica 1988; 177:159–164.
64. Nazzaro-Porro M, Passi S, Boniforti L, Belsito F. Effects of aging on fatty acids in skin surface lipids. J Invest Dermatol 1979; 73:112–117.
65. Wilhelm KP, Cua AB, Maibach HI. Skin aging: effect on transepidermal water loss, stratum corneum hydration, skin surface pH, and casual sebum content. Arch Dermatol 1991; 127:1806–1809.
66. Jacobsen E, Billings JK, Frantz RA, Kinney CK, Stewart ME, Downing DT. Age-related changes in sebaceous wax ester secretion in men and women. J Invest Dermatol 1985; 85:483–485.
67. Yamamoto A, Serizawa S, Ito M, Sato Y. Effect of aging on sebaceous gland activity and on the fatty acid composition of wax esters. J Invest Dermatol 1987; 89:507–512.
68. Hara M, Kikuchi K, Watanabe M, Denda M, Koyama J, Nomura J, Horii I, Tagami H. Senile xerosis: functional, morphological, and biochemical studies. J Geriatr Dermatol 1993; 1:111–120.
69. Ghadially R, Brown BE, Sequeira-Martin SM, Feingold KR, Elias PM. The aged epidermal permeability barrier. J Clin Invest 1995; 95:2281–2290.
70. Ghadially R, Brown BE, Hanley K, Reed JT, Feingold KR, Elias PM. Decreased

epidermal lipid synthesis accounts for altered barrier function in aged mice. J Invest Dermatol 1996; 106:1064–1069.
71. Haratake A, Ikenaga K, Katoh N, Uchiwa H, Hirano S, Yasuho H. Topical mevalonic acid stimulates de novo cholesterol synthesis and epidermal permeability barrier homeostasis in aged mice. J Invest Dermatol 2000; 114:247–252.
72. Denda M, Kitamura K, Elias PM, Feingold KR. trans-4-(aminomethyl)cyclohexane carboxylic acid (t-AMCHA), an anti-fibrinolytic agent, accelerates barrier recovery and prevents the epidermal hyperplasia induced by epidermal injury in hairless mice and humans. J Invest Dermatol 1997; 109:84–90.
73. Mauro T, Grayson S, Gao WN, Man MQ, Kriehuber E, Behne M, Feingold KR, Elias PM. Barrier recovery is impeded at neutral pH, independent of ionic effects: implications for extracellular lipid processing. Arch Dermatol Res 1998; 290:215–222.
74. Denda M, Katagiri C, Hirao T, Maruyama N, Taahashi M. Some magnesium salts and a mixture of magnesium and calcium salts accelerate skin barrier recovery. Arch Dermatol Res 1999; 291:560–563.
75. Lee SH, Elias PM, Proksch E, Menon GK, Man MQ, Feingold KR. Calcium and potassium are important regulators of barrier homeostasis in murine epidermis. J Clin Invest 1992; 89:530–538.
76. Denda M, Hosoi J, Ashida Y. Visual imaging of ion distribution in human epidermis. Biochem Biophys Res Commun 2000; 272:134–137.
77. Menon GK, Elias PM, Lee SH, Feingold KR. Localization of calcium in murine epidermis following disruption and repair of the permeability barrier. Cell Tis Res 1992; 270:503–512.
78. Watt FM. Terminal differentiation of epidermal keratinocytes. Curr Opin Cell Biol 1989; 1:1107–1115.
79. Watanabe R, Wu K, Paul P, Marks DL, Kobayashi T, Pittelkow MR, Pagano RE. Up-regulation of glucosylceramide synthase expression and activity during human keratinocyte differentiation. J Biol Chem 1998; 273:9651–9655.
80. Menon GK, Price LF, Bommannan B, Elias PM, Feingold KR. Selective obliteration of the epidermal calcium gradient leads to enhanced lamellar body secretion. J Invest Dermatol 1994; 102:789–795.
81. Vicanova J, Boelsma E, Mommas AM, Kempenaar JA, Forslind B, Pallon J, Egelrund T, Koerten HK, Ponec M. Normalization of epidermal calcium distribution profile in reconstructed human epidermis is related to improvement of terminal differentiation and stratum corneum barrier formation. J Invest Dermatol 1998; 111:97–106.
82. Seabra MC, Goldstein JL, Sudhof TC, Brown MS. Rab geranylgeranyl transferase. J Biol Chem 1992; 267:14497–14503.
83. Novick P, Brennward P. Friends and family: the role of the Rab GTPases in vesicular traffic. Cell 1993; 75:597–601.
84. Song HJ, Rossi A, Ceci R, Kim IG, Anzano MA, Jang SI, DeLaurenzi V, Steinert PM. The genes encoding geranylgeranyl transferase α-subunit and transglutaminase I are very closely linked but not functionally related in terminally differentiating keratinocytes. Biochem Biophys Res Commun 1997; 235:10–14.
85. Imakado S, Bickenbach JR, Bundman DS, Rothnagel JA, Atter PS, Wang XJ, Wal-

czak VR, Wisniewski S, Pote J, Gordon JS, Heyman RA, Evans RM, Roop DR. Targeting expression of a dominant-negative retinoic acid receptor mutant in the epidermis of transgenic mice results in loss of barrier function. Genes Dev 1995; 9:317–329.
86. Hanley K, Jiang Y, Crumrine D, Bass NM, Appel R, Elias PM, Williams ML, Feingold KR. Activators of the nuclear hormone receptors PPARα and FXR accelerate the development of the fetal epidermal permeability barrier. J Clin Invest 1997; 100: 705–712.
87. Hanley K, Komuves LG, Bass NM, He SS, Jiang Y, Crumrine D, Appel R, Friedman M, Bettencourt J, Min K, Elias PM, Williams ML, Feingold KR. Fetal epidermal differentiation and barrier development in vivo is accelerated by nuclear hormone receptor activators. J Invest Dermatol 1999; 113:788–795.
88. Komuves LG, Hanley K, Jiang Y, Elias PM, Williams ML, Feingold KR. Ligands and activators of nuclear hormone receptors regulate epidermal differentiation during fetal rat skin development. J Invest Dermatol 1998; 111:429–433.
89. Komuves L, Hanley K, Lefebvre AM, Man MQ, Williams M, Elias PM, Auwerx J, Feingold KR. PPARα-activation promotes epidermal keratinocyte differentiation in vivo. J Invest Dermatol (abstr) 1999; 112:550.
90. Hanley K, Komuves LG, Ng DC, Schoonjans K, He SS, Lau P, Bikle DD, Williams ML, Elias PM, Auwerx J, Feingold KR. Farnesol stimulates differentiation in epidermal keratinocytes via PPARα. J Biol Chem 2000; 275:11484–11491.
91. Hanley K, Jiang Y, He SS, Friedman M, Elias PM, Bikle DD, Williams ML, Feingold KR. Keratinocyte differentiation is stimulated by activators of the nuclear hormone receptor PPARα. J Invest Dermatol 1998; 110:368–375.
92. Feingold KR. Role of nuclear hormone receptors in regulating epidermal differentiation. Program and preprints of annual scientific seminar, Society of Cosmetic Chemists 1999; 50:30–31.
93. Komves LG, Hanley K, Man MQ, Elias PM, Williams ML, Feingold KR. Keratinocyte differentiation in hyperproliferative epidermis: topical application of PPARα activators restores tissue homeostasis. J Invest Dermatol 2000; 115:361–367.
94. Ashida Y, Denda M, Hirao T. Histamine H1 and H2 receptor antagonists accelerate skin barrier repair and prevent epidermal hyperplasia induced by barrier disruption in a dry environment. J Invest Dermatol 2001; 116:261–265.
95. Halkier-Sorensen L, Menon GK, Elias PM, Thestrup-Pedersen K, Feingold KR. Cutaneous barrier function after cold exposure in hairless mice: a model to demonstrate how cold interferes with barrier homeostasis among workers in the fish-processing industry. Br J Dermatol 1995; 132:391–401.

5
Investigating Human Skin Barrier Lipids with In Vitro Skin Models

Annie Black and Odile Damour
Hôpital Edouard Herriot, Lyon, France

Kordula Schlotmann
Henkel KGaA, Düsseldorf, Germany

I. INTRODUCTION

Advances in technology and a movement toward reduced animal testing have spurred the development of alternative methods for determining skin irritation, sensitization, and efficacy of cosmetic products. The first generation of tissue-engineered epidermal equivalents are simple cultures that are easy to produce and permit efficient screening of the irritation potential of large quantities of molecules. However, it was recently demonstrated that the response of skin-equivalent systems to topically applied consumer products is influenced not only by epithelium but also by epithelial-mesenchymal interactions [1].

Thus, skin-equivalent models, comprising a reconstructed epithelium together with dermal equivalent, have been designed to meet the research and testing needs of the cosmetics, pharmaceutical, and chemical industries as well as for mechanistic studies of skin physiology and clinical applications for acute or chronic wounds. Unfortunately, the incomplete barrier function and short life span of these constructs are considered major limitations preventing their extensive use. The successful application of tissue-engineered skin equivalents requires the correct organization and function of the reconstructed tissue.

This chapter gives a brief overview of keratinocyte and fibroblast cell culture techniques. The different epidermis and skin-equivalent models are discussed as well as the characterization of a skin-equivalent model; finally, skin barrier lipid analysis of these different epidermis and skin models are discussed.

II. KERATINOCYTE AND FIBROBLAST MONOLAYER CULTURES

As a result of the work of Rheinwald and Green, serial passage of keratinocytes has become accessible and permits the production of large quantities of cells [2]. Keratinocytes can be isolated form skin biopsies obtained from breast or abdomen skin as well as foreskin by different enzymatic digestion techniques with trypsin, dispase, a combination of the two [3], or a thermolysin-trypsin combination [4]. Epidermal cell suspensions then are seeded on a feeder layer of lethally irradiated 3T3 mouse fibroblasts or human fibroblasts that produce both extracellular matrix and growth factors. These fibroblasts support keratinocyte attachment, stimulate proliferation [5], and inhibit the growth of contaminating fibroblasts.

For culturing keratinocytes, various culture media are used, the most common being a mixture of Dulbecco's Modified Eagle's Medium (DMEM) and Ham's F12 medium (3:1), supplemented with 10% bovine serum, hydrocortisone, insulin, transferrin, cholera toxin, 1,25-dihydroxyvitamin D3, ascorbic acid, and epidermal growth factor (EGF). EGF stimulates multiplication and increases the life span of keratinocytes [6]. Medium and culture conditions have also evolved and permit the growth of the cells without a feeder layer and in the absence of serum [7,8].

Cultures of dermal fibroblasts can be generated from separated dermis of foreskin, abdomen, or breast or any other type of skin sample. The tissue can be digested by a collagenase treatment, after which the cell suspension can be seeded in a mixture of DMEM with 10% serum. Fibroblasts can also be obtained from explant cultures.

A single skin specimen can generate keratinocytes and fibroblasts and also melanocytes [9] endothelial cells [10], and dendritic cells such as Langerhans cells [11,12] and dermal dendrocytes [13,14] (Fig. 1).

Conventional monolayer cell cultures present many advantages for toxicity and efficacy testing. They provide a large supply of material, in simple culture conditions and in a relatively short time, which permits large-scale screening. However, they have a limited capacity to organize into a physiological structure and exhibit in vivo characteristics. Moreover, testing material is limited to water-soluble compounds, and the lack of epidermal differentiation and stratum corneum can result in an overestimation of toxicity of the tested compound.

Despite the fact that the use of two-dimensional cultures has resulted in considerable progress in the understanding of keratinocyte biology, tissue-engineered epidermal and skin equivalents have been developed in order to im-

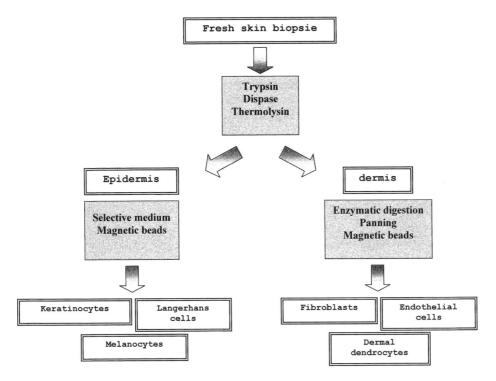

Figure 1 Schematic representation of the extractions of dermal and epithelial cells from fresh skin biopsies. First, epidermis and dermis can be separated by enzymatic digestion by different enzymes. From the resulting cell suspension, melanocytes and Langerhans cells can be isolated by selective medium, magnetic beads, or cell sorting. Dermal cell suspensions (mainly fibroblasts) can be obtained after incubation with collagenase or by explants. Endothelial cells can be obtained by selective medium or panning technique.

prove our understanding of cell-cell, cell-matrix, and dermal-epidermal interactions.

III. TISSUE-ENGINEERED EPIDERMIS AND SKIN MODELS

Epidermis and skin models are three-dimensional systems produced by seeding single cell suspensions of keratinocytes on an appropriate support. The keratinocytes attach to the substrate and proliferate. The culture is kept submerged for a few days to ensure coverage of the substrate. Subsequently, the support covered by a layer of keratinocytes is elevated at the air-liquid interface to induce terminal differentiation [15–17]. Growth and differentiation of keratinocytes is also affected by calcium and epidermal growth factor (EGF) concentration. These factors and a retinoic acid gradient created by the air-liquid interface is crucial for the appearance of a multilayered differentiated tissue [17–21].

Various types of substrates can be used to support keratinocyte growth and differentiation into reconstructed epidermis. They can be synthetic or biological, which can be either cellular or acellular.

A. Epidermis Models

Keratinocytes can be seeded either on a culture insert made of polycarbonate [22] or cellulose acetate [23] or on a layer of collagen type I and III overlayed by collagen type IV [24].

Epidermalization can also be achieved on human de-epidermalized dermis (DED) obtained from cadaver skin. The epidermis is removed and the dermal cells are killed by multiple freeze-thaw cycles. Only the dermal extracellular matrix and basement membrane components remain [25–27], which facilitates the adhesion of keratinocytes and the reconstruction of the dermal-epidermal junction.

These epidermal equivalents provide a local environment similar to that of in vivo tissues in regard to the effect of compounds in contact with the epidermis. The response of reconstructed tissues to these compounds may be normalized compared to monolayer cultures. Moreover, the effect of keratinocytes on the physiological organization and function of melanocytes and Langerhans cells can be studied [28–30].

B. Skin-Equivalent Models

Skin-equivalent models consist of a living dermal equivalent associated with an overlaying epidermis. Besides fibroblast-populated collagen gels, three-dimensional dermal equivalents can be generated by synthesis of extracellular matrix by fibroblasts inside a scaffold or on inert filters before keratinocyte seeding (Fig. 2).

Three-dimensional dermal equivalents have been also generated using combinations of fibroblasts and endothelial cells either in a collagen gel [31] or in a collagen sponge where the endothelial cells reorganize themselves into capillary-like structures [32,33] (Fig. 3).

The inclusion of melanocytes [34–37], melanocytes in combination with Langerhans cells [30], and melanocytes with endothelial cells compose the many variations that have been developed (Fig. 4). These cells not only share the environment of the keratinocytes and fibroblasts but interact with them in a manner that can influence their activities.

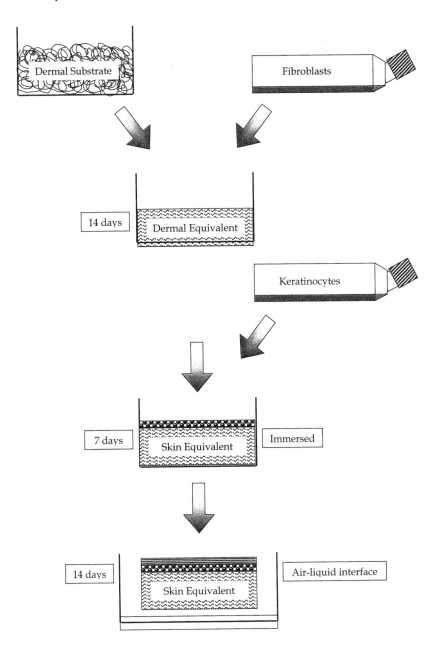

Figure 2 Human skin equivalent culture. Fibroblasts are seeded onto a dermal substrate and cultured for 14 days to obtain a dermal equivalent. The dermal equivalent is then seeded with keratinocytes and cultured for 7 days immersed in culture medium and 14 days at the air-culture medium interface for a total of 35 days.

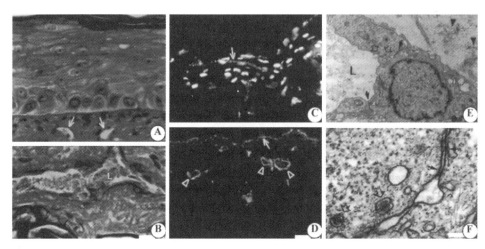

Figure 3 Tissue-engineered endothelialized skin equivalents. Biopolymers were seeded with HUVEC and fibroblasts and keratinocytes and cultured for 35 days. Sections were stained with Masson's trichrome. White arrows show the capillary-like structures with a lumen (L). Scale: (A) 1 cm = 16 µm; (B) 1 cm = 9 µm. Frozen sections were immunostained for human vWF (C) and laminin (D). Note the typical granular staining of vWF in the cells implicated in tubular structures (arrow) (C). Nuclei are stained with Hoechst 33258 (C). Basement membrane constituent, laminin (D), is observed surrounding tubular structures (arrowheads). Deposition of laminin was also detected at the dermal-epidermal junction (arrow) (D). Scale: (C) 1 cm = 27 µm; (D) 1 cm = 53 µm. Transmission electron microscopy of a portion of a capillary-like structure observed. An internal lumen (L), intercellular junctions (arrow), and Weibel-Palade bodies were seen (open arrows) (E, F). Scale: (E) 1 cm = 0,8 µm; (F) 1 cm = 10 µm.

1. Collagen-based Dermal Analogues

With the knowledge that collagen is the essential compound of connective tissue, Yannas was the first to use it to reconstitute an in vitro dermal analogue [38]. This acellular Dermal Substrate (DS) is obtained by lyophilization of a composite matrix made by coprecipitating bovine collagen type I and III and chondroitin sulfate. The cross-linking of collagen by glutaraldehyde is the most common procedure [39–44]. However, glutaraldehyde is toxic and therefore requires a fastidious elimination protocol. In addition, it can induce calcification of the sponges, which can have harmful consequences [45,46]. This is why other methods of cross-linking have been developed, such as periodate [47], acyl azide [46], or a dehydrothermic physical treatment [48]. This treatment consists of dehydrating the sponge to draw the constituent molecules closer together and thus permitting the creation of amide links. Last, in order to avoid the use of chemical prod-

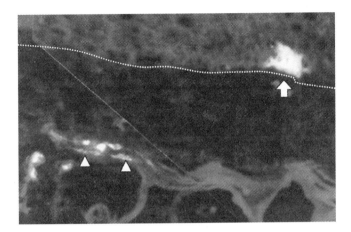

Figure 4 Melanocyte and endothelial cells are shown on serial frozen sections. Melanocytes are stained with MEL-5 (white arrow) and a capillary-like structure is stained with EN4 (arrowheads) in a pigmented endothelialized skin equivalent. The dotted line shows the basement membrane zone. Magnification: 40×.

ucts, a collagen sponge, made from bovine collagen and chondroitin sulfate, has been insolubilized by chitosan [49]. Chitosan, obtained by deacetylation of chitin, creates ionic bonds with carboxyl groups of collagen and sulfate groups of chondroitin sulfate. After lyophilization, these collagen-glycosaminoglycan-chitosan (C-GAG-C) polymers form an alveolar structure between 50 and 150 μm. The ionic bonds formed are sufficiently strong to give this sponge good mechanical properties [50,51].

2. Noncollagenous Dermal Analogues

Many synthetic biocompatible polymers, biodegradable or not, can be used as supports for fibroblast culture in the production of dermal equivalents. They differ by their structure—either porous, mesh, or layer—or by their composition—nylon, polyglycolic acid [52–54], poly-L-lactid, polyethylene oxide, or polybutylene terephthalate [55,56]. When these dermal analogue fibroblasts are seeded, they multiply, synthesize, and deposit new ECM.

3. Collagen Gel

Another way to obtain a dermal equivalent is by using a collagen gel [57,58]; this was proposed by Karasek and Charlton [59] and developed by Bell [60]. The resistance and insolubility of collagen is obtained by retraction of the gel by fibroblasts. This living DE has a final size proportional to the number of cells it contains and inversely proportional to the collagen concentration. In this model, cell proliferation and collagen synthesis are inhibited, probably due to biochemical confinement, the fibroblasts being restricted in an environment of retracted collagen.

The collagen gel model was also completed by the addition of hypodermis composed of pre-adipocytes and mature adipocytes cultured in the inferior part of a gel [61]. In addition to these cell populations, this model can also support the growth of hair follicles [62,63].

4. Self-Assembly Approach

The self-assembly approach comprises the multiplication of fibroblasts and the synthesis of extracellular matrix by the cells to obtain a three-dimensional tissue without the support of any artificial scaffold [64]. The dermal equivalent is composed of stacked fibroblast sheets where keratinocytes can be seeded. Complete pilosebaceous units were also integrated [64].

IV. CHARACTERIZATION OF A TISSUE-ENGINEERED COLLAGEN-GLYCOSAMINOGLYCAN-CHITOSAN SKIN EQUIVALENT

Previous studies with a skin equivalent based on a collagen-glycosaminoglycan-chitosan dermal substrate have demonstrated the importance of (1) fibroblasts in the epithelial morphogenesis [65], (2) fibroblast newly synthetized ECM in keratinocyte adhesion and the formation of a complex dermal-epidermal junction [66], (3) fibrillar collagens in the re-expression of collagen types XII and XIV [67], and (4) keratinocytes in the maturation and organization of elastin [68].

In all of these studies, the culture conditions favored fibroblast and keratinocyte proliferation and extracellular matrix synthesis but hindered keratinocyte differentiation. To improve keratinocyte differentiation, which is necessary for optimal pharmacotoxicological studies, it is necessary to control the proliferation-differentiation balance.

Despite the similarities in tissue architecture, the reconstructed epidermis in serum-containing media still exhibits some deviations from normal epidermis. It lacks a stratum granulosum, and the stratum corneum contains a high number of nuclear remnants. Many factors seem to be responsible for modulating epithelial differentiation [69–73]. The development of optimal culture conditions, eliminating serum and ill-defined supplements from medium, should improve reproducibility and keratinocyte differentiation.

After 14 days of culture at the air-liquid interface in serum-free medium, the epidermis of the skin equivalent is morphologically similar to normal human skin [74] (Fig. 5).

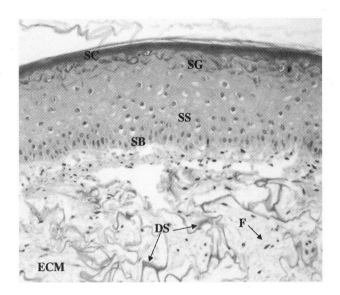

Figure 5 Histological analysis of the skin equivalent. Keratinocytes seeded on the dermal equivalent multiply to form a stratified, differentiated epidermis with stratum basal (**SB**), stratum spinosum (**SS**), stratum granulosum (**SG**), and stratum corneum (**SC**). The fibroblasts (**F**) are numerous and have filled the pores of the dermal substrate (**DS**) with newly synthesized extracellular matrix (**ECM**). Magnification: 40×.

The epidermis is stratified, well-organized, and uniform. A layer of cuboidal basal keratinocytes is firmly attached to the dermal equivalent underlying the stratum spinosum. The stratum granulosum is well developed, with large numbers of keratohyalin granules and a stratum corneum.

Hemidesmosome structures can be found in the basal surface of basal keratinocytes of the epidermis in contact with the dermal equivalent (Fig. 6).

Keratinocytes, joined by large numbers of desmosomes throughout the epidermis, contain abundant cytoplasmic keratin filaments and synthetic organelles (Fig. 7).

In the stratum granulosum, many keratohyalin granules are present (Fig. 7). The stratum corneum is six to seven layers thick, interspaced with desmosomal remnants (Fig. 7). The borders of the cornified cells overlap, interdigitate, and insert into adjacent cells (Fig. 7), and keratin filaments become more randomly oriented (Fig. 7).

Lamellar granules (LG) were found in the cytoplasm of spinous keratinocytes. LG were partially filled with stacks of lamellae (Fig. 8). This fully developed skin equivalent expresses a number of protein products typically found in

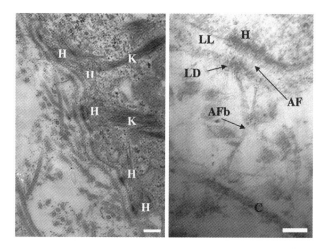

Figure 6 Ultrastructural analysis of the basement membrane. The basement membrane zone contains many hemidesmosomes (**H**) linked to keratin filaments (**K**) and characteristic normal human skin structures such as lamina lucida (**LL**) containing anchoring fibrils (**AF**). Underneath the lamina densa (**LD**) are anchoring fibers (**AFb**). These anchoring fibers are linked to collagen fibers (**C**). Left: Bar indicates 200 nm. Right: Bar indicates 100 nm.

normal human skin. Keratin 14 is present in the basal layer and gradually disappears in the suprabasal layers, whereas keratin 10 is found exclusively in suprabasal keratinocytes (Fig. 9).

The absence of serum-containing media for culture at the air-liquid interface improved keratinocyte differentiation. This could be explained by the absence of retinoids, which are contained in serum and impede terminal differentiation in vitro [75–77]. Another explanation of the differences in terminal differentiation obtained in previous studies with this model is the presence of epidermal growth factor. Chih-Shan et al. have demonstrated that in the continuing presence of 10–20 ng/ml of epidermal growth factor, the epidermis is less organized, thinner, and less proliferative [78]. EGF also depressed several indicators of differentiation, such as keratohyalin granules, membrane-coating granules, and filaggrin expression, and frequent nuclear retention was noted in the stratum corneum [78,71]

The presence of basement membrane proteins and their correct structural arrangement are critical elements for the stability of the dermal-epidermal junction. Contrary to findings in a collagen gel model cultured in defined conditions [72], the absence of serum does seem to affect the synthesis of basement membrane components in the equivalent grown on the collagen-glycosaminoglycan-

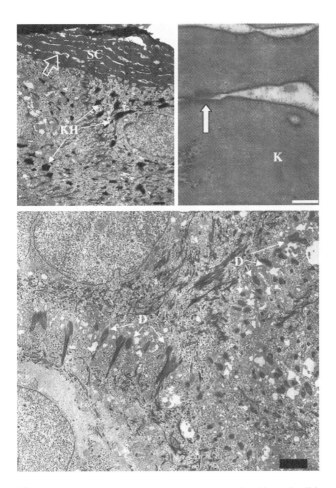

Figure 7 Ultrastructural analysis of suprabasal epidermal cell layers. The stratum spinosum shows numerous desmosomes (**D**), and the stratum granulosum contains keratohyalin granules (**KH**). The stratum corneum is composed of six to seven layers interspaced with desmosomal remnants (large white arrow). The borders of the cornified cells overlap, interdigitate, and insert into adjacent cells (clear arrow) and keratin filaments become more randomly oriented (**K**). Top left: Bar indicates 100 nm. Top right: Bar indicates 200 nm. Bottom: Bar indicates 200 nm.

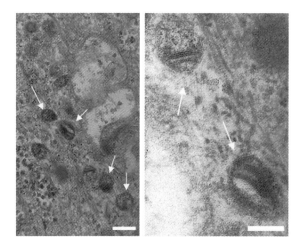

Figure 8 Lamellar granule ultrastructure. Lamellar granules are present in the spinous layer of the epidermis but they are not completely filled with lamellae. Left: Bar indicates 200 nm. Right: Bar indicates 100 nm.

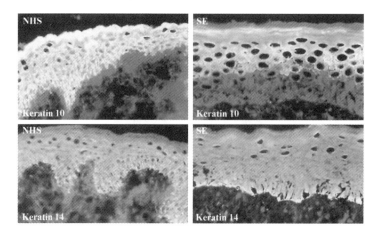

Figure 9 Immunohistochemical analysis of keratins. A layer of organized and adhering basal cells expressing keratin 14 and suprabasal keratinocytes showing differentiation-associated keratin 10 characterizes the epidermis of the skin equivalent. Similar staining patterns are seen in normal human skin (NHS).

chitosan dermal substrate [74]. Immunohistochemistry has demonstrated the presence of major proteins of the dermal-epidermal junction such as collagen type VII, laminin, and hemidesmosome-associated integrin α6. The structural integrity of the basement membrane zone ensured by series of linked extracellular structures including anchoring filaments and anchoring fibrils is also observed as demonstrated before in serum-containing conditions [66]. These results suggest that basement membrane stability is independent of serum [74].

Fibroblasts seeded in the dermal substrate proliferate in the pores of the sponge and synthesize their own extracellular matrix in which collagen and protein synthesis is abundant even in reduced serum conditions. The histological and immunohistochemical results show that fibroblasts express some of their morphogenic potential: reorganization of the matrix with synthesis de novo of matrix constituents. The composition of this neosynthesized extracellular matrix is close to that of normal human dermis; type I, III, and V collagens and fibronectin (Fig. 10) as well as the reconstruction of elastic tissue has been demonstrated (Fig. 11). Fibrillin-1 is detected in the dermis, with short fibers appearing perpendicular to the basement membrane; deeper in the dermis, fibers are parallel. Elastin is located close to the basement membrane zone in a parallel fashion.

Transmission electron microscopy has revealed a structured organization of the neosynthesized extracellular matrix in which the collagen quarter-staggered fibrils are regrouped into a network of bundles indicating complex dermal reconstruction, and in between these collagen fibrils is an abundant microfibrillar material (Fig. 12).

The absence of serum did not seem to affect the expression or stability of these extracellular matrix proteins. Duplan-Perrat et al. demonstrated that keratinocytes influence fibroblast migration, proliferation, and extracellular matrix synthesis [68]. Thus, keratinocytes could sustain this complex extracellular matrix organization in the absence of serum.

Thirty-five days are required to obtain a mature skin equivalent in these conditions. The equivalents were cultured for 14 additional days for a total of 49 days. This was an attempt to determine if the skin equivalents could be cultured longer with an optimal differentiation. In these conditions, both dermal and epidermal cells maintained their histotypical morphology. Basal cells are still proliferating and the epidermis still shows the histotypical morphology of normal human skin. Differentiation markers filaggrin and transglutaminase are still expressed in their respective positions, keratin 10 is expressed in the suprabasal layers, and Ki67 demonstrates proliferating cells in the basal layer of the epidermis (Fig. 13), showing optimal differentiation after 4 weeks at the air-liquid interface.

The complexity of the neosynthesized extracellular matrix provides a good support for long-term keratinocyte culture survival. The long-term survival of this collagen-glycosaminoglycan-chitosan–based skin-equivalent culture could

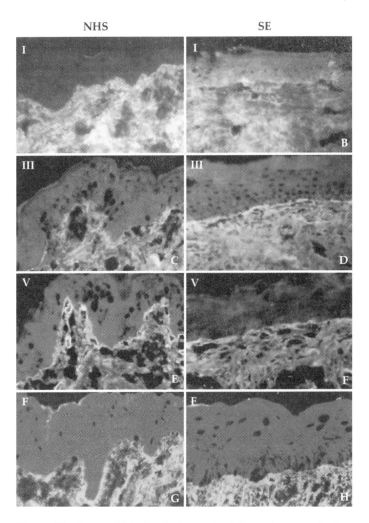

Figure 10 Immunohistochemical analysis of the newly synthesized extracellular matrix. Frozen sections were immunostained for human type I collagen, type III collagen, type V collagen, and fibronectin (F). Note that all these dermal extracellular matrix proteins are synthesized by fibroblasts in the skin equivalent (SE) as seen in normal human skin (NHS).

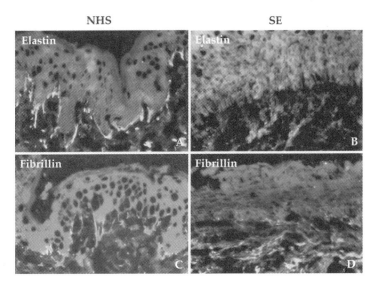

Figure 11 Elastic tissue reconstruction. Elastin and fibrillin-1 constitute the two major components of the elastic system in normal human skin (NHS). In the skin equivalent (SE), fibrillin-1 is detected in the dermis and shows an organization into microfibrils appearing perpendicularly to the BMZ, and deeper in the dermis, fibers are parallel. Elastin is deposited onto this microfibrillar network. Elastin staining seems to be more intense in the lower part of the reconstructed dermis.

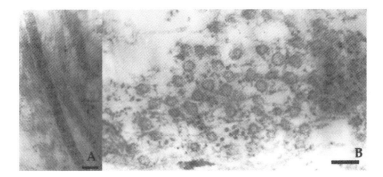

Figure 12 Ultrastructural analysis of neosynthetized extracellular matrix. Quarter-staggered collagen fibrils are organized into mature collagen fibers, regrouped in cross and longitudinal sections. Abundant microfibrillar material is also observed. Bars indicate 100 nm.

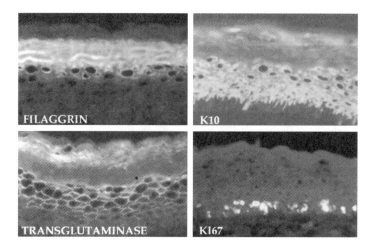

Figure 13 Long-term culture of the skin equivalent. Differentiation markers are still expressed in their respective positions and Ki67 demonstrates proliferating cells in the basal layer of the epidermis.

be explained by two reasons. First, the keratinocyte extraction and culture techniques permit the transfer of stem cells into the skin equivalent [74]. Second, the human extracellular matrix synthesized by the fibroblasts in conjunction with soluble factors secreted by these cells could help maintain the keratinocytes in their proper differentiated state by maintaining the balance between proliferation and differentiation. The maintenance of the in vivo characteristics will permit a more thorough investigation of the mechanisms involved in skin homeostasis. An increase in life span will allow the investigation of mechanisms that require longer exposure to active molecules or simply favor a more flexible management of pharmacotoxicology tests.

V. SKIN BARRIER LIPID COMPOSITION OF IN VITRO MODELS

The stratum corneum consists of protein-enriched corneocytes embedded in a lipid-enriched, intercellular matrix [79,80]. These extracellular lipids are organized in bilayers originating from lamellar bodies. The lipid and proteolytic enzyme content is discharged and reorganized into broad lamellar bilayers composed mainly of cholesterol, ceramides, and free fatty acids [81]. The composition and the organization of epidermal lipids are thought to contribute substantially to the permeability barrier of the stratum corneum.

Regarding the use of reconstructed human skin equivalents for fundamental and applied research, for efficacy testing and toxicity screening, and as models for drug transport and metabolism studies, the barrier function should resemble normal human skin as closely as possible. Thus, the ordered extrusion of lamellar bodies, the correct organization of the lipids in the lamellar bilayers as well as a physiological lipid composition, has to be achieved. Different techniques can be used to examine these phenomena—e.g., electron microscopy visualization of intercorneocyte lipid bilayers, using ruthenium tetroxide as an additional fixative. Various spectroscopy techniques such as small angle x-ray scattering and ATR-FTIR spectroscopy can help to determine the molecular structure of the lipids present in reconstructed skin models [82,83]. The lipid content and profile can be evaluated by the extraction of lipids from total epidermis or stratum corneum and analysis by high-performance thin-layer chromatography (HPTLC) [84–87], whereas the functionality of reconstructed skin equivalents can also be examined by techniques such as transepidermal water loss (TEWL) and percutaneous absorption.

Thirteen years ago, Ponec et al. [21] and Boddé et al. [20] investigated the intercellular lipid structures and the lipid content of keratinocytes grown at the air-liquid interface on de-epidermized dermis (DED). They found that the epidermal architecture was close to the in vivo situation, including the formation of lamellar bodies and intercellular lipid lamella. However, unusual lipid structures were observed locally between the corneocytes, and lamellar body structures were abnormal. Although all lipid species that are present in native skin can be produced, the lipid composition, analyzed by thin-layer chromatography, differed from that of human skin: Cultures contained significantly less linoleic acid (18:2) and significantly more oleic acid (18:1) and palmitoleic acid (16:1), and the triglyceride content was higher, whereas the content of glycosylceramides and ceramides was lower compared with normal human skin [88]. These deviations in lipid composition may induce the observed local imperfections and may contribute to an impaired barrier function. The suggestion was that an improvement of the lipid composition and a resulting improvement of the stratum corneum barrier may be achieved by changing the culture conditions. This hypothesis could be confirmed by different researchers. Mak et al. demonstrated that the water permeability of reconstructed skin could be lowered by reduction of the relative humidity in the environment, which possibly leads to a modulation in lipid biosynthesis in the stratum corneum [88].

Ponec and her group examined whether the triglyceride accumulation in the air-exposed cultures may be a result of an excessive supplementation of cells with glucose. Lowering of the glucose content in the medium resulted in a decrease in lactate production and triglyceride synthesis, but the triglyceride content remained still higher than in vivo, and the stratum corneum barrier function still remained impaired [89]. Later, it could be shown that the high triglyceride content is a consequence of EGF in the medium [71].

A clear improvement of lipid composition and an improved structural organization of the stratum corneum lipids could be obtained by using a fully defined serum- and EGF-free medium and culturing the epidermis models on DED at 33°C [70]. Under these conditions, not only was the balance between cell proliferation and differentiation improved, resulting in an extension of the life span of the cultures, but also the triglyceride content was low and similar to human skin, and the amounts of ceramides and free fatty acids (FFA) were increased. This may lead to a better solubilization of cholesterol and a reduction in the formation of crystalline cholesterol in the intercellular spaces. Additionally, as seen by small-angle x-ray scattering (SAXS), the structural organization of stratum corneum lipids was markedly improved. Similar results with medium without serum and no epidermal growth factor were seen using ATR-FTIR spectroscopy [83].

Although the described attempts led to an improvement of the lipid profile, the content of glycosylceramides still remained low, and the proportions of ceramide 4 to 7 were reduced, the largest reduction occuring in ceramides 6 and 7. Starting from the point that the presence of ascorbate as a cofactor may be required for the hydroxylation of sphingoid bases and fatty acids, and antioxidants may also be required to prevent lipid peroxidation during the enzymatic hydroxylation steps, supplementation of the medium with vitamin C and vitamin E was performed [90]. As observed in the normalization of the lipid profile by high-performance thin-layer chromatography (HPTLC), vitamin C plays a crucial role in lipogenesis. Both glucosphingolipids (GSL) and the ceramide profiles were normalized. The different GSL fractions (including acylglycosylceramides) and ceramides 6 and 7 were synthesized in significant amounts in vitro. Vitamin E alone did not show this effect. Adding vitamin C to the medium, the dermal substrate used for the reconstruction of the epidermis was also not important regarding the lipid composition: Whether using DED, inert filter fibroblast-populated collagen matrix, or DED populated with fibroblasts, the overall lipid profiles and the relative amount of ceramides were similar and close to those in native epidermis [90].

Vitamin C improved additionally the epidermal morphology and the ultrastructure of the stratum corneum as well as the stratum corneum lipid organization (SAXS analysis). As in native epidermis, numerous lamellar bodies were present in the stratum granulosum and excreted at the stratum granulosum/stratum corneum interface. Lipid lamellae appeared with multiple alternating electron-dense and electron-lucent bands. The number of intracellulary located lipid droplets within the corneocytes was very low. Although great similarities in structure between reconstructed and native skin was observed by electron microscopy, there were still some differences in the SAXS pattern that were difficult to explain: only the long-range lipid lamellar phase was present; the short lamellar phase, usually present in native skin, was missing. It was suggested that the desquamation process is involved in the regulation of stratum corneum lipid organization. Because this process is still impaired in reconstructed human skin, this may lead to the deviations observed in SAXS analysis [90].

Up to now, the information concerning lipids in reconstructed skin is obtained mainly from ''in-house'' models of reconstructed skin from the group of Maria Ponec; the question is, therefore, whether the barrier function of other

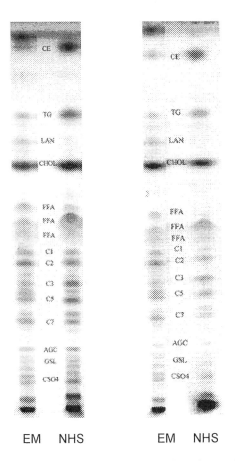

EM: epidermis model
NHS: normal human skin
CSO4: cholesterol sulfate,
GSL: glucosphingolipids,
AGC: acylglucosylceramides,
C 1-7: ceramides, **FFA:** free fatty acid, **CHOL:** cholesterol, **DG:** diglycerides, **LAN:** lanosterol, **TG:** triglycerides esters, **CE:** cholesterol esters

Figure 14 Analysis of lipid profiles of normal human epidermis (HE) and Skinethic epidermis model (with kind permission from Skinethic®, Nice, France).

epidermis models is also close to the in vivo situation. Lipid content and profile as well as the ultrastructure of the epidermis models EpiDerm™, SkinEthic™, and Episkin™ were compared in a recently published study [87].

Lipid analysis revealed that the major skin barrier lipids are synthesized in vitro but not in the same proportions as found in native skin (Fig. 14). The content of free fatty acids, one of the major lipid fractions in the skin, was mostly lower in reconstructed skin than in native skin. Lanosterol and cholesterol ester were also found in lower amounts in reconstructed tissues, whereas the triglyceride content was variable. The profile of glycosphingolipids and ceramides was incomplete. The content of ceramides 5 and 6 was markedly reduced and the most polar ceramide 7 was missing (Table 1). This may be the result of a lack of ascorbic acid in the medium. Regarding the ultrastructure examined by electron microscopy using ruthenium tetroxide as an additional fixative, all culture models showed the unique organization of the characteristic alternating electron-dense and electron-lucent lipid lamellae. However, this pattern did not exist throughout the whole intercellular space. The stratum corneum in some Skinethic™ and in most Episkin™ cultures was very compact, and multiple lipid droplets were seen in the corneocytes, which were found only occasionally in Epiderm™ cultures [87].

In conclusion, knowledge of the barrier function of reconstructed skin and of lipids in the stratum corneum of these models has improved markedly during the past years. Modulation of the culture conditions (lowering of temperature and humidity) and of the medium conditions (no serum in the air-liquid interphase, lowering of EGF and glucose, addition of vitamin C) has led to a nearly physiological lipid profile. Although some deviations still exist in comparison to normal

Table 1 Ceramide Profile in Different Skin Models

% of total ceramides	Epidermis models			Native epidermis
	EpiDerm (n = 5)	SkinEthic (n = 4)	Episkin (n = 3)	(n = 6)
1	10.5 ± 3.6	15.4 ± 3.2	9.8 ± 2.9	9.2 ± 1.9
2	45.9 ± 2.8	53.7 ± 5.5	42.6 ± 3.6	24.7 ± 5.7
2a	12.6 ± 3.9	2.4 ± 2.4	16.7 ± 4.2	2.7 ± 2.3
3	15.1 ± 2.6	18.8 ± 3.7	23.8 ± 2.6	21.7 ± 0.5
4	4.5 ± 0.6	3.5 ± 1.6	3.0 ± 1.4	4.8 ± 1.1
5	8.9 ± 1.8	4.7 ± 1.3	3.4 ± 1.8	20.9 ± 0.3
6	2.4 ± 0.6	1.4 ± 0.5	0.6 ± 0.4	5.7 ± 0.5
7	0.0 ± 0.0	0.0 ± 0.0	0.0 ± 0.0	10.3 ± 0.8

Source: From Ref. 87.

human skin, reconstructed skin models are of excellent use in applied and fundamental research—e.g., as a model system for the study of the modulation of lipid metabolism [91] as well as for pharmatoxicology and efficacy testing in cosmetics.

VI. CONCLUSION

Tissue-engineered skin equivalents as alternatives to animal experimentation offer not only a way to meet the demands of regulatory agencies, animal welfare organizations, consumers, and scientists but also to provide a means to improve and extend our knowledge of biological processes in the skin. This is particularly important when the purpose of these studies is to extrapolate the results to humans. Progress in three-dimensional cell culture techniques has provided coculture models for growing keratinocytes, fibroblasts, melanocytes, endothelial cells, and Langerhans cells. At this time, no validated and accepted model can accurately replace animal experimentation, but alternative methods have evolved very rapidly in the past few years, being a high scientific priority. In this respect, human tissue engineered skin equivalents form promising alternatives, because cells are of human origin and the equivalents closely resemble the tissue of origin. Especially in regard to the stratum corneum, the lipid content and profile, and the barrier function, the goal to achieve skin models that closely resemble human skin has been nearly achieved by optimization of the culture medium and culture conditions. Nevertheless, it has to be kept in mind that the lipid profile and the organization of lipids in the stratum corneum of skin equivalents deviate in some ways from human skin.

REFERENCES

1. Bernhofer LP, Seiberg M, Martin KM. The influence of the response of skin equivalent systems to topically applied consumer products by epithelial-mesenchymal interactions. Toxicol In Vitro 1999; 13:219–229.
2. Rheinwald JG, Green H. Serial cultivation of strains of human keratinocytes: the formation of keratinizing colonies from single cells. Cell 1975; 6:331–344.
3. Prunieras M, Delescluse C, Régnier M. The culture of skin: a review of theories and experimental methods. J Invest Dermatol 1976; 67:58–65.
4. Germain L, Rouabhia M, Guignard R, Carrier L, Bouvard V, Auger FA. Improvement of human keratinocyte isolation and culture using thermolysin. Burns 1993; 19:99–104.
5. Alitalio K, Kuismanen E, Myllyla R, Kiistala U, Asko-Selijavaara S. Vaheri A. Extracellular matrix proteins of human epidermal keratinocytes and feeder 3T3 cells. J Cell Biol 1982; 94:497–505.

6. Rheinwald JG, Green H. Epidermal growth factor and the multiplication of cultured human epidermal keratinocytes. Nature 1977; 265:421–424.
7. Peehl DM, Ham RG. Growth and differentiation of human keratinocytes without a feeder layer or conditioned medium. In Vitro Cell Dev Biol 1980; 16:516–525.
8. Boyce ST, Ham RG. Cultivation of frozen storage and clonal growth of normal human epidermal keratinocytes in serum-free medium. J Tissue Cult Meth 1985; 9: 83–93.
9. Eisinger M, Marko O. Selective proliferation of normal human melanocytes in vitro in the presence of phorbol ester and cholera toxin. Proc Natl Acad Sci USA 1986; 79:2015–2022.
10. Swerlick RA, Garcia-Gonzalez E, Kubota Y, Xu YL, Lawley TJ. Studies of the modulation of MHC antigen and cell adhesion molecule expression on human dermal microvascular endothelial cells. J Invest Dermatol 1991; 97:190–196.
11. Romani N, Lenz A, Glassl H, Stossel H, Stanzl U, Majdic O, Fritsh P, Schuler G. Cultured Langerhans cells resemble lymphoid dendritic cells in phenotype and function. J Invest Dermatol 1989; 93:600–609.
12. Teunissen MBM, Wormeester J, Krieg SR, Peters PJ, Vogel IMC, Kapsenberg ML, Bos JD. Human epidermal Langerhans cells undergo profound morphological and phenotypical changes during in vitro culture. J Invest Dermatol 1990; 94:166–173.
13. Lenz A, Heine M, Schuler G, Romani N. Human and murine dermis contain dendritic cells. J Clin Invest 1993; 92:2587–2596.
14. Nestle FO, Zheng XG, Thompson CB, Turka LA, Nickoloff BJ. Characterization of dermal dendritic cells obtained from normal human skin reveals phenotypic and functionally distinctive subsets. J Immunol 1993; 151:6535–6545.
15. Régnier M, Pruniéras M, Woodley D. Growth and differentiation of adult human epidermal cells on dermal substrates. Front Matrix Biol 1981; 9:4–35.
16. Pruniéras M, Régnier M, Woodley D. Methods for cultivation of keratinocytes with an air-liquid interface. J Invest Dermatol 1983; 81:28s–33s.
17. Régnier M, Schweizer J, Michel S, Bailly C, Pruniéras M. Expression of high molecular weight (67kD) keratin in human keratinocytes cultured on dead de-epidermized dermis. Exp Cell Res 1986; 165:63–72.
18. Asselineau D, Bailly C, Bernard BA, Darmon M. Epidermal morphogenesis and induction of the 67K keratin polypeptide by cultures of human keratinocytes at the air liquid interface. Exp Cell Res 1985; 159:536–539.
19. Asselineau D, Bernard BA, Bailly C, Darmon M. Retinoic acid improves epidermal morphogenesis. Dev Biol 1989; 133:322–335.
20. Boddé HE, Holman B, Spies F, Weerheim A, Kempenaar J, Mommaas M, Ponec M. Freeze-fracture electron microscopy of in vitro reconstructed human epidermis. J Invest Dermatol 1990; 95:108–116.
21. Ponec M, Weerheim A, Kempenaar J, Mommaas A, Nugteren DH. Lipid composition of cultured human keratinocytes in relation to their differentiation. J Lipid Res 1988; 29:949–961.
22. Rosdy M, Clauss L. Terminal epidermal differentiation of human keratinocytes grown in chemically defined medium on inert filter substrates at the air-liquid interface. J Invest Dermatol 1990; 96:409–414.
23. Zhao JF, Zhang YJ, Kubilus J, Jin XH, Santella RM, Athar M, Wang ZY, Bickers

DR. Reconstituted 3-dimensional human skin as a novel in vitro model for studies of carcinogenesis. Biochem Biophys Res Commun 1999; 8:49–53.
24. Tinois E, Tiollier J, Gaucherand M, Dumas H, Tardy M, Thivolet J. In vivo and post-transplantation differentiation of keratinocytes grown on the human type IV collagen film of a bilayered dermal substitute. Exp Cell Res 1991; 193:310–319.
25. Freeman AE, Igel HJ, Herman BJ, Kleinfield KL. Growth and characterization of human skin epithelial cell cultures. In Vitro 1976; 12:352–362.
26. Ponec M, Weerheim A, Kempenaar J, Mommaas A, Nugteren DH. Lipid composition of cultured human keratinocytes in relation to their differentiation. J Lipid Res 1988; 29:949–961.
27. Régnier M, Asselineau D, Lenoir MC. Human epidermis reconstructed on dermal substrate in vitro: an alternative to animals. Skin Pharmacol 1990; 3:70–85.
28. Bertaux B, Morlière P, Moreno G, Courtalon A, Masse JM, Dubertret L. Growth of melanocytes in a skin equivalent model in vitro. Br J Dermatol 1988; 119:503–512.
29. Bessou S, Surleve-Bazeille JE, Sorbier E, Taieb A. Ex vivo reconstruction of the epidermis with melanocytes and the influence of UVB. Pigment Cell Res 1995; 8: 241–249.
30. Régnier M, Staquet MJ, Schmitt D, Schmidt R. Integration of Langerhans cells into a pigmented reconstructed human epidermis. J Invest Dermatol 1997; 109:510–512.
31. Smola H, Stark HJ, Thiekotter G, Mirancea N, Krieg T, Fusenig NE. Dynamics of basement membrane formation by keratinocyte-fibroblast interactions in organotypic skin culture. Exp Cell Res 1998; 15:399–410.
32. Black AF, Berthod F, L'Heureux N, Germain L, Auger FA. In vitro reconstruction of a human capillary-like network in a tissue-engineered skin equivalent. FASEB J 1998; 12:1331–1340.
33. Black AF, Hudon V, Damour O, Germain L, Auger FA. A novel approach for studying angiogenesis: a human skin equivalent with a capillary-like network. Cell Biol Tox 1999; 15:81–90.
34. Regnier M, Duval C, Galey JB, Philippe M, Lagrange A, Tuloup R, Schmidt R. Keratinocyte-melanocyte co-cultures and pigmented reconstructed human epidermis: models to study modulation of melanogenesis. Cell Mol Biol (Noisy-le-grand) 1999; 45:969–980.
35. Bessou S, Surleve-Bazeille JE, Sorbier E, Taieb A. Ex vivo reconstruction of the epidermis with melanocytes and the influence of UVB. Pigment Cell Res 1995; 8: 241–249.
36. Boyce ST, Medrano E, Abdul Malek Z, Supp A, Dodick J, Nordlund J, Warden G. Pigmentation and inhibition of wound contraction by cultured skin substitute with adult melanocytes after transplantation to athymic mice. J Invest Dermatol 1993, 4: 360–365.
37. Nakazawa K, Nakazawa H, Sahuc F, Lepavec A, Collombel C, Damour O. Pigmented human skin equivalent—new method of reconstruction by grafting an epithelial sheet onto a non-contractile dermal equivalent. Pigment Cell Res 1997; 10:382–390.
38. Yannas IV, Burke JF. Design of an artificial skin: I. Basic design principles. J. Biomed Mat Res 1980; 14:65–68.

39. Oliver RF, Barker H, Cooke A, Grant RA. Dermal collagen implants. Biomaterials 1982a, 3:38–40.
40. Oliver RF, Barker H, Cooke A, Stephen L. 3H-collagen turnover in non-cross-linked and aldehyde-cross-linked dermal collagen grafts. Br J Exp Path 1982b, 63:13–17.
41. Doillon CJ, Silver FH. Collagen-based wound dressing: effects of hyaluronic acid and fibronectin on wound healing. Biomaterials 1986; 7:3–8.
42. Doillon CJ, Silver FH, Berg RA. Fibroblast growth on a porous collagen sponge containing hyaluronic acid and fibronectin. Biomaterials 1987; 8:195–200.
43. Boyce ST, Christianson DJ, Hansbrough JF. Structure of a collagen-GAG dermal skin substitute optimized for cultured epidermal keratinocytes. J Biomed Mat Res 1988; 22:939–957.
44. Suzuki S, Matsuda K, Isshiki N, Tamada Y, Ikada Y. Experimental study of a newly developed bilayer artificial skin. Biomaterials 1990; 11:356–360.
45. Levy RJ, Schoen FJ, Sherman FS, Nichols J, Hawley MA, Lund SA. Calcification of subcutaneously implanted type I collagen sponges. Am J Pathol 1986; 122:71–92.
46. Petite H, Rault I, Huc A, Menashe P, Herbage D. Use of the acyl-azide method for cross-linking collagen-rich tissues as pericardium. J Biomed Mat Res 1990; 24:179–187.
47. Tiollier J, Dumas H, Tardy M, Tayot JL. Fibroblast behavior on gels of type I, III and IV human placental collagens. Exp Cell Res 1990; 191:95–104.
48. Koide M, Osaki K, Konishi J, Oyamada K, Katakura T, Takahashi A, Yoshizato K. A new type of biomaterial for artificial skin: dehydrothermally cross-linked composites of fibrillar and denatured collagens. J Biomed Mater Res 1993; 27:79–87.
49. Collombel C, Damour O, Gagnieu C, Marichy J, Poinsignon F. Biomaterials with a base of collagen, chitosan and glycosaminoglycans; process for preparing them and their application in human medicine. French patent 8708252, 1987. European patent 884101948, 1988. US patent PCT/FR/8800303, 1989.
50. Berthod F, Damour O, Collombel C. Collagen synthesis by fibroblasts cultured within a collagen sponge. Biomaterials 1993; 14:749–754.
51. Damour O, Gueugniaud PY, Berthin M, Rousselle P, Berthod F, Collombel C. A dermal substrate made of collagen-GAG-chitosan for deep burn coverage: first clinical uses. Clin Mat 1994; 15:273–276.
52. Landeen LK, Zeigler FC, Halberstadt C, Cohen R, Slivka SR. Characterization of a human dermal replacement. Wounds 1992; 4:167–175.
53. Cooper ML, Hansbrough JF, Spielvogel RL. In vivo optimization of a living dermis substitute employing cultured human fibroblasts on a biodegradable polyglycolic acid mesh. Biomaterials 1991; 12:243–248.
54. Contard P, Bartel R, Jacobs L, Perlich J, Mac Donald D, Handler L, Cone D, Fleishmajer P. Culturing keratinocytes and fibroblasts in a three dimensionnal mesh results in epidermal differentiation and formation of a basal lamina anchoring zone. J Invest Dermatol 1993; 100:35–39.
55. Beumer G. In vitro evaluation of biodegradable polymers as constituents of a potential living skin equivalent. J Invest Dermatol 1990; 42:462.

56. Beumer G, Van Blitterswijk C, Ponec M. Biocompatibility of a biodegradable matrix used as a skin substitute: an in vivo evaluation. J Biomed Mat Res 1994; 28:454–552.
57. Ehrmann RL, Gey GO. The growth of the cells on a transparent gel of reconstituted rat tail collagen. J Natl Cancer Inst 1956; 16:1375–1386.
58. Elsdale TR, Bard JB. Collagen substrata for studies on cell behavior. J Cell Biol 1972; 42:298–311.
59. Karasek MA, Charlton ME. Growth of post-embryonic skin epithelial cells on collagen gels. J Invest Dermatol 1971; 56:205–210.
60. Bell E, Ivarsson B, Merril C. Production of a tissue-like structure by contraction of collagen lattices by human fibroblasts of different proliferative potential in vitro. Proc Natl Acad Sci USA 1979; 76:1274–1278.
61. Sugihara H, Toda S, Miyabara S, Kusaba Y, Minami Y. Reconstruction of the skin in three-dimensional collagen gel matrix culture. In Vitro Cell Dev Biol 1991; 27A:142–146.
62. Yuspa SH, Wang Q, Weinberg WC, Goodman L, Ledbetter S, Dooley T, Lichti U. Regulation of hair follicle development: an in vitro model for hair follicle invasion of dermis and associated connective tissue remodeling. J Invest Dermatol 1993; 101:27s–32s.
63. Jahoda CAB, Reynolds AJ. Dermal-epidermal interaction follicle-derived cell populations in the study of hair-growth mechanisms. J Invest Dermatol 1993; 101:33s–38s.
64. Michel M, L Heureux N, Pouliot R, Xu W, Auger FA, Germain L. Characterization of a new tissue-engineered human skin equivalent with hair. In Vitro Cell Dev Biol Anim 1999; 35:318–326.
65. Saintigny G, Bonnard M, Damour O, Collombel C. Reconstruction of epidermis on a chitosan cross-linked collagen-GAG lattice: effects of fibroblasts. Acta Derm Venereol 1993; 73:175–180.
66. Sahuc F, Nakazawa K, Berthod F, Damour O, Collombel C. Mesenchymal-epithelial interactions regulate gene expression of type VII collagen and kalinin in keratinocytes and dermal-epidermal junction formation in a skin equivalent model. Wound Rep Reg 1996; 4:93–102.
67. Berthod F, Germain L, Guignard R, Lethias C, Garrone R, Damour O, Van der Rest M, Auger FA. Differential expression of collagens XII and XIV in human skin and in reconstructed skin. J Invest Dermatol 1997; 108:737–742.
68. Duplan-Perrat F, Damour O, Montrocher C, Peyrol S, Grenier G, Jacob MP, Braye FM. Keratinocytes influence the maturation and organization of the elastin network in a skin equivalent. J Invest Dermatol 2000; 114:365–370.
69. Bouwstra JA, Gooris GS, Dubbelaar FER, Weerheim AM, Ponec M. pH, cholesterol sulfate and fatty acids affect the stratum corneum lipid organization. J Invest Dermatol 1998; 3:69–74.
70. Gibbs S, Vicanovà J, Bouwstra J, Valstar D, Kempenaar J, Ponec M. Culture of reconstructed epidermis in a defined medium at 33°C shows a delayed epidermal maturation, prolonged lifespan and improved stratum corneum. Arch Dermatol Res 1997; 289:585–595.

71. Ponec M, Gibbs S, Weerheim A, Kempenaar J, Mulder A, Mommaas AM. Epidermal growth factor and temperature regulate keratinocyte differentiation. Arch Dermatol Res 1997; 289:317–326.
72. Stark HJ, Baur M, Breitkreutz D, Mirancea N, Fusenig NE. Organotypic keratinocyte cocultures in defined medium with regular epidermal morphogenesis and differentiation. J Invest Dermatol 1999; 112:681–691.
73. Vicanova J, Boelsma E, Mommaas AM, Kempenaar JA, Forslind B, Pallon J, Egelrud T, Koerten HK, Ponec M. Normalization of epidermal calcium distribution profile in reconstructed human epidermis is related to improvement of terminal differentiation and stratum corneum barrier formation. J Invest Dermatol 1998; 111:97–106.
74. Black AF, Schlotmann K, Kaeten M, Foerster T, Auger FA, Damour O. Characterization of a human skin equivalent model used for pharmacotoxicology testing. Submitted.
75. Watt F. Terminal differentiation of epidermal keratinocytes. Curr Opin Cell Biol 1989; 1:1107–1115.
76. Fuchs E. Epidermal differentiation. Curr Opin Cell Biol 1990; 2:1028–1035.
77. Rosdy M, Clauss LC. Terminal epidermal differentiation of human keratinocytes grown in chemically defined medium on inert filter substrates at the air-liquid interface. J Invest Dermatol 1990; 95:409–414.
78. Chih-Shan JC, Lavker RM, Rodeck U, Risse B, Jensen PJ. Use of a serum-free epidermal culture model to show the deleterious effects of epidermal growth factor on morphogenesis and differentiation. J Invest Dermatol 1995; 104:107–112.
79. Schürer NY, Elias PM. The biochemistry and function of stratum corneum lipids. In: Elias PM, ed. Skin Lipids, Advances in Lipid Research. Vol 24. San Diego: Academic Press; 1991:27–56.
80. Wertz PW, Downing DT. Ceramides of pig epidermis: structure determination. J Lipid Res 1983; 24:1135–1139.
81. Gray GM, Yardley HJ. Different populations of pig epidermal cells: isolation and lipid composition. J Lipid Res 1975; 16:441–447.
82. Bouwstra JA, Gooris GS, van der Spek JA, Bras W. The structure of human stratum corneum as determined by small angle X-ray scattering. J Invest Dermatol 1991; 96:1006–1014.
83. Pouliot R, Germain L, Auger FA, Tremblay N, Juhasz J. Physical characterization of an in vitro human skin equivalent produced by tissue engineering and its comparison with normal human skin by ATR-FTIR spectroscopy and thermal analysis (DSC). Biochim Biophys Acta 1999; 1439:341–352.
84. Bligh EG, Dyer WJ. A rapid method of total lipid extraction and purification. Can P Biochem Physiol 1959; 37:911–917.
85. Ponec M, Weerheim A. Retinoids and lipid changes in keratinocytes. In: Packer L, ed. Methods in Enzymology. San Diego: Academic Press; 1990:30–41.
86. Bowser PA, White RJ. Isolation, barrier properties and lipid analysis of stratum compactum, a discrete region of the stratum corneum. Br J Dermatol 1985; 112:1–14.
87. Ponec M, Boelsma E, Weerheim A, Mulder A, Bouwstra J, Mommaas M. Lipid and ultrastructural characterization of reconstructed skin models. Int J Pharm 2000; 203:211–225.

88. Mak VHW, Cumpstone MB, Kennedy AH, Harmon CS, Guy RH, Potts RO. Barrier function of human keratinocyte cultures grown at the air-liquid interface. J Invest Dermatol 1991; 96:323–327.
89. Ponec M, Kempenaar J, Weerheim A, de Lannoy L, Kalkman I, Jansen H. Triglyceride metabolism in human keratinocytes cultured at the air-liquid interface. Arch Dermatol Res 1995; 287:723–730.
90. Ponec M, Weerheim A, Kempenaar J, Mulder A, Gooris GS, Bouwstra J, Mommaas M. The formation of competent barrier lipids in reconstructed human epidermis requires the presence of vitamin C. J Invest Dermatol 1997; 109:348–355.
91. Rivier M, Castiel I, Safonova I, Ailhaud G, Michel S. Peroxisome proliferator-activated receptor-α enhances lipid metabolism in a skin equivalent model. J Invest Dermatol 2000; 114:681–687.

6
Analytical Techniques for Skin Lipids

Kristien De Paepe and Vera Rogiers
Vrije Universiteit Brussel, Brussels, Belgium

I. INTRODUCTION

The lipid lamellae of the intercellular spaces of the stratum corneum (SC) are essential components of an adequate barrier function [1–3]. In combination with the hydrolipidic layer—a water and oil mixture thought to cover part of the body surface—the barrier lipids (i.e., ceramides, cholesterol, and free fatty acids [FFAs]) prevent excessive transepidermal water loss (TEWL) and skin penetration of irritants and potentially harmful xenobiotics [4–6]. Because aged skin displays an increased susceptibility to irritants and the water retention capacity of surfactant-induced scaly skin is decreased, it has been hypothesized that changes in barrier lipid composition or their physical structure may account for these observations [7–11]. Furthermore, several skin disorders, such as atopic dermatitis [12–14], lamellar ichthyosis [15–17], severe xerosis [18,19], and psoriasis [20–23], most likely can be attributed to abnormalities of the intercellular lipids in the SC, which could be responsible for the functional disorders.

However, it still remains questionable whether the observed barrier function impairment is directly linked with changes in barrier lipid composition. A general problem with reports in this field is the lack of background information with respect to the many variables that could play a crucial role in the regulation and modulation of the barrier lipid composition. Also, one should critically analyze the techniques used, because these often are complex, are not standardized, and only provide semiquantitative results. In addition, skin sampling is not well standardized, meaning that a variety of methods is used, including blister techniques, skin strippings, surgical skin removal and in vivo lipid extractions. Factors affecting the barrier composition of the SC are given below.

II. FACTORS AFFECTING THE BARRIER COMPOSITION OF THE SC

A. Age and Sex

Individual factors have been reported to play a role in barrier lipid composition, such as age [9,11,24,25] and sex [9]. Although Denda et al. [9] showed sex differences in sphingolipid composition are a function of aging, it is now generally acknowledged that there are no sex-related differences in barrier integrity or barrier recovery [26,27].

With respect to age, conflicting data have been reported. These may result from differences in sampling of SC lipids or from an incomplete assessment of the lipid profile. Differences above the age of 50 [12,18,28] or 80 years have been described [11]. Arguments for age-related differences are found in the observation that aged skin displays an altered drug permeability and an increased susceptibility to irritant contact dermatitis (ICD), suggesting an impaired epidermal barrier function with an increased TEWL [11]. After solvent treatment or tape stripping, the barrier recovered more slowly in aged than in young volunteers. Also, xerosis is very common among the elderly. Its appearance changes according to the season (winter xerosis), and it is more severe in a low-humidity environment [25].

In various reports, decreases as a function of aging of the total SC lipid content and, more specifically, of the ceramides, have been described [9–12, 19,24,25]. In one report, no correlation could be found between the SC lipid content and age or sex of the subjects [27]. Questionable, however, is that none of these parameters was specified for the volunteers involved.

B. Anatomical Site

The measurement site on the body also is important because it has been shown that regional anatomic variations in skin permeability are directly related to SC lipid concentrations [19,25,29]. The differences in lipid weight found at four sites (abdomen, leg, face and sole, see Table 1) were inversely proportional to the permeability; also, a relationship with TEWL measurements was observed [29,30]. In this respect, palms had the highest TEWL and the lowest lipid content, whereas face samples revealed the opposite.

Table 1 Regional Variation in Lipid Weight Percentage (wt %) of Human SC Samples (Results expressed as mean values ± SEM)

Site	Abdomen	Leg	Face	Plantar
Lipids (wt %)	6.5 ± 0.5	4.3 ± 0.8	7.2 ± 0.4	2.0 ± 0.6

SEM: standard error of the mean.
Source: Ref. 29.

C. Racial Factors

Skin type [26] and race [31,32] also seem to be reflected in the barrier lipid concentrations. In a report published in 1959 [31], higher amounts of barrier lipids in black than in white people were found, but these results were never confirmed. Till today, data in the literature about race and phenotype were very confusing and sometimes even conflicting. It is generally thought that racial differences do exist at the level of the epidermal structure and function, but clear evidence is hardly given. It is suggested that darkly pigmented skin displays a more resistant barrier, which also recovers faster after mechanical perturbation (tape stripping) [26,32]. However, when combining these observations with noninvasive bioengineering measurements, such as TEWL, this type of conclusion becomes problematic [33].

D. Dietary and Hormonal Factors

Except for essential fatty acid deficiencies (EFAD) [34,35], no hard data have been found with respect to dietary influences or hormonal effects. Denda et al. [9] suggested hormonal effects on epidermal lipid synthesis because the age-related differences observed in SC lipid content were limited to women. Other groups could not find any effect of the menstrual cycle on barrier integrity [26]. However, both Agner et al. [36] and Elsner et al. [37] reported about the changes in skin sensitivity to irritation in relation to the menstrual status of women.

E. Medical Therapy

In some reports, skin effects by systemic treatment with hypocholesterolemic drugs are discussed. Although local application of an inhibitor (e.g., lovastatin) of hydroxymethylglutaryl CoA-reductase—the rate-limiting enzyme in cholesterol synthesis—has been shown to delay barrier repair on murine skin [38], no effect on the water permeability barrier of human skin was seen when statins were taken orally [39,40].

Table 2 Summary of Weight Percentages of the Different SC Lipids of Healthy Human Skin as Found by Different Research Groups (The numbering of the ceramide subgroups is according to Wertz et al. [47]. If not indicated otherwise, values are expressed as a percentage of the total lipids [mean ± SD].)

Lipids	Author: Wertz et al. 1985 [47] Forearm skin	Melnik et al. 1989 [48] Plantar skin	Lavrijsen et al. 1997 [49] Forearm skin a	b	Wertz and Downing 1991 [50] NI	Wertz 1996 [3] NI	Brod 1991 [51] NI^d	Yamamoto et al. 1991 [13] Forearms	Lampe et al. 1983 [29]; Elias 1983 [52] Abdominal^d	Leg^d	Face^d	Plantar^d
Sphingolipids							30 ± 3.5		18.2 ± 2.8	25.9 ± 1.3	26.5 ± 0.9	34.8 ± 2.1
Glycosylceramides			0.15	0	0.0		Traces		2.6	3.4	6.7	6.4
Ceramides		35.0 ± 3.2	18.0	38.6			30 ± 3.0	16.8 ± 3.7	15.5	22.6	19.9	28.4
Cer 1	7.0 ± 3.2		11.8	7.9	3.2	4.4		9.9 ± 3.5				
Cer 2	21.0 ± 4.9		22.4	28.7	8.9	18.0		12.2 ± 4.3				
Cer 3	13.4 ± 4.3		21.4	18.3	4.9	7.4		20.5 ± 4.3				
Cer 4	22.2 ± 4.5		7.19	6.4	6.1	4.7		8.9 ± 7.9				
Cer 5			19.6	20.3	5.7	4.9		26.3 ± 9.9				
Cer 6I	9.8 ± 1.1		6.4	5.9	12.3	8.2		21.9 ± 4.0				
Cer 6II	13.6 ± 4.5		10.9	12.5								
Phospholipids	5.0	3.2 ± 1.1	1.95	0	ND		2.0 ± 0.5	5.0 ± 1.4	4.9 ±1.6	5.2 ± 1.1	3.3 ± 0.3	3.2 ± 0.9
Cholesterol sulfate	3.2	1.8 ± 0.7	1.4	1.1	1.9		1.5 ± 0.4	3.2 ± 0.6	1.5 ± 0.2	6.0 ± 0.9	2.7 ± 0.3	3.4 ± 1.2
Neutral lipids	35.1						66.0 ± 2.8		77.7 ± 5.6	65.7 ± 1.8	66.4 ± 1.4	60.4 ± 0.9
Free sterols/CHOL	14.9	28.9 ± 3.6	10.9	16.8	26.9	28.0	32 ± 2.5	14.9 ± 2.8	14.0 ± 1.1	20.1 ± 2.0	17.3 ± 0.5	32.8 ± 1.6
FFA	8.0	19.2 ± 3.5	7.1	13.0	9.1	14.1	20 ± 2.0	8.0 ± 4.9	19.3 ± 3.7	13.9 ± 1.8	19.7 ± 0.6	9.0 ± 1.7
DG/TG	9.1	3.5 ± 1.4	50.4	13.4	0.0		2 ± 0.5	9.1 ± 2.1	25.2 ± 4.6	20.1 ± 1.0	13.5 ± 1.0	5.9 ± 0.6
Cholesterol esters	3.1		2.7	4	10.0		10 ± 1.0	3.1 ± 0.5				
Sterol/wax esters		6.5 ± 1.5						11.0 ± 3.0	6.1 ± 0.6	4.6	6.2 ± 0.7	7.1 ± 0.7
Squalene		0.2 ± 0.1					2 ± 0.5	21.3 ± 6.2	6.5 ± 2.0	3.6	6.9 ± 0.3	2.9 ± 1.8
n-Alkanes		1.7 ± 1.3	3.3	7.0					3.7 ± 0.5	3.0	2.8 ± 0.3	2.9 ± 1.0
Others					11.1			7.6				

F. Environmental Factors

Environmental variables, including seasonal variations [25,41], exposure to UV light [42], and the presence of sebaceous lipids [3,43] (see Sec. III.B.4), could also affect the composition of the SC barrier lipids.

Winter xerosis is characterized by a scaly, rough, dry, and often itching skin. In dermatology, the term is used for those conditions in which skin dryness is more severe than commonly observed. Xerosis is especially frequent on the lower legs. Only small changes in total SC lipid concentrations have been reported, and no correlation has been found between cholesterol sulfate nor ceramides and the degree of non-eczematous dry skin [18,24,44]. However, in more recent reports, a reduction in ceramides, and more precisely in ceramide 1 esterified with linoleate, was demonstrated in the winter months compared with the summer period [25,41]. These results suggest that the SC is functionally inferior

Table 2 (Continued)

									Fulmer and		
Imokawa et al. 1991 [12]	Paige et al. 1994 [16]	Lavrijsen et al. 1994 [53]	Wertz 1992 [54]	Jass and Elias 1991 [30]	Bonté et al. 1997 [55]		Long et al. 1985 [56]		Kramer 1986 [57]	Zellmer and Lasch 1997 [27]	Bonté et al. 1995 [58]
					Forearm skin		Forearm skin				
		Forearm							Lower		
Arms	Lower legs[c]	skin	Leg skin	Abdominal	f	g	Intact SC	Squamated	legs[h]	Plantar skin	j	k
				24.4 ± 3.8								
		0–1.3	0	traces	ND	ND					ND	ND
20.0	22.3 ± 2.80	5.3–36.4	38	24.4 ± 3.8	18.8	31.7			41.8 ± 3.5	20.25 ± 0.67	19.5	17.9
1.7	2.67 + 1.13						6.2 ± 2.8	5.3 ± 1.7	5.0 ± 0.6			
4.20	8.24 ± 2.03						18.5 ± 4.3	19.7 ± 6.2	8.2 ± 0.7			
3.96	5.99 ± 1.44						11.8 ± 3.8	8.2 ± 1.4	8.0 ± 0.8			
5.24	7.14 ± 1.65						19.5 ± 4.0	25.6 ± 7.6	10.9 ± 0.9			
									9.2 ± 0.9			
4.88	7.84 ± 2.22						8.6 ± 1.0	14.2 ± 2.5				
							23.8 ± 5.4	23.6 ± 3.1				
		0–6.6	0	6.6 ± 2.2	7.3	12.2				Traces	12.6	7.0
		0.3–2.9	0	2.0 ± 0.3	1.5	2.4	12.0 ± 4.0	3.5 ± 1.5		2.06 ± 0.13	1.2	1.7
				66.9 ± 4.8								
		4.8–23.6	21.6	18.9 ± 1.5	7.3	12.2			23.7 ± 1.4	43.53 ± 3.04	13.5	5.9
		2.4–12.3	23	26.0 ± 5.0	18.8	31.7			25.3 ± 2.3	20.16 ± 1.12	16.1	6.8
		10.5–75.9	8.1	variable	11.6	NI			8.9 ± 1.5	4.56 ± 0.54		
		1.9–11.9	9.3		5.8	9.8			10.2 ± 0.9	9.44 ± 0.67	8.1	6.5
				7.3 ± 1.2	21.7	NI					11.0	47.3
		0–0.2		6.5 ± 2.7	7.3	NI					18.0	6.9
		1.3–5.1		8.2 ± 3.5								

CHOL: cholesterol.
FFA: free fatty acids.
DG/TG: di-/triglycerides.
NI: not indicated.
ND: not detected.
[a] Integral lipid extraction of SC with CHCl$_3$/MeOH (1:2) (median values).
[b] Topical lipid extraction with acetone/diethyl ether (1:1).
[c] Together they account for ±10% of the total lipid amount.
[d] Mean values ± SEM.
[e] μg Ceramide/mg SC ± SD.
[f] % Total lipids with sebum.
[g] % Total lipids without sebum.
[h] ng Lipid/μg protein ± SEM.
[i] Detected but not quantified.
[j] Mammary SC.
[k] In vivo extraction of forearm skin.

during the winter season. This type of xerosis is not always correlated with a significant decrease in the SC water content. It is thus better to talk about "rough skin" instead of "dry skin," especially since it is known that a decrease of the sebum function occurs [18,35]. Moreover, Saint-Léger et al. [44] found that a decreased amount of neutral lipids (sterol esters and triglycerides) together with higher amounts of FFA (increased esterase activity) was associated with the severity of rough skin.

G. UV Irradiation

As long as UV irradiation occurred at suberythemal doses, increases in the amount of SC lipids and improvement of barrier function were observed [42]. This explains the beneficial effects of phototherapy on atopic skin. Lamellar body content extrusion was improved, as was their conversion into lipid lamellae in the SC [45]. However, high UV doses above the suberythemal level caused barrier damage characterized by inflammation and scaliness of the skin with a concomitant increase in TEWL [42].

H. Expression of Results Obtained from Lipid Analysis

Table 2 gives an overview of the different SC lipids of healthy human skin as found by various research groups. As already mentioned, the large variation found between the different lipid components is most likely due to differences in extraction procedure and efficiency, lipid analysis protocols, and sampling location. Not all authors give a complete concentration profile of the main SC lipid components (Table 2). The lack of total percentage of sphingolipids or neutral lipids makes comparison between different publications even more difficult.

It is also important to notice that the quantitative expression of the results is not uniform. Individual concentrations of skin lipids may be expressed as a percentage of the total lipids or skin proteins. Expression as a percentage of the epidermal or SC weight (dry versus wet weight) also is an option. In the case of in vivo extractions, the total amount of lipids yielded after evaporation of the extraction solvents may even be expressed versus the skin surface (in square centimeters) involved [46].

III. ANALYTICAL TECHNIQUES TO DETERMINE SKIN BARRIER LIPIDS

A. Sources of Skin Samples

Most skin sampling techniques are invasive. Apart from in vivo skin extractions (see Sec. III.B.2) [46,47,53,58], skin samples can be obtained from biopsies after

Analytical Techniques for Skin Lipids 155

surgery [21,22,29,57,59,60], tape stripping [14,19,53,61], or induced skin blisters [62–66]. Also, SC sample collection by repeated scraping of the skin with a single-edged razor blade [27,53] and a standardized shaving technique [48,57,67] have been described.

When human skin samples are collected, the investigator has the responsibility to ask the local ethics committee or other authorized instances for approval of the protocols involved. Guidelines such as those developed for skin irritation and contact allergy testing should be followed [68]. It is also important that all volunteers provide signed informed consent. When healthy human skin is studied, only volunteers with no skin pathologies, as diagnosed by a dermatologist supervising the experiments, should enter the study. Exclusion criteria should be stated, and variables, including age, sex, medical history, and every parameter of potential importance for the study under consideration, should be noted. In several cases, it is advisable to work with a homogeneous test population.

1. Ex Vivo Biopsies

When obtained during cosmetic or general surgery (e.g., breast, abdominal or facial skin), full-thickness skin samples (*ex vivo*) should be transported as soon as possible to the analyzing laboratory. This can be done in sterile medium (T-medium) containing 10 µg gentamycin/ml PBS (phosphate buffered saline) solution [69]. When extraction is done directly, dry transportation is also possible. In the latter case, skin samples should be properly spread out in appropriate recipients (e.g., Petri dish) to avoid contamination of the upper skin layers by lipids from the adipose tissue.

Attention should be paid that, before handling, appropriate safety tests with respect to bacterial and viral infections have been carried out. When not otherwise specified in the protocol, skin samples should be obtained from areas free from diseases or infections. It is also important that no scars or abrasions are present and that the samples are relatively hairless [29].

2. Tape Stripping

Tape stripping techniques have gained popularity during the past few years. Current protocols can be divided into two groups. The first one makes use of lipid-free glass slide strips with one or two droplets of cyanoacrylate resin (e.g. LocTite® Superglue-5, Kreglinger-Loctite NV, Antwerp, Belgium or Bison® Super-colle, Perfecta Chemie, Goes, The Netherlands) [12,14,70].

In the second group, a number of adhesive tapes (Tesafilm™-Leukoflex™, Beiersdorf, Germany; Scotch tape 3M or D-Squame®, CuDerm Corporation, Dallas, TX, USA) are consecutively applied to and removed from the skin [19,41,53,55,71]. For both techniques, it seems important that a constant, reproducible pressure is used when the strips are applied on the skin.

In the case of cyanoacrylate resin, a single strip (applied for 45 to 60 sec) should be sufficient to remove the complete SC ($\pm$ 10 µm); for adhesive tapes, 15 to 20 strippings are needed (see also Sec. III.B.3) [53]. Imokawa and coworkers [12] showed by histological biopsies that only 5 to 10 SC layers were stripped when cyanoacrylate resin was used. Although under appropriate conditions both techniques can remove the same amount of SC layers and lipid material, the cyanoacrylate resin protocol seems to be less acceptable to the volunteers.

The amount of SC removed by stripping is not linearly proportional to the number of strips removed [58,72]. This lack of correlation is even more pronounced when inter- and intra-individual differences in anatomical regions are studied.

Because tape stripping methods lead to significant increases in water loss through the upper layers of the skin, transepidermal water loss (TEWL) measurements can be used for the assessment of the degree of barrier perturbation [33]. TEWL, however, does not give any idea about the amount of SC that has been removed. Barrier disruption is followed by a cascade of metabolic responses in the viable epidermis, including the secretion of newly formed lamellar bodies. Within 1 hr to 6 days, the physiological barrier repair process is completed [73,74].

To quantify the stripped material, weighing of each single strip is commonly performed [55] (see also Sec. III.B.3) but it is rather labor intensive and time-consuming. Some commercially available tapes are hygroscopic, making proper weighing procedures even more complex. Therefore, Marttin and coworkers [72] proposed an easier and faster method using spectrophotometric examination (at 278 nm) of the proteins present on the tape. Although less reliable (larger variability), this technique revealed that light absorption correlated to the weight of the SC material (see also Sec. III.C.4). An advantage of the latter method is that the homogeneity of the individual strips can be examined.

3. Skin Blisters

An alternative to tape stripping, especially when the whole epidermis ($\pm$ 100 µm thick) needs to be collected, is the blister technique. Originally, the blister method was developed to separate epidermis from dermis by mechanical force only, avoiding chemical and thermal damage [62,63].

A negative pressure gradient is applied, provoking a multidirectional extension leading to a separation of the two skin compartments at the dermo-epidermal junction. This basal membrane zone is attached with hemidesmosomes to the epidermis and with a network of fibers to the dermis, resulting in a tight junction responsible for several interactions between both skin layers (reviewed in [75]). By electron microscopy, three structured layers are visible between epidermis and dermis: the lamina lucida, the lamina densa, and finally the sublamina densa.

In vivo blistering can also be induced by exposing the skin to liquid nitrogen for 20 to 25 sec. This method is useful when small samples are required—for instance, from the dorsal side of the hands or fingers [76].

An example of a commercially available device with a special dome-shaped cap, made of Plexiglass, is Dermovac® (Instrumentarium Corp., Helsinki, Finland.) In a typical in vivo blister experiment, a suction of about 200 mmHg (0.263 Pa) below the atmospheric pressure is applied consistently. The diameter of the blister depends on the holes in the cap that is attached to the skin. The applied pressure induces a suction power that increases the tension on the cell junctions, which eventually will—in normal individuals—tear off the epidermis at the level of the lamina lucida. The basal cell layer of the epidermis is still intact, allowing self-regeneration of the epidermis. Normally, it takes 2 to 3 hr to induce blister formation. Around 50 to 150 µl suction blister fluid is present, and the SC as well as the underlying viable epidermis can be harvested. Clear blister fluid roughly corresponds to interstitial fluid. Suction blister fluid is free of white cells for the first 5 hr. Thereafter, few polymorphonuclear leukocytes appear. The protein content of suction blisters is 60–70% of the corresponding serum value. In diseased skin, blisters cannot be raised properly and, as seen in some skin pathologies, externally applied suction may induce basal cell cytolysis [65]. Consequently, secretion of fluid into the capsule with and without bursting of the blister roof is observed. Standardization of the blister technique in diseased skin has, as far as we know, not been established.

In Table 3, a chronological overview of the mechanical blister techniques performed on human skin is described. The blister technique is based on an old methodology, and therefore the basic literature goes back to the early 1960s. More recent publications still use those basic principles and describe the applicability of the blister technique in various biochemical and grafting studies [64,80]. Also, the epidermal blister roof, which is not contaminated by fibroblasts and other dermal cells, provides cell material for epidermal cell cultures [62].

Epidermal recovery follows blister formation. This occurs without the use of external applied products; however, the latter could improve faster skin repairment [82]. After removal of the blister roof, the existing disruption needs to heal and partial epithelization is reported after 4 or 5 days [81]. The healing period is very person-dependent, but usually epithelization is completed within 2 to 3 weeks [80,81]. Although blister techniques are mainly described as being painless and without permanent scar formation [80,81], skin pigmentation disturbances may occur and remain visible for 2 to 12 months [64,80,83]. This is one of the reasons why volunteers need to be properly informed before they give their written consent. Individuals with a known medical history of bad wound healing may not be included in this type of experiment.

Kiistala [63] investigated a number of parameters that could be important for the induction of blisters. He reported that the induction time—for the same

Table 3 Overview of Variables Occurring During In Vivo Blister Induction in Humans (Suction blister time indicates the period between the application of the pressure gradient and the appearance of the first vesicles.)

Authors	Anatomical site	φ (mm)	mmHg	Time	Apparatus	Recovery	Advantages	Disadvantages
Kiistala & Mustakallio 1964 [77]	Chest	20	200	<2 h	Angiosterrometer	NI	NI	NI
Kiistala 1968 [62]	Chest, Abdomen	5–70	150–200	<3 h	Dermovac®	NI	NI	NI
Kiistala 1972 [63]	Ventral forearm							
*Karimieni et al. 1980 [78]	Abdomen	5	200	2 h	NI	NI	NI	NI
Fernandes et al. 1985 [79]	Ventral forearm	8	400	105 min	Gelman vacuum pump	NI	NI	Intradermal bleedings
Suvanprakorn et al. 1985 [80]	Different body sites (Hand, foot, neck)	NI	200–500 (Site depended)	1–2 h	NI	Re-pigmentation After 7–14 days, completed after 28–95 days	Painless, no scars	Hyper-pigmentation
Michiyuki 1988 [64]	Abdomen, tights	20	200	3–4 h	Oil rotary Vacuum pump	Normalization of pigmentation after 6 months	No scars	Risk of loss of pigmentation, or hyper-pigmentation
Willsteed et al. 1990 [76]	Ventral forearm	5	200	90–120 min	NI	NI	NI	NI
Taylor et al. 1993 [65]	Ventral forearm	6	200	±90 min	Eschmann VP50	NI	NI	NI
†Kallioinen et al. 1995 [81]	Abdomen	20	200	<2 h	Angiosterrometer	Partial epithelization after 4–5 days, completed after 9 days	NI	NI
Grönneberg et al. 1996 [82]	Ventral forearm	9	300 at 39°C	2–3 h	Negative pressure Suction system	NI	NI	NI
*Haapasaari et al. 1997 [66]	Abdomen	5–70	150–200	<3 h	Dermovac®	NI	NI	NI
Shukuwa et al. 1997 [83]	Ventral forearm	8	1520	1½–2 h	Hand vacuum pump (Meward Enterprise)	NI	No scars, painless	Hyper-pigmentation

NI = not indicated
* according to Kiistala 1968 [62]
† according to Kiistala and Mustakallio 1964 [77]

pressure applied—was significantly decreased above the age of 30 years. The latter probably was due to the age-dependent decrease in cohesion between epidermis and dermis. From that age on, no sex-related differences were observed. However, in younger volunteers, females showed more resistance on mechanical epidermal rupture than males did. This was explained as women having a more elastic skin compared to age-matched male subjects. Nowadays, this explanation is doubtful, and clear evidence is still lacking, especially because the elastic properties of the skin are mainly provided by the underlying dermis and not by the epidermis nor SC (reviewed in [75]). Kiistala [63] also reported differences in blister induction as a function of anatomical body site (chest, abdomen, ventral forearm). Abdominal skin was shown to exhibit the lowest blister time.

4. In vitro Models

Lipid extracts can also be obtained from cultures of human epidermal keratinocytes [84,85]. Cultured skin equivalents are being used not only for transplantation and treatment of severe skin ulcers but also for toxicity studies and research of potential skin irritants [86].

Conventional monolayer cultures of human skin cells (keratinocytes, fibroblasts) have their origin in the late 1970s [87,88]. Keratinocytes were plated on plastic or on collagen-coated dishes. Disadvantages were an incomplete differentiation and the absence of SC [89]. In the past two decades, successful attempts have been made to develop keratinocyte cultures that are morphologically equivalent to the native epidermis. These skin equivalents are three-dimensional structures based on the recombination of cultured cells to approach the skin tissue more closely. Skin equivalents are generated by seeding normal human keratinocytes on an appropriate dermal substrate and culturing them at the air-liquid interface [90]. Dermal substrates may consist of different materials of biological or artificial origin and can be either cell-free or populated with fibroblasts. Under the air-exposed culture conditions, the keratinocytes undergo complete terminal differentiation that also results in the formation of a corneous layer. Only air-lifted skin cultures have been proposed as having a functional permeability barrier [90,91].

Commercially available three-dimensional human skin models comprising a reconstructed epidermis with a functional SC are SkinEthic™ (SkinEthic Laboratories, Nice, France) and EpiDerm™ (MatTek Corporation, Ashland, MA, USA). Both models are lacking the dermis and consist of keratinocytes cultured on an inert filter (e.g., cellulose acetate). The type of reconstructed skin experimentally used depends on the requirements of the protocol under investigation. Also EpiSkin™ (owned by L'Oréal, Paris, France) is an epidermis model reconstructed on a collagen matrix. In addition to keratinocytes, the main cell types in the epidermis, melanocytes [92] and Langerhans cells [93], can be incorporated.

An actual problem of these laboratory-made skin models, which do not contain any blood vessels, is their short life span. Most of the time they can not be kept alive for longer than one month [86,90].

In these so-called air-exposed keratinocyte cultures, all typical in vivo differentiation markers can be observed at the ultrastructural level, including desmosomes, keratohyalin granules, lamellar bodies, and the intercellular lipid structures [91,94].

Some abnormalities in the expression of several specific protein markers and disturbances at the level of the lipid composition still reflect deficiencies in the quality and barrier properties of various in vitro skin equivalents (reviewed in [89,94]). Lower contents of certain sphingolipids and the free fatty acids (FFAs), linoleic and arachidonic acid, were reported [95]. Also, normal desquamation did not take place, revealing the occurrence of impaired cell shedding [91,96,97].

B. Lipid Extraction

Various extraction protocols have been reported and there is not yet a clear agreement which technique should be preferably used. Out of our own experience there is, however, no doubt that the composition of the solvent mixtures, the number of extraction steps, the polarity of the solvents, and many other factors will have a tremendous effect on the quantitative outcome of the lipid analysis.

1. Ex vivo Biopsies and in vitro Skin Models

When lipids are extracted from biological samples (biopsies or blistered skin) or from reconstructed skin, the classic approach is the use of a mixture of chloroform ($CHCl_3$) and methanol (MeOH) according to a modified Folch et al. [98] or Bligh-Dyer [95] method [16,27,29,48,53,54,57,59,60,84,100,101]. In Figure 1, the extraction procedure used in our laboratory and partially based on the method published by Ponec and Weerheim [95], is given [102].

Other solvent mixtures used for the extraction of skin lipids from tissue samples are n-hexane/MeOH 2:3 (v/v) [44,58].

It is important that all solvents used for extraction are of chromatographic grade (e.g., pro analysis (pa) solvents). The recipients used to collect the extracted lipids should be made of glass with Teflon inlays for the screw caps. The supernatants need to be transferred with glass pipettes to avoid any possible contamination with components extracted from various plastic materials.

After evaporation of the solvent mixture, the total amount of the extracted lipids can be weighed on a micro balance (e.g., Type M5 SA, MT5, AT20 (2–20 µg), Mettler Instrument AG, Zürich, Switzerland). To allow inter-person comparisons, the mass of each lipid fraction, as determined after lipid separation, can

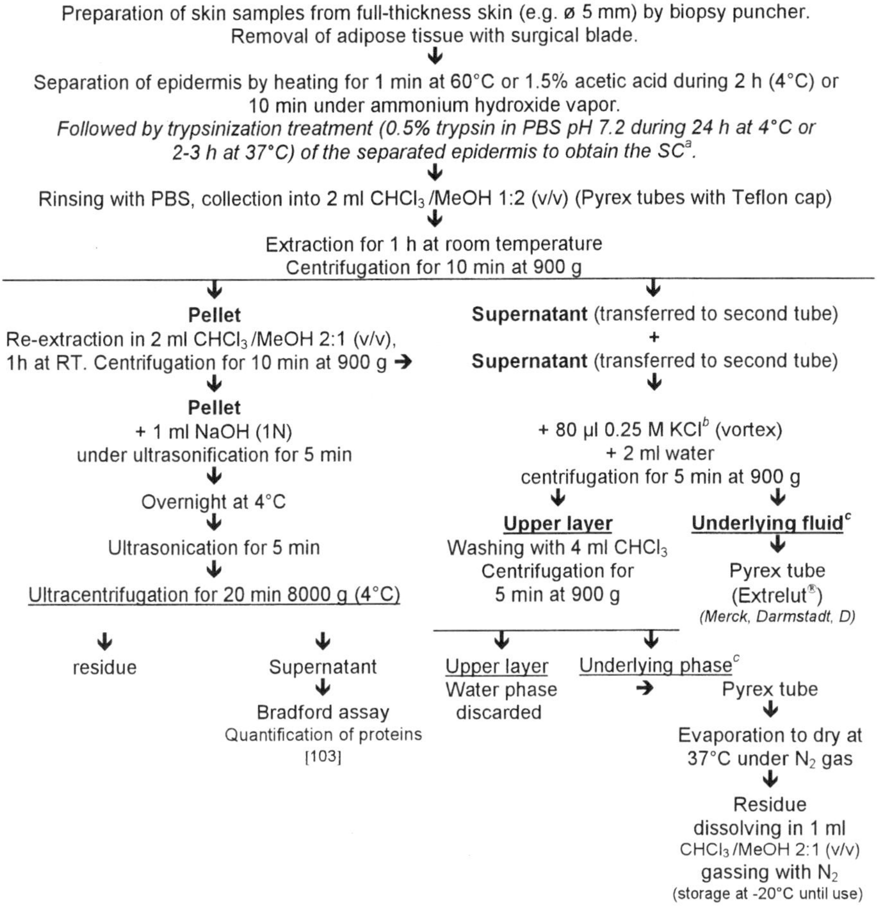

Analytical Techniques for Skin Lipids 161

Preparation of skin samples from full-thickness skin (e.g. ø 5 mm) by biopsy puncher.
Removal of adipose tissue with surgical blade.
↓
Separation of epidermis by heating for 1 min at 60°C or 1.5% acetic acid during 2 h (4°C) or 10 min under ammonium hydroxide vapor.
Followed by trypsinization treatment (0.5% trypsin in PBS pH 7.2 during 24 h at 4°C or 2-3 h at 37°C) of the separated epidermis to obtain the SC[a].
↓
Rinsing with PBS, collection into 2 ml $CHCl_3$/MeOH 1:2 (v/v) (Pyrex tubes with Teflon cap)
↓
Extraction for 1 h at room temperature
Centrifugation for 10 min at 900 g

Pellet	Supernatant (transferred to second tube)
Re-extraction in 2 ml $CHCl_3$/MeOH 2:1 (v/v), 1h at RT. Centrifugation for 10 min at 900 g →	+ Supernatant (transferred to second tube)

Pellet
+ 1 ml NaOH (1N)
under ultrasonification for 5 min
↓
Overnight at 4°C
↓
Ultrasonication for 5 min
↓
Ultracentrifugation for 20 min 8000 g (4°C)

↓   ↓
residue   Supernatant
                ↓
           Bradford assay
           Quantification of proteins
           [103]

+ 80 µl 0.25 M KCl[b] (vortex)
+ 2 ml water
centrifugation for 5 min at 900 g
↓   ↓
Upper layer   Underlying fluid[c]
Washing with 4 ml $CHCl_3$ ↓
Centrifugation for   Pyrex tube
5 min at 900 g   (Extrelut®)
                    (Merck, Darmstadt, D)
↓   ↓
Upper layer   Underlying phase[c]
Water phase   →   Pyrex tube
discarded   ↓
              Evaporation to dry at
              37°C under N_2 gas
                       ↓
                  Residue
              dissolving in 1 ml
              $CHCl_3$/MeOH 2:1 (v/v)
                gassing with N_2
              (storage at -20°C until use)

Figure 1 Flow diagram for sequential extraction of human skin lipids.
[a] For the extraction of the total epidermal lipids this preparative step is not carried out.
[b] Addition of KCl is done to ensure extraction of the polar cholesterol sulfate into the organic phase.
[c] Filtered through Whatman n°43.

be normalized to the total lipid amount of either the epidermis or the SC. The pellet is used for the determination of the protein concentration [103].

2. In vivo Extractions

Due to their toxicological profile and skin aggression, $CHCl_3$ and MeOH cannot be used for the in vivo collection of SC lipids [104]. Even by using light microscopy, it could be shown that $CHCl_3$/MeOH induces cell damage in the living epidermal cell layers [53].

Also n-hexane/MeOH 2:3 (v/v), as used by Saint-Léger et al. [44], was not an option because exhaustive lipid extractions went together with severe skin irritation. Cyclohexane/ethanol 2:8 (v/v) was alternatively used [58].

To overcome this problem, various in vivo sampling extraction procedures have been developed, including the use of pure acetone [31], ethanol 95% [13,47,56], n-hexane/isopropanol 3:2 (v/v) [105,106], ethanol/diethyl ether 3:1 (v/v) [42,67], and a mixture of acetone/diethyl ether 1:1 (v/v) [12,46,53]. The latter solvent mixture was used during two consecutive extraction steps of 5 and 25 minutes, respectively. Proper validation showed that it was a suitable extraction protocol to study SC lipid profiles, even in various skin disorders [53].

To bring extraction solvents in contact with the skin, a precleaned glass or stainless metal cylinder [46,47,53] is applied and held properly in position to prevent lateral leakage. The extraction can be split up in two consecutive steps, having the advantage that with the first short step, the exogenous lipids as well as the lipids of the hydrolipidic mixture can be largely removed. The second step then collects the barrier lipids [53]. Sebaceous lipid contamination can also be avoided by washing the skin with a cotton swab drenched in cyclohexane [106] or by removal of the upper two horny cell layers by tape stripping [42].

Sequential application of either acetone or petroleum ether, utilizing precleaned (delipidized) cotton swabs followed by an extraction of the pooled swabs by a Bligh-Dyer method, has been described only for murine skin [1]. Extraction of human skin lipids by pure acetone [31] dissolves the hydrolipidic mixture and, being a nonpolar solvent, selectively removes only a limited amount of some superficial skin lipids [46]. After acetone treatment of the skin, no irregularities have been reported at the level of the lamellar bodies, and physiological barrier repair occurred without delay [107].

3. Extraction of Tape Strippings

$CHCl_3$/MeOH mixtures cannot be used for tape stripping extraction because of contamination by the adhesive material. For cyanoacrylate resin, a smear of the lipid bands occurs during chromatographic separation, making identification and quantitative determination of the samples impossible [55,70].

Extraction with n-hexane/ethanol 95:5 (v/v) under ultrasonication, followed by filtration through a solvent-resistant filter (e.g., Millex-SR, Millipore 0.2–0.5 µm, Molsheim, France or Whatman filters n°1 or 43, Whatman Ltd, UK), has been reported by Imokawa et al. [7,12] and Di Nardo et al. [14] for cyanoacrylate strippings. After solvent evaporation under N_2, the samples were redissolved in $CHCl_3$/MeOH 2:1 (v/v) and kept stored at −20°C before use. When extracted lipids (µg) are expressed per milligram of SC, the residue on the glass slide, composed of SC and cyanoacrylate resin, can be ultrasonicated with dimethyl formamide (DMF) [12,14]. Consequently, the cyanoacrylate resin is solubilized, while the dispersed SC can be separated from the DMF solution by filtration. Finally, the residue on the filter is washed (DMF / MeOH), dried under vacuum, and weighed.

Other research groups have used different extraction protocols, which can be summarized as follows:

A pre-extraction step with MeOH (sonication), followed by a conventional $CHCl_3$/MeOH extraction [19,41]
A two-step extraction with cyclohexane/ethanol 2:8 and 5:5 (v/v) [55,58]
A one-hour extraction in ethyl acetate/MeOH 60:40 (v/v) [61].

4. Factors Affecting SC Lipid Profiles During Extraction

As already emphasized, the collection method of the lipid samples can have a tremendous effect on the outcome of the lipid concentrations measured, making comparison of reported data difficult. Indeed, in some methods whole epidermis is collected (blisters, *ex vivo* biopsies); in other methods, only surface lipids or superficial SC lipids are harvested. Also, the SC is not uniform across its entire thickness, which can be at the origin of the collection of a non–properly defined part of the SC in the case of tape stripping.

From results shown in Table 4 it is clear that, depending on the location in the epidermis, a different lipid profile in weight percentage (wt%) is found as a function of epidermal differentiation and cornification [52,59]. Also, Bonté et al. [55] showed that the lipid gradient observed in the upper SC could be linked with the number of tape strippings. For the assessment of the main barrier lipids (i.e., ceramides, cholesterol, FFAs), collection of the SC layers is sufficient, implying a tape stripping procedure of 15 to 25 strips or a trypsinization treatment of epidermal samples (biopsies, skin blisters) [53,55].

Apart from the origin of the lipid samples, one cannot avoid that some contamination occurs by exogenous lipids [108] and lipids originating from the sebaceous glands [3]. A counter argument for the latter was reported by Lampe et al. [29], who showed measurable quantities of squalene and wax esters in plantar SC (Table 2). The sole is depleted of sebaceous glands, so these nonpolar lipids could not be of sebaceous origin. A consecutive study of Lampe et al. [59] showed that squalene and n-alkanes are evenly distributed in the viable epidermal layers (Table 4), suggesting that they are not simply of sebaceous or environmental origin. Also, Williams and Elias [109] excluded all possible sources of contaminations and still found representative levels of n-alkanes. High quantities of squalene, however, could reflect a high level of cholesterol metabolism in the epidermis. So, one still has to be very critical when analyzing the lipid profiles obtained.

Table 4 Variations in Lipid Composition During Epidermal Differentiation and Cornification (Samples obtained from abdominal skin.)

Lipids	Stratum basale/ spinosum (wt %)[a]	Stratum granulosum (wt %)[a]	Stratum corneum (wt %)[a]
Sphingolipids	7.3 ± 1.0	11.7 ± 2.7	18.2 ± 2.8
Glycosylceramides	3.5 ± 0.3	5.8 ± 0.2	Traces
Ceramides	3.8 ± 0.2	8.8 ± 0.2	18.1 ± 0.4
Phospholipids	44.5 ± 3.4	25.3 ± 2.6	4.9 ± 1.6
Cholesterol sulfate	2.6 ± 3.4	5.5 ± 1.3	1.5 ± 0.2
Neutral lipids	51.0 ± 4.5	56.5 ± 2.8	77.7 ± 5.6
Free sterols/CHOL	11.2 ± 1.7	11.5 ± 1.1	14.0 ± 1.1
FFA	7.0 ± 2.1	9.2 ± 1.5	19.3 ± 3.7
DG/TG	12.4 ± 2.9	24.7 ± 4.0	25.2 ± 4.6
Sterol/wax esters[b]	5.3 ± 1.3	4.7 ± 0.7	6.1 ± 0.6
Squalene	4.9 ± 1.1	4.6 ± 1.0	6.5 ± 2.0
n-Alkanes	3.9 ± 0.3	3.8 ± 0.8	3.7 ± 0.5
TOTAL	99.1	101.1	99.3

[a] Weight percentage (wt%) mean ± SEM.
[b] Sterol/wax esters present in approximately equal quantities as determined by acid hydrolysis (12% borontrichloride in MeOH for 30 min at 90°C).
Source: Refs 52, 59.

An overall finding was that the total amount of lipids, extracted by using topical in vivo extraction procedures, was lower than that found by integral extraction of the SC [53] (see also Table 2). Questions to be answered included: "To what depth does in vivo extraction with mainly acetone/diethyl ether remove lipids? What proportion of the SC lipids is extracted? Do the extracted lipid classes reflect the overall SC lipid profile?" These questions were tackled by Lavrijsen and coworkers [53], and they showed indeed that in vivo extraction procedures did not remove significant amounts of lipids from the lower layers of the SC or the epidermis, but that it was possible to collect a representative ceramide profile. Thus, although the total lipid profile was not completely recovered, the assessment of the main barrier lipids was satisfactory.

Although studies were sometimes based on only a few samples, several authors have shown a great inter-individual variation in the overall lipid composition of the skin barrier [71,86]. The latter was irrespective of the extraction protocol used. However, ceramide profiles from different skin samples taken from the same subject were similar. This profile uniformity was also found between different subjects [53]. In addition, Norlén et al. [106] showed that the long-chain

FFAs of the SC were a stable, uniform lipid group, with a low inter-individual variation, dominated by saturated lignoceric (C24:0) and hexacosanoic acid (C26:0).

To control the inter-individual variations, it is of utmost importance to standardize the site and area of extraction in a homogeneous population (age, sex, phototype, lifestyle). In case of tape stripping, time and applied pressure should also be controlled (see Sec. III.A.2) [86].

In extraction studies, TEWL measurements have often been included to assess the amount of water loss through the upper skin layers [21]. A direct relationship between the efficiency of the barrier to water loss and the total lipid weight percentage has been shown [1]. Consequently TEWL can be used as a parameter to reflect the degree of barrier impairment after solvent extraction [46] or tape stripping [53].

C. Lipid Analysis

In order to study epidermal barrier physiology, skin disorders, and barrier recovery, a rapid and accurate analysis technique of the epidermal lipids is necessary. The latter can be derived either from human biopsies, in vivo extractions, and tissue culture specimens [48]. The lipid amounts in SC—especially in damaged skin—are low, so sensitive techniques must be developed.

The most commonly used techniques are chromatographic methods such as thin-layer chromatography for total lipid determination and ceramide profiles (C.1) followed by gas chromatography, more specific for the quantification of FFA, combined with mass spectrometry (C.2). Recently, Norlén and coworkers were able to develop a high-performance liquid-chromatography (HPLC)/light scattering detection (LSD)–based analysis method for the quantitative determination of skin lipids (C.3) [71,106]. Also spectrophotometric techniques (C.4) can be used for individual lipid components and overall skin barrier characterization.

An overview of the various analytical techniques for qualitative and quantitative lipid determination is available in a number of comprehensive books [110–112]. This chapter does not have the intention to repeat this work, but for some of the techniques (TLC, HPLC, GLC/MS), attention will be focused on their relevance for extracted skin lipids. Apart from a short summary of the required methodology, emphasis will be put on some pitfalls, and practical hints will be given for experimental protocols and study designs.

1. Thin-Layer Adsorption Chromatography (TLC)

For a complete introduction to TLC and a full description of the analytical technique, the standard works of Jork and coworkers [113,114] are recommended. One-dimensional high-performance thin-layer chromatography (HPTLC) is frequently used. It is a suitable and easy-handling method for the separation of individual epidermal lipid classes, based on their differences in mobility in a system of two phases [34,47,48,95]. Besides a rapid screening, this technique also enables an accurate quantitative determination of the different lipid classes.

Two-dimensional HPTLC has the advantages of exhibiting a higher resolution and allowing a faster identification of a great variety of individual lipid fractions. For quantitative purposes, however, it is not very practical [95].

Standard lipids. Pure lipid standards, all of the highest purity available are needed (e.g., Sigma, St. Louis, MO, USA). They usually include Cer type III, Cer type IV, cerebrosides, trioleine representative for the triglycerides (TG), oleic or stearic acid as FFA, cholesterol (CHOL), cholesterol oleate as cholesterol ester (CE), and cholesterol sulfate (CSO_4). Pentacosane, as a marker for n-alkanes (ALK), and squalene (SQ) are often incorporated.

These lipids are dissolved in $CHCl_3$/MeOH 2:1 (v/v) and are used to produce calibration curves necessary to identify and quantify unknown skin samples. Lipid standard mixtures mostly contain 100 ng to 2 µg of each lipid fraction. However, for preparative HPTLC development, much higher individual amounts are applied, ranging from 5 to 15 µg [115].

Because large differences exist in the mass of individual epidermal fractions and peak areas are not always a linear function of lipid weight, several increasing quantities of the standard mixture are applied [19,27,95].

Thin-layer Plates. The most commonly used TLC plates are precoated silicagel 60 HPTLC plates without concentration zone (size 10 × 10 cm, 20 × 10 cm, 20 × 20 cm; e.g., Merck, Darmstadt, Germany). The advantages of HPTLC plates is that much lesser amounts of total lipid extracts (10–20 µg) are required compared to ordinary TLC plates (2–3 mg), still allowing proper separation of the different lipid fractions [48].

In order to remove impurities that may interfere with lipid separation, the plates are first washed in $CHCl_3$/MeOH/H_2O/glacial acetic acid 120:70:9:1 (v/v/v/v) [21] or in MeOH/ethyl acetate 60:40 (v/v) [95], sometimes followed by a second wash with $CHCl_3$/ethyl acetate/diethyl ether 30:20:50 (v/v/v). After evaporation of all solvents, the HPTLC plates are activated for 10 min at 130°C. Heating can be done on commercially available heating plates (e.g., TLC Plate Heater III, Camag, Muttenz, Switzerland or Heat Plate, Desaga, Wiesloch, Germany).

Application of Samples and Lipid Standards. Starting at a height of approximately 5 mm above the bottom edge of the plate, increasing volumes of the lipid extracts are applied as narrow bands (3 mm or broader) at a constant distance of about 5 mm from each other. For this purpose, automatic appliers can be used (e.g., Linomat IV or Automatic TLC Sampler 4 (ATS4), Camag, Muttenz, Switzerland or TLC-Applicator AS 30, Desaga, Wiesloch, Germany) or appropriate manual syringes. The advantages of those appliers are easy-handling and an accurate dot- or line-application, according to the spray-on technique. For calibration reasons, it is essential to apply the standard lipids on the same plate as the one used for the samples [13].

Lipid Separation. After application of both lipid extracts and standard lipids, the plate is placed in a developing chamber containing the eluting solvent. Horizontal developing chambers, in which the silica gel side of the plate is facing down (e.g., Desaga, Wiesloch, Germany or Camag, Muttenz, Switzerland), are preferred. For proper use, the chambers must be precisely horizontally leveled and covered by a glass plate in order to avoid interference from the environment. By capillary action the solvent moves up the plate, taking the various lipid components with it at different rates, depending on their adsorption. TLC analyses are typically performed at room temperature, and temperature fluctuations must be avoided.

Skin lipids contain a variety of individual lipids differing in chemical structure and polarity [52]. This is the reason why it is virtually impossible to separate all lipid classes by TLC using one solvent or even one solvent mixture. To allow optimal separation of all barrier lipids on one plate, a multistep HPTLC technique was developed [95]. According to this technique, HPTLC plates are sequentially developed with series of different solvent mixtures (e.g., pro analysis). These mixtures vary in polarity, allowing migration of only a limited number of lipids during each development step; others remain adsorbed by the silica gel layer.

Different solvent mixtures for the development of total lipid extracts or, more specific, for the determination of the ceramides profile, have been reported. A brief overview of how lipids can be fractionated by using different development systems is given in Table 5. After each development step, the plate needs to be dried on a plate heater at 40°C for 3 min.

Development systems more complex than those shown in Table 5 have been reported. Bonté and coworkers [58] described a method for the major SC lipids by using an automated multiple development system (e.g., AMD Camag, Muttenz, Switzerland or TLC-MAT, Desaga, Wiesloch, Germany). The technique consisted of a single isocratic step with toluene to remove and separate the sebum lipids, followed by a 25-step gradient with a decreasing polarity range from $MeOH/H_2O$ 97:3 (v/v) to n-hexane. The drying procedures between each step were performed automatically. Two years later, this method was improved by Zellmer and Lasch [27], resulting in a better separation between FFA and triglycerides.

Staining and Charring. After the development is completed, the plate is removed from the chamber and dried at 40°C. Subsequently, the plate is heated at 100°C for 3 minutes, mainly in order to evaporate the ethyl- or hexyl acetate. Finally, visualization of the separated lipid fractions is done. To detect certain

Table 5 Sequential Development Systems for the Separation of Individual Lipid Fractions by HPTLC

Lipid extracts obtained from:	
↓ Skin samples (biopsies/blisters) or in vivo extraction	↓ Tape stripping (cyanoacrylate resin or adhesive tapes)
1) petroleum ether/diethyl ether/acetic acid (35:15:0.5) → 50 mm 2) toluene/n-hexane (40:10) → 100 mm 3) heptane → 100 mm ↳ for separation of non-polar lipids 1) $CHCl_3$/MeOH/H_2O (40:10:1) → 65 mm 2) $CHCl_3$/MeOH/H_2O/acetic acid (40:10:1:1) → 35 mm 3) petroleum ether/diethyl ether/acetic acid (15:35:0.5) → 100 mm ↳ for separation of polar lipids [57][b]	1) $CHCl_3$/MeOH/H_2O (40:10:1) → 100 mm 2) $CHCl_3$/MeOH/glacial acetic acid (190:9:1) → 160 mm 3) n-hexane/diethyl ether/acetic acid (70:30:1) → 200 mm [14][b]
	1) CH_2Cl_2/ethyl acetate/aceton/ 2-propanol (80:16:4:1) → 30 mm 2) $CHCl_3$/acetone/MeOH (76:8:16) → 20 mm 3) n-hexane/$CHCl_3$/hexyl acetate/ acetone/MeOH (6:80:0.10:4) → 95 mm [95][s]
1) MeOH/$CHCl_3$/H_2O (20:95:1) → 55 mm 2) petroleum benzine/diethyl ether/acetic acid (80:20:10) → 90 mm ↳ for separation of ceramides [42][s]	1) MeOH/$CHCl_3$/H_2O (20:95:1) → 60 mm 2) n-hexane/diethyl ether/acetic acid (80:20:10) → 85 mm 3) petroleum benzine → 100 mm [70][b]

[b] distance from the bottom of the plate
[s] distance from the spot line

types of lipids or functional groups, spraying with a specific chemical reagent can be carried out; suitable nondestructive dyes—allowing preparative HPTLC—include aqueous (0.25%) 8-anilino-1-naphthalene sulfonic acid (lipophilic fluorochrome) for glycosphingolipids and neutral lipids [21,29,59,115] and 2′,7′-dichlorofluorescein [56], both visualized under UV light. Alternatively, the plates can be sprayed with a phosphoric or sulfuric acid solution combined with $CuSO_4$ after which the lipids are charred by heating the plates at 160°C for 10 min [14,48,61,95]. Before increasing the temperature of the heat block to 160°C, the plates are preheated at 60°C until the staining mixture is evaporated. The separated lipids become visible as discrete bands of black carbon spots.

Analytical Techniques for Skin Lipids

To improve the quantification of various lipid fractions, a first identification can be performed when the lipids are charred at 80°C during 10 min. At that temperature, the sterol-containing components, as well as the most polar ceramides, appear as blue/violet bands [61]. Subsequently, the temperature can be further increased to 160°C.

Determination of Lipid Profiles. Photodensitometric scanning of the bands allows quantification. For this purpose a Camag TLC Scanner 3 (Muttenz, Switzerland) equipped with a computerized image analyzer [22] or a BioRad Scanning Videodensitometer (GS 710 or 800, BioRad, Hercules, CA, USA) [14] can be used. Desaga (Wiesloch, Germany) and Shimadzu (Kyoto, Japan) have, respectively, the Densitometer CD 60 and the Photodensitometer CS-9301PC, on the market.

Examples of commercially available software for densitometric evaluation are winCATS® (Camag, Muttenz, Switzerland), ProQuant® (Desaga, Wiesloch, Germany), or BioRad Quantity One® (BioRad, Hercules, CA, USA).

After choosing the appropriate conditions (slit width and height = beam dimensions, measuring wavelength and light source, absorption/reflection mode, etc.), these methods are suitable for automatic measurements, integration of samples, and evaluation of standards. After video integration, individual peak lists are reported containing the name of the component, Y-position, peak height, and area.

Photodensitometric scanning also allows recording of spectra. This technique is useful for a quick identification of unknown substances and is indispensable for determining the optimum wavelength for which increased sensitivity and small peaks identification are characteristic. Background spectra are used for the correction of the substances spectra. Current densitometers for direct photometric evaluation cover a wavelength range from 190 to 800 nm.

Research groups do not always provide all technical details about their methodology. We know of scannings at 420 nm, 450 nm, or 550 nm in the absorbance/reflectance mode after degrative charring, by the reports of Rawlings et al. [19], Bonté et al. [58], and Zellmer and Lasch [27], respectively. A wavelength of 632.8 nm was used by Melnik et al. [48]. However, this technique was questioned by Röpke et al. [70], and they repeated the same protocol followed by individual recording of the spectra of the different lipid components. They found an absorption maximum at 325 nm. This was exactly the same wavelength as found by our group when lipid extracts were separated according to the protocol described in Figure 2. [102]. As an example, the separation of an epidermal (abdomen) skin sample is shown.

The ratio to the front (Rf) of each lipid fraction can be calculated by dividing the distance of the migrated lipid component by the total distance of the solvent front. Compared to the individual Rf values of the co-migrated lipid stan-

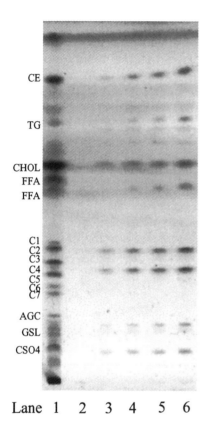

Figure 2 HPTLC lipid profile of a human epidermal skin sample obtained after the extraction protocol given in Figure 1. *Lane 1*: Sample, human abdominal epidermis. *Lane 2*: Blank solvent (for background subtraction). *Lanes 3–6*: Lipid standards in increasing amounts. Cer type III and Cer type IV (Sigma, St. Louis, MO, USA) were used as standards for cer 1 and cer 2, and for cer 3 to 7, respectively. Lipid separation was done according to the sequential ceramide development system. (modified from Refs. 17,95).

1. 40 mm n-hexane/CHCl$_3$/acetone (8:90:2)
2. 10 mm CHCl$_3$/acetone/MeOH (76:8:16)
3. 70 mm n-hexane/CHCl$_3$/hexyl acetate/acetone/MeOH (6:80:0.1:10:4)
4. 15 mm CHCl$_3$/acetone/MeOH (76:4:20)
5. 75 mm n-hexane/CHCl$_3$/hexyl acetate/ethyl acetate/MeOH (8:80:0.1:6:6)
6. 95 mm n-hexane/diethyl ether/ethyl acetate (78:18:4)

The plate was sprayed with acetic acid/H$_2$SO$_4$/H$_3$PO$_4$/H$_2$O 5:1:1:95 (v/v/v/v) and 0.5% CuSO$_4$ after which the lipids were charred by heating the plates at 160°C for 10 min.

Cholesterol esters (CE), Rf 0.87; triglycerides (TG), Rf 0.75; cholesterol (CHOL), Rf 0.64; free fatty acids (FFAs), Rf 0.55 and 0.58; ceramides (cer), Rf 0.27, 0.29, 0.32, 0.34, 0.36, 0.40, 0.42 (numbering according to Ref. 116); acylglucosphingolipids (AGC), Rf 0.18; glucosphingolipids (GSL), Rf 0.16; cholesterol sulfate (CSO$_4$), Rf 0.11.

dards, qualitative determination of the unknown lipid fractions is allowed. From the list of peak surfaces of the standards, calibration functions are calculated. Consequently, because the weight of the applied standards (in µg) and the optical density of both the standard lipids and the sample fractions are known, the sample lipid weight—by linear, but mostly polynome regression—can be calculated.

As already emphasized, the degree of charring varies for different lipid classes. This means that for each HPTLC plate, individual calibration curves are required for every lipid class [27,95]. In addition, one should realize that most of the standard lipids used are rather representative for—and not identical to—the various skin lipid components recovered after extraction. Consequently, quantification of unknown skin samples relatively depends on the charring behavior of the selected standard lipids. The latter usually are different in the reports published [16,29,48,95], making comparisons of results even more difficult.

Another possibility for quantifying samples is to include an internal standard in the standard lipid mixture. For this purpose, methyl oleate, an ester that is not present in human skin, can be used (0.25 mg/ml solvent) [14].

In order to allow correct quantification, solvent blanks should be applied along with the lipid samples and standards to bring the background profile and possible solvent contamination into account [109]. The latter is of particular importance when tape strips are analyzed, as tape-stripping contaminants co-migrate in different chromatographic solvent systems.

Nowadays, chromatogram evaluation by video scanning technology, image acquisition, and quantitative image analysis (intensities of the image pixels) is gaining field. Although the principles of planar chromatography remain unchanged, it is probably only a matter of some years before video technology—which is faster, is more flexible, and allows electronical storage of chromatograms—will take over the classic densitometry (e.g., Video-Technology ProViDoc® with ProResult® program, Desaga, Wiesloch, Germany or Video Imaging System, Shimadzu, Kyoto, Japan).

2. Gas-Liquid Chromatography/Mass Spectrometric Techniques (GLC/MS)

Gas-liquid chromatography (GLC) with flame ionization detection (e.g., GC-17A, Shimadzu, Kyoto, Japan) is very suitable for the determination of the methyl esters (FAME) of FFA or of FA liberated after chemical hydrolysis [29,55,117]. This method is also useful to establish the structural analysis of individual ceramides and the composition of their FAs and long chain bases [25,41].

After preparative HPTLC (i.e., sprayed with an aqueous 8-anilino-1-naphthalene solution), the neutral and polar lipid bands are excised followed by a re-extraction of the lipid-containing silica gel by the extraction solvent (e.g., $CHCl_3$/MeOH 2/1 (v/v)). The residue is then separated from the supernatant, and after a second wash, the organic phase is dried under nitrogen and weighed. Dissolved in MeOH/n-hexane 2:0.5 (v/v), FAs can be converted to FAMEs by a complex of (12%) borontrichloride (or -fluoride) in MeOH at 90–100°C for 1–1½ hours followed by extraction of methyl esters in n-hexane and fractionation in the solvent system petroleum ether/diethyl ether/glacial acetic acid 80:20:1 (v/v/v) to recover purified FAMEs [29,59,105]. Besides ester preparation, it is important that chromatographic parameters including resolution, peak asymmetry, and total analysis time are optimized as well as the linearity of the flame ionization detection and the injection technique [106].

Glass capillary columns with polar polyester liquid stationary phases are most suitable for FAME analysis as they allow a clear separation of the esters of the same chain length, unsaturated components eluting after the related saturated ones. A chromatogram is obtained in which the individual FAME peaks are characterized by their retention time and area. Identification of the individual FAMEs can be provided by direct comparison of their retention times with those of known standards [13]. Tricosanoic acid (C23:0) is normally not present in the human SC and is often used as an internal standard [106,118]. Determination of the areas of the chromatographic peaks can be made by a computing integrator (e.g., Shimadzu, Kyoto, Japan).

In the study of Lampe et al. [59], GLC combined with mass spectrometry (MS) was used to identify cholesterol sulfate (CSO_4), which was believed to be present in trace amounts. By using both desorption and fast atom bombardment (FAB) sources, and by identification of free sterols after solvolysis, it has been shown that CSO_4 was present in all viable epidermal layers, with the highest concentration in the stratum granulosum (see also Table 4) [59,119]. From Table 4, it is clear that significantly lower concentrations of CSO_4 are present in the SC, providing the evidence that its desulfation by the enzyme steroid sulfatase is required for the physiological desquamation of the SC [119,120]. Current HPTLC separation and determination techniques have sufficiently been improved and refined so that even small amounts of CSO_4 can be determined with planar chromatography, without the use of expensive equipment [61].

In the study of Norlén et al. [106], special attention was focused on the FFA fraction because the unsaturated FFAs exert a key function in the regulation of the skin barrier properties. They also showed that short-chain FFAs (less than C20), including the unsaturated species, are mainly contaminants from sebaceous glands and the environment.

3. Liquid Chromatography

Although planar chromatographic methods are easy-handling and offer quick qualitative screenings of various lipid classes, relevant criticisms include the limited densitometric sensitivity and the influences of temperature and humidity as

Analytical Techniques for Skin Lipids 173

well as the higher degree of variability in solvent mixtures [121]. Also, day-to-day variability of Rf values and resolution cannot be ignored. To avoid those problems, some research groups have tried to develop HPLC methods allowing direct quantification of the major lipid classes using one or more detectors [71,106,118,121].

Because GLC methods are not able to separate complex mixtures of neutral lipids of varying polarity, Murphy and coworkers [121] standardized a HPLC method for the separation of triacylglycerols, cholesterol and cholesterol esters. In addition, Flamand et al. [122] focused on the concentration and distribution of the free long-chain sphingoid bases (sphinganine, sphingosine and phytosphingosine) in the SC, whereas Schäfer and Kragball [118] analyzed phospholipids.

Whenever a complete determination of all individual SC lipids is required in conventional TLC, multiple development steps are necessary (see Sec. III.C.1). Therefore, Norlén and his group developed a new HPLC-based method for the quantitative analysis of the inner SC lipids, meaning the intact skin barrier, unperturbed by degradation processes [71,106]. They were able to separate and quantify all main lipid classes by HPLC and light scattering detection (LSD), while the FFA fractions were further analyzed by GLC. MS was used for peak identification and flame ionization detection for quantification (see Sec. III.C.2). However, it must be noted that with the single-step HPLC method, a complete separation between different skin ceramide subtypes was not obtained [106].

For this type of analysis, more specialized laboratory equipment is needed including a HPLC pump, column oven, injector (manual or automatic), gradient system controller, detector (i.e., spectrofluorimeter or light scattering detection), integrator, etc. (e.g., Shimadzu, Kyoto, Japan). All solvents used need to be of HPLC grade. To ensure high accuracy during long-term experiments, calibration curves for all lipid components need to be run daily to control the variations in detector response [106]. Coefficients of correlation must be calculated for all calibration curves, because only within the linear detectable range can reliable concentration measurements be performed. Consequently, direct quantification of several lipid classes is carried out by peak area analysis. In the case of tape stripping, an appropriate blank can be obtained by making a chromatographic analysis of the derivatized extract of an identical piece of adhesive tape.

4. Spectrophotometric Techniques

As described in Sec. III.A.2, spectrophotometric measurements of SC tape strips are useful for the inspection of the homogeneity of single strips. They also allow a rapid screening of the amount of SC material stripped. For this purpose, a Shimadzu UV 190 double-beam spectrophotometer equipped with a SC-3 Gel Scanning device was used (Shimadzu, Kyoto, Japan). Untouched, nonused tapes served as blanks. A drawback of this method, however, is the light scattering of the adhesive tapes used, which could overshadow the absorption of the proteins present in the SC material [72].

For the various UV/VIS systems, software packages are offered by different specialized companies (e.g., Shimadzu, Kyoto, Japan), allowing data processing as well as good laboratory practice (GLP)-specific functions.

Whereas individual lipid fractions are commonly fractionated by quantitative, sequential HPTLC techniques, various lipid classes can also be spectrophotometrically determined after the formation of a specific colored complex with a known absorbance. After extraction and drying (see Sec. III.B.), the total lipid content is redissolved in $CHCl_3$ and can consequently be determined using the sulfo-phosphovanillin-reaction. The colored complex obtained (pink) is then measured at 530 nm [69]. Two of a number of colorimetric reactions that can be carried out for some of the major SC lipid classes are as follows:

> Phospholipids can be determined by the molybdate-vanadate reaction (yellow color) and measured at 405 nm.
> Cholesterol dissolved in acetic acid can be determined after conversion by H_2SO_4 into cholestadiene, which reacts with a second molecule to form bischolestadiene. This dimer is sulfonated (catalyzed by Fe^{3+}) to its mono- or disulfonic acid derivative, which gives a maximum absorption at 550 nm.

The assays described are simple and quick; due to a co-prepared standard series, calibration curves can be calculated, allowing quantitative determinations. In colorimetric techniques, plain reagent solutions serve as blanks.

IV. CONCLUDING REMARKS

It is acknowledged that the composition and organization of the intercellular lipids play an important role in the barrier function of human SC. Differences in SC lipid profiles between healthy and diseased skin exist, but can only be characterized by applying accurate methods for collection and quantification of epidermal and SC lipids. Therefore, the analytical techniques currently applied for the determination of skin lipids need to be evaluated critically.

For the validation of both in vivo and in vitro experiments, it is important to standardize the many variables that play a crucial role in the composition of the SC lipid fractions. Person-linked factors such as age, sex, race, and anatomical site have been reported to influence the lipid composition. Also, environmental factors and UV irradiation may affect the outcome of the lipid concentrations measured. These variables must be controlled in a reproducible and standardized way in order to be able to show a clear and direct relationship between various skin conditions, barrier function alterations, and barrier lipid composition.

Various sources for sampling epidermal or SC lipids (in vivo, ex vivo, in vitro) are described, including skin biopsies after surgery, tape stripping, and skin blisters. Both the tape stripping and blister techniques have the advantage that skin samples can be obtained from different parts of the body, allowing comparative studies. This is not always the case for skin samples coming from plastic surgery, which are mostly restricted to breast, abdominal, or facial skin. A major advantage of the stripping method is that the lipid composition can be studied as a function of SC depth. However, these sampling techniques are invasive and should be applied only under approved ethical conditions.

Although various extraction protocols can be compared, a clear agreement on which technique should be preferably used is still doubtful. Attention needs to be focused not only on conventional extraction protocols of biological samples but also on human in vivo techniques. A prerequisite for direct in vivo sampling of SC lipids is the application of noninvasive techniques. In this respect, the use of a mixture of acetone/diethyl ether 1:1 (v/v) is proposed. Many factors, such as sample sources and contaminations, may affect the SC lipid profiles during extraction and should therefore be controlled.

For chemical analysis of epidermal lipids, chromatographic methods are most widely used. Quantitative analysis of the different lipid fractions after sequential separations is commonly performed by high-performance thin-layer chromatography (HPTLC). Individual lipids are separated on HPTLC plates and subsequently quantified by computerized densitometry. Although lipid separation is quickly obtained, disadvantages still are the time-consuming quantification of the different lipid fractions and the appropriate calibration procedure. However, as an answer to the relevant criticisms of planar chromatographic methods, high-performance liquid-chromatographic (HPLC) techniques are making their way up.

V. ACKNOWLEDGMENTS

The authors thank Dr. Maria Ponec and Mr. Arij Weerheim from the Skin Research Laboratory, Leiden University Medical Center (Leiden, The Netherlands) for their kind advice and critical review of the manuscript.

SUPPLIERS

BIO-RAD LABORATORIES Headquarters, ✉ 2000, Alfred Nobel Drive, Hercules, CA 94547, USA
☎ ++510 724 1000; ⊇ ++510 741 5800
🖥 www.biorad.com

BOEHRINGER MANNHEIM YAMANOUCHI K.K., ✉ 3-10-11 Toranomon MF bldg n°10, Minato-Ku, Tokyo 105, Japan
☎ ++81 (0) 3 34326211; 📠 ++81 (0) 3 34326219
💻 http://biochem.boehringer-mannheim.com; bmkkbio@cet.co.jp

CAMAG, ✉ Sonnenmattstrasse 11, CH-4132 Muttenz, Switzerland
☎ ++41 61 4673434; 📠 ++41 61 4610702
💻 www.camag.de; info@camag.ch

DESAGA GmbH, ✉ Postfach/P.O. Box 1280, D-69153 Wiesloch, Germany
☎ ++49 (0) 6222 92880; 📠 ++49 (0) 6222 928892
💻 sales@desaga-gmbh.de

EPIDERM MatTek Corporation, ✉ 200, Homer Avenue, Ashland, MA 01721, USA
☎ ++508 881 6771; 📠 ++508 879 1532
💻 www.mattek.com; information@mattek.com

INSTRUMENTARIUM CORPORATION, ✉ Kuartaneenkatu 2, Helsinki, P.O. Box 100, Fin-00031, Finland
☎ ++358 1039411; 📠 ++358 9144172
💻 www.instrumentarium.com; ir@instrumentarium.fi

SHIMADZU Head Office, ✉ 1 Nishinokyo Kuwabaracho, Nakagyou-ku, Kyoto 604-8511, Japan
☎ ++81 (0) 75 8230077 or 75 8231111; 📠 ++81 (0) 75 8231361
💻 www.shimadzu.com

SKINETHIC Laboratories, ✉ 45, Rue Saint-Philippe, 06000 Nice, France
☎ ++33 493 97 77 27; 📠 ++33 493 97 77 28
💻 www.skinethic.com; infos@skinethic.com

REFERENCES

1. Grubauer G, Feingold KR, Harris RM, Elias PM. Lipid content and lipid type as determinants of the epidermal permeability barrier. J Lipid Res 1989; 30:89–96.
2. Holleran WM, Mao-Qiang M, Gao WN, Menon GK, Elias PM, Feingold KR. Sphingolipids are required for mammalian epidermal barrier function: Inhibition of sphingolipid synthesis delays barrier recovery after acute perturbation. J Clin Invest 1991; 88:1338–1345.
3. Wertz PW. The nature of the epidermal barrier: biochemical aspects. Advan Drug Delivery Rev 1996; 18:283–294.

4. Schürer NY, Plewig G, Elias PM. Stratum corneum lipid function. Dermatologica 1991; 183:77–94.
5. Downing DT. Lipids: their role in epidermal structure and function. Cosmet Toiletries 1991; 106:63–69.
6. Elias PM, Feingold KR. Lipids and the epidermal water barrier: metabolism, regulation and pathophysiology. Semin Dermatol 1992; 11:176–182.
7. Imokawa G, Akasaki S, Minematsu Y, Kawai M. Importance of intercellular lipids in water-retention properties of the stratum corneum: induction and recovery study of surfactant dry skin. Arch Dermatol Res 1989; 281:45–51.
8. Wilhelm KP, Cua AB, Maibach HI. Skin aging: Effect on transepidermal water loss, stratum corneum hydration, skin surface pH, and casual sebum content. Arch Dermatol 1991; 127:1806–1809.
9. Denda M, Koyama J, Horii I, Takahashi M, Hara M, Tagami H. Age- and sex-dependent change in stratum corneum sphingolipids. Arch Dermatol Res 1993; 285: 415–417.
10. Rawlings AV, Rogers J, Mayo AM. Changes in lipids in the skin aging process. Biocosmet Skin Aging 1993; 1:31–45.
11. Ghadially R, Brown BE, Sequeira-Martin SM, Feingold KR, Elias PM. The aged epidermal permeability barrier: structural, functional, and lipid biochemical abnormalities in humans and a senescent murine model. J Clin Invest 1995; 95:2281–2290.
12. Imokawa G, Abe A, Jin K, Higaki Y, Kawashima M, Hidano A. Decreased level of ceramides in SC of atopic dermatitis: an etiologic factor in atopic dry skin? J Invest Dermatol 1991; 96:523–526.
13. Yamamoto A, Serizawa S, Ito M, Sato Y. Stratum corneum lipid abnormalities in atopic dermatitis. Arch Dermatol Res 1991; 238:219–223.
14. Di Nardo A, Wertz PW, Gianetti A, Seidenari S. Ceramide and cholesterol composition of the skin of patients with atopic dermatitis. Acta Derm Venereol (Stockh) 1998; 78:27–30.
15. Yardley HJ. Epidermal lipids. Int J Cosmet Sci 1987; 9:13–19.
16. Paige DG, Morse-Fischer N, Harper JI. Quantification of stratum corneum ceramides and lipid envelope ceramides in the hereditary ichthyoses. Br J Dermatol 1994; 131:23–27.
17. Lavrijsen APM, Bouwstra JA, Gooris GS, Weerheim A, Boddé HE, Ponec M. Reduced skin barrier function parallels abnormal stratum corneum lipid organization in patients with lamellar ichtyosis. J Invest Dermatol 1995; 105:619–624.
18. Akimoto K, Yoshikawa N, Higaki Y, Kawashima M, Imokawa G. Quantitative analyses of stratum corneum lipids in xerosis and asteatotic eczema. J Dermatol 1993; 20:1–6.
19. Rawlings AV, Watkinson A, Rogers J, Mayo AM, Hope J, Scott IR. Abnormalities in stratum corneum structure, lipid composition, and desmosome degradation in soap-induced winter xerosis. J Soc Cosmet Chem 1994; 45:203–220.
20. Elias PM, Menon GK. Structural and lipid biochemical correlates of the epidermal permeability barrier. Adv Lipid Res 1991; 24:1–26.
21. Motta S, Monti M, Sesana S, Mellesi L, Ghidoni R, Caputo R. Abnormality of water barrier function in psoriasis: role of ceramide fractions. Arch Dermatol 1994; 130:452–456.

22. Motta S, Sesana S, Monti M, Giuliani A, Caputo R. Interlamellar lipid differences between normal and psoriatic stratum corneum. Acta Derm Venereol (Stockh) 1994; suppl 186:131–132.
23. Ilzuka H, Ishida-Yamamoto A, Honda H. Epidermal remodelling in psoriasis. Br J Dermatol 1996; 135:433–438.
24. Saint-Léger D, François AM, Lévêque JL, Stoudemayer TL, Grove GL, Kligman AM. Age-associated changes in stratum corneum lipids and their relation to dryness. Dermatologica 1988; 177:159–164.
25. Rogers J, Harding CR, Mayo AM, Banks J, Rawlings AV. Stratum corneum lipids: the effect of ageing and the seasons. Ach Dermatol Res 1996; 288:765–770.
26. Reed JT, Ghadially R, Elias PM. Skin type, but neither race nor gender, influence epidermal permeability barrier function. Arch Dermatol 1995; 131:1134–1138.
27. Zellmer S, Lasch J. Individual variation of human plantar SC lipids, determined by automated multiple development of high-performance thin-layer chromatography plates. J Chromatogr 1997; 691:321–329.
28. Jin K, Higaki Y, Takagi Y, Higuchi K, Yada YK, Kawashima M, Imokawa G. Analysis of beta-glucocerebrosidase and ceramidase activities in atopic and aged dry skin. Acta Derm Venereol 1994; 74:337–340.
29. Lampe MA, Burlingame AL, Whitney J, Williams ML, Brown BE, Roitman E, Elias PM. Human SC lipids: characterization and regional variations. J Lipid Res 1983; 24:120–130.
30. Jass HE, Elias PM. The living stratum corneum: implications for cosmetic formulation. Cosmet Toiletries 1991; 106:47–53.
31. Reinertson RP, Wheatley VR. Studies on the chemical composition of human epidermal lipids. J Invest Dermatol 1959; 32:49–59.
32. Hood HL, Wickett RR. Racial differences in epidermal structure and function. Cosmet Toiletries 1992; 107:47–48.
33. Kompaore F, Marty JP, Dupont C. In vivo evaluation of the stratum corneum barrier function in blacks, Caucasians and Asians with two noninvasive methods. Skin Pharmacol 1993; 6:200–207.
34. Wertz PW, Swartzendruber DC, Abraham W, Madison KC, Downing DT. Essential fatty acids and epidermal integrity. Arch Dermatol 1987; 123:1381–1384.
35. Brooks G, Idson B. Skin lipids. Int J Cosmet Sci 1991; 13:103–113.
36. Agner T, Damm P, Skouby SO. Menstrual cycle and skin reactivity. J Am Acad Dermatol 1991; 24:566–570.
37. Elsner P, Wilhelm D, Maibach HI. Effect of low-concentration sodium lauryl sulfate on human vulvar and forearm skin: age-related differences. J Reprod Med 1991; 36:77–81.
38. Feingold KR, Mao-Qiang M, Proksch E, Menon GK, Brown BE, Elias PM. The lovastatin-treated rodent: a new model of barrier disruption and epidermal hyperplasia. J Invest Dermatol 1991; 96:201–209.
39. Ramsing D, Agner E, Malinowski J, Meibom J, Agner T. Effect of systemic treatment with cholesterol-lowering drugs on the skin barrier function in humans. Acta Derm Venereol (Stockh) 1995; 75:198–201.
40. Brazzelli V, Distante F, Perani G, Berardesca E. Effects of systemic treatment with

statins on skin barrier function and stratum corneum water-holding capacity. Dermatology 1996; 192:214–216.
41. Conti A, Rogers J, Verdejo P, Harding CR, Rawlings AV. Seasonal influences on stratum corneum ceramide 1 fatty acids and the influence of topical essential fatty acids. Int J Cosmet Sci 1996; 18:1–12.
42. Wefers H, Melnik BC, Flür M, Bluhm C, Lehmann P, Plewig G. Influence of UV irradiation on the composition of human stratum corneum lipids. J Invest Dermatol 1991; 96:959–962.
43. Stewart ME. Sebaceous gland lipids. Semin Dermatol 1992; 11:100–105.
44. Saint-Léger D, François AM, Lévêque JL, Stoudemayer TJ, Kligman A, Grove G. Stratum corneum lipids in skin xerosis. Dermatologica 1989; 178:151–155.
45. Fartasch M, Bassukas ÍD, Diepgen T. Disturbed extruding mechanism of lamellar bodies in dry non-eczematous skin of atopics. Br J Dermatol 1992; 127:221–227.
46. Imokawa G, Akasaki S, Hattori M, Yoshizuka N. Selective recovery of deranged water holding properties by stratum corneum lipids. J Invest Dermatol 1986; 87:758–761.
47. Wertz PW, Miethke MC, Long SA, Strauss JS, Downing DT. The composition of the ceramides from human stratum corneum and from comedones. J Invest Dermatol 1985; 84:410–412.
48. Melnik B, Hollmann J, Erler E, Verhoeven B, Plewig G. Microanalytical screening of all major stratum corneum lipids by sequential high-performance thin-layer chromatography. J Invest Dermatol 1989; 92:231–234.
49. Lavrijsen APM, Weerheim A, Bouwstra JA, Ponec M. Differences in stratum corneum barrier properties between hairless mouse skin and human skin. In: Thesis Lavrijsen APM. Stratum Corneum Barrier in Healthy and Diseased skin. ISBN 90-9010205-1 Leiden: Labor Vincit, 1997:26–38.
50. Wertz PW, Downing DT. Epidermal lipids. In: Goldsmith LA, ed. 2nd ed, Physiology, Biochemistry and Molecular Biology of the Skin. Oxford, Oxford University Press, 1991:205–235.
51. Brod J. Characterization and physiological role of epidermal lipids. Int J Dermatol 199; 30:84–90.
52. Elias PM. Epidermal lipids, barrier function and desquamation. J Invest Dermatol 1983; 80:44s–49s.
53. Lavrijsen APM, Higounec IM, Weerheim A, Oestmann E, Tuinenburg EE, Boddé HE, Ponec M. Validation of an in vivo extraction method for stratum corneum ceramides. Arch Dermatol Res 1994; 286:495–503.
54. Wertz PW. Epidermal lipids. Sem Dermatol 1992; 11:106–113.
55. Bonté F, Saunais A, Pinguet P, Meybeck A. Existence of a lipid gradient in the upper stratum corneum and its possible biological significance. Arch Dermatol Res 1997; 289:78–82.
56. Long SA, Wertz PW, Strauss JS, Downing DT. Human stratum corneum polar lipids and desquamation. Arch Dermatol Res 1985; 277:284–287.
57. Fulmer AW, Kramer GJ. Stratum corneum lipid abnormalities in surfactant-induced dry scaly skin. J Invest Dermatol 1986; 86:598–602.
58. Bonté F, Pinguet P, Chevalier J, Meybeck A. Analysis of all stratum corneum lipids

by automated multiple development high performance thin-layer chromatography. J Chromatogr B 1995; 664:311–316.
59. Lampe MA, Williams ML, Elias PM. Human epidermal lipids: characterizations and modulations during differentiation. J Lipid Res 1983; 24:131–140.
60. Friberg SE, Kayali I, Rhein LD, Simion FA, Cagan RH. The importance of lipids for water uptake in stratum corneum. Int J Cosmet Sci 1990; 12:5–12.
61. Ponec M, Weerheim A. Determination of stratum corneum lipid profile by tape stripping in combination with high performance thin-layer chromatography. In: Brain KR, Walters KA, eds. Conference Proceedings Perspectives in Percutaneous Penetration 2000, Cardiff: STS Publishing, 2000:37.
62. Kiistala U. Suction blister device for separation of viable epidermis from dermis. J Invest Dermatol 1968; 50:129–137.
63. Kiistala U. The influence of age, sex and body region on suction blister formation in human skin. Ann Clin Res 1972; 4:236–246.
64. Michiyuki K. Epidermal grafting using the tops of suction blisters in the treatment of vitiligo. Arch Dermatol 1988; 124:1656–1658.
65. Taylor G, Vennig V, Wojnarowska F, Millard PR. Suction-induced basal cell cytolysis in the Weber-cockayne variant of epidermolysis bullosa simplex. J Cutaneous Pathol 1993; 20:389–392.
66. Haapasaari K-M, Risteli J, Karvonen J, Oikarinen A. Effect of hydrocortisone, methylprednisolone aceponate and mometasone furoate on collagen synthesis in human skin in vivo. Skin Pharmacol 1997; 10:261–264.
67. Melnik B, Hollmann J, Hofmann U, Yuh MS, Plewig G. Lipid composition of outer SC and nails in atopic and control subjects. Arch Dermatol Res 1990; 282:549–551.
68. SCCNFP/0003/98 and SCCNFP/0068/98final. Guidelines on the use of human volunteers in compatibility testing of finished cosmetic products, European Union, 1998.
69. Vanden Bossche H, Willemsens G, Schreuders H, Coene M-C, Van Hove C, Cools W. Use of human skin models in the study of retinoic acid metabolism and lipid synthesis, effects of liarozole. In: Rogiers V, Sonck W, Shephard E, Vercruysse A, eds. Human Cells in *in vitro* Pharmaco-Toxicology: Present status within Europe. Brussels: VUBpress, 1993:61–76.
70. Röpke E-M, Augustin W, Gollnick H. Improved method for studying skin lipid samples from cyanoacrylate strips by high-performance thin-layer chromatography. Skin Pharmacol 1996; 9:381–387.
71. Norlén L, Nicander I, Lundh-Rozell B, Ollmar S, Forslind B. Inter and intra individual differences in human stratum corneum lipid content related to physical parameters of skin barrier function in-vivo. J Invest Dermatol 1999; 112:72–77.
72. Marttin E, Neelissen-Subnel MTA, De Haan FHN, Boddé HE. A critical comparison of methods to quantify SC removed by tape stripping. Skin Pharmacol 1996; 9:69–77.
73. Kompaore F, Dupont C, Marty JP. In vivo evaluation in man by two noninvasive methods of the stratum corneum barrier function after physical and chemical modification. Int J Cosmet Sci 1991; 13:293–302.

74. Lee SH, Elias PM, Feingold KR, Mauro T. A role for ions in barrier recovery after acute perturbation. J Invest Dermatol 1994; 102:976–979.
75. Schaefer H, Redelmeier TE. Structure of the skin & composition and structure of the stratum corneum. In: Schaefer H, Redelmeier TE, eds. Skin Barrier: Principles of Percutaneous Absorption. Basel: Karger, 1996:1–20, 43–87.
76. Willsteed EM, Bhogal BS, Das A, Bekir SS, Wojnarowska F, Black MM, Mckee PH. An ultrastructural comparison of dermo-epidermal separation techniques. J Cutaneous Pathol 1990; 18:8–12.
77. Kiistala U, Mustakallio KK. In vivo separation of epidermis by production of suction blisters. Lancet 1964; 27:1444–1445.
78. Kariniemi A-L, Kousa M, Asko-Seljavaara S. Dissociation of suction blister roof epidermis with trypsin and desoryribonuclease into viable single cells. Acta Derm Venereol (Stockh) 1980; 61:58–61.
79. Fernandes AC, Anderson R, Ras GJ. An objective filter-based, enzymatic method for the in vivo measurement of the migration of human polymorphonuclear leucocytes. J Immunol Methods 1985; 83:259–271.
80. Suvanprakorn P, Dee-Ananlap S, Pongsomboon C, Klaus SN. Melanocyte autologous grafting for treatment of leukoderma. J Am Acad Dermatol 1985; 13:968–974.
81. Kallioinen M, Koivukangas V, Jarvinen H, Oikarinen A. Expression of cytokeratins in regenerating human epidermis. Br J Dermatol 1995; 133:830–835.
82. Grönneberg R, van Hage-Hamsten M, Halldén G, Hed J, Raud J. Effects of salmeterol and terbutaline on IgE-mediated dermal reactions and inflammatory events in skin chambers in atopic patients. Allergy 1996; 51:610–646.
83. Shukuwa T, Kligman AM, Stoudemayer TJ. A new model for assessing the damaging effects of soaps and surfactants on human stratum corneum. Acta Derm Venereol 1997; 77:29–34.
84. Takami J, Abe A, Matsuda T, Shayman JA, Radin NS, Walter RJ. Effect of an inhibitor of glucosylceramide synthesis on cultured human keratinocytes. J Dermatol 1998; 25:73–77.
85. Ponec M, Boelsma E, Weerheim A, Mulder A, Bouwstra J, Mommaas M. Lipid and ultrastructural characterization of reconstructed skin models. Int J Pharm 2000; 203:211–225.
86. Boelsma E, Tanoja H, Boddé HE, Ponec M. An in vitro-in vivo study to the use of a human skin equivalent for irritancy screening of fatty acids. Toxic in Vitro 1997; 11:365–376.
87. Rheinwald JG, Green H. Serial cultivation of strains of human keratinocytes: the formation of keratinizing colonies from single cells. Cell 1975; 6:331–344.
88. Prunieras M, Delescluse C, Regnier M. The culture of skin: a review of theories and experimental methods. J Invest Dermatol 1976; 67:58–65.
89. Ponec M. In vitro cultured human skin cells as alternatives to animals for skin irrtancy screening. Int J Cosmet Sci 1992; 14:245–264.
90. Boelsma E, Gibbs S, Faller C, Ponec M. Characterization and comparison of reconstructed skin models: morphical and immuno-histochemical evaluation. Acta Derm Venereol 2000; 80:82–88.

91. Ponec M, Weerheim AM, Kempenaar JA, Mommaas AM, Nugteren DH. Lipid composition of cultured human keratinocytes in relation to their differentiation. J Lipid Res 1988; 29:949–961.
92. Regnier M, Duval C, Galey JP, Philippe M, Lagrange A, Tuloup R, Schmidt R. Keratinocyte-melancoyte co-cultures and pigmented reconstructed human epidermis: models to study modulation of melanogenesis. Cell Mol Biol 1999; 45:969–980.
93. Regnier M, Patwardhan A, Scheynius A, Schmidt R. Reconstructed human epidermis composed of keratinocytes, melanocytes and langerhans cells. Med Biol Eng Comput 1998; 36:821–824.
94. Ponec M. Reconstructed human epidermis in vitro: An alternative to animal testing. ATLA 1995; 23:97–110.
95. Ponec M, Weerheim A. Retinoids and lipid changes in keratinocytes. In: Packer L (ed.). Methods in Enzymology, vol. 190, Retinoids, part B, Academic Press, San Diego, 1990:30–41.
96. Ponec M, Weerheim A, Kempenaar J, Mulder A, Gooris G, Bouwstra J, Mommaas AM. The formation of competent barrier lipids in the reconstructed human epidermis requires the presence of vitamin C. J Invest Dermatol 1997; 109:348–355.
97. Vicanová J, Mommaas AM, Mulder AA, Koerten HK, Ponec M. Impaired desquamation in the in vitro reconstructed human epidermis. Cell Tissue Res 1996; 286:115–122.
98. Folch J, Lees M, Stanley SGH. A simple method for the isolation and purification of total lipids from animal tissues. J Biol Chem 1956; 226:497–509.
99. Bligh EG, Dyer WJ. A rapid method of total lipid extraction and purification. Can J Biochem Physiol 1959; 37:911–917.
100. Hatfield RM, Fung W-M. Molecular properties of a stratum corneum model lipid system: large unilamellar vesicles. Biophys J 1995; 68:196–207.
101. Norlén L, Emilson A, Forsling B. Stratum corneum swelling. Biophysical and computer assisted quantitative assessments. Arch Dermatol Res 1997; 289:506–513.
102. De Paepe K, Weerheim A, Ponec M, Houben E, Roseeuw D, Rogiers V. Analysis of the epidermal lipids of healthy human skin: Study concerning the different factors affecting the design of a control population. Skin Pharmacol 2001, submitted.
103. Bradford MM. A rapid and sensitive method for the quantitation of microgram quantities of protein utilizing the principle of protein-dye binding. Anal Biochem 1976; 72:248–254.
104. de Groot AC, Weyland JW, Nater JP. eds. Unwanted effects of cosmetics and drugs used in dermatology. 3rd ed, Amsterdam: Elsevier, 1994:770.
105. Ghyczy M, Nissen H-P, Biltz H. The treatment of acne vulgaris by phosphatidylcholine from soybeans with a high content of linoleic acid. Sci Cosmetol 1995; 4:5–15.
106. Norlén L, Nicander I, Lundsjö A, Cronholm T, Forslind B. A new HPLC-based method for the quantitative analysis of inner stratum corneum lipids with special reference to the free fatty acid fraction. Arch Derm Res 1998; 290:508–516.
107. Fartasch M. The nature of the epidermal barrier: structural aspects. Adv Drug Deliv Rev 1996; 18:273–282.

108. Bortz JT, Wertz PW, Downing DT. The origin of alkanes found in human skin surface lipids. J Invest Dermatol 1989; 93:723–727.
109. Williams ML, Elias PM. N-Alkanes in normal and pathological human scale. Biochem Biophys Res Commun 1982; 107:322–328.
110. Christie WW. ed. Lipid analysis. 2nd ed. Oxford: Pergamon Books, 1982:207.
111. Christie WW. ed. Gas chromatography and lipids. Glasgow: The Oily Press, 1989: 307.
112. Hamilton RJ, Hamilton S. eds. Lipid analysis: A practical approach. Oxford: Oxford University Press, 1991:22–23.
113. Jork H, Funk W, Fischer W, Wimmer H. eds. Thin-layer chromatography: Reagents and detection methods. Vol 1a: Physical and chemical detection methods: fundamentals, reagents I. Weinheim: VCH Verlagsgesellschaft, 1990:464.
114. Jork H, Funk W, Fischer W, Wimmer H. eds. Thin-layer chromatography: reagents and detection methods. Vol 1b: Physical and chemical detection methods: Activation reactions, reagent sequences, reagents II. Weinheim: VCH Verlagsgesellschaft, 1994:496.
115. Müthing J, Kemminer S. Nondestructive detection of neutral glycosphingolipids with lipophilic anionic fluorochromes and their employment for preparative high-performance TLC. Anal Biochem 1996; 238:195–202.
116. Robson KJ, Stewart ME, Michelsen S, Lazo ND, Downing DT. 6-Hydroxy-4-sphingosine in human epidermal ceramides. J Lipid Res 1994; 35:2060–2068.
117. Vicanová J, Tvrzická E, Stulik K. Capillary gas chromatography of underivatized fatty acids with a free fatty acid phase column and a programmed temperature vaporizer injector. J Chromatogr B 1994; 656:45–50.
118. Schäfer L, Kragball K. Abnormalities in epidermal lipid metabolism in patients with atopic dermatitis. J Invest Dermatol 1991; 96:10–15.
119. Williams ML, Elias PM. Stratum corneum lipids in disorders of cornification. Increased cholesterol sulfate content of stratum corneum in recessive X-linked ichthyosis. J Clin Invest 1981; 68:1404–1410.
120. Downing DT, Stewart ME, Wertz PW, Colton SW, Abraham W, Strauss JS. Skin lipids: an update. J Invest Dermatol 1987; 88:2–6.
121. Murphy EJ, Rosenberger TA, Horrocks LA. Separation of neutral lipids by high-performance liquid chromatography: quantification by ultraviolet, light scattering and fluorescence detection. J Chromatogr B 1996; 685:9–14.
122. Flamand N, Justine P, Bernaud F, Rougier A, Gaetani L. In vivo distribution of free long-chain sphingoid bases in the human stratum croenum by high-performance liquid chromatographic analysis of strippings. J Chromatogr B 1994; 656: 65–71.

7
Biophysical Methods for Stratum Corneum Characterization

Hans Lambers and Hans Pronk
Sara Lee Household and Body Care Research, The Hague, The Netherlands

I. INTRODUCTION

A. General Background

There is no doubt that the application of cosmetic lipids has many positive effects on the structure and function of the skin. These effects are pleiotropic, caused either by direct interaction with the epidermis, particularly the stratum corneum, or indirectly, by influencing the physiologic, homeostatic condition of the skin.

Generally, the overall result may be a range of skin benefits such as:

Improved moisturization
Enhanced barrier function
Smoothing of the microrelief
Improved mechanical properties
Reduction in scaliness
Cosmetic lipids have also more general benefits, such as improvement in skin pH, temperature, and color.

Special strategies with regard to penetration and targeting of cosmetic lipids may be considered in order to enhance various skin properties; for instance, whether to increase the superficial stratum corneum emolliency, to enforce the stratum corneum lipid barrier, or even to realize trans-(epi)dermal transport all need different product formulation approaches. This will be covered in other chapters of this book (see, for example, Chapter 9).

Many biophysical methods have been developed during the past 50 years, all of which involve the application of physical principles and methods to study and explain the structure and function of the skin and to assess the above-mentioned properties. These properties of the skin are somehow interlinked and related to each other, which make proper and accurate assessment of biophysical parameters often a complex matter. For instance, a change in moisturization is intimately linked to changes in mechanical properties such as elasticity, but there also exists a relation to the degree of desquamation, barrier function, and so on.

Existing biophysical methods are maturing and improving all the time; also, the development of new biophysical methods is a continuous challenge to the cosmetic scientist. Although, in general, this will improve the possibilities to characterize the structure and function of the skin, it must be realized that proper standardization of measuring devices and protocols is often still lacking; this may give rise to different results in different laboratories, making inter-laboratory comparison sometimes a difficult task.

In order to perform biophysical evaluations, a number of general conditions should be met for optimal results.
Important for almost all of the methods that are described in this chapter are the following:

> All examinations should be done in an air-conditioned laboratory that guarantees constant room temperature (preferably between 20 and 23°C) and humidity (preferably between 40% and 60% relative humidity) [1].
> The volunteers should sit in this air-conditioned laboratory at least 30 minutes before the assessment and throughout the test procedure. Short-term conditions affecting the assessment (e.g., journey to the test institute by bicycle or car) are equalized in this way. Standardized conditions are essential for physiological examinations of the skin. However, this procedure does not compensate for climatic conditions.
> Subjects should be in a relaxed physiological and psychological state and be equilibrated fully to the room conditions.
> A suitable body test site for measuring influences of topical treatments should have an even surface and few hair follicles. The volar aspect of the forearm meets most of the requirements.
> Preferably, the assessor should be skilled in the art and possess a large amount of experience.

B. Scope of the Chapter

This chapter focuses on the description of the most popular, easy-to-handle, and frequently used noninvasive biophysical methods—c.q., devices to characterize in vivo human skin, in particular the stratum corneum. In some cases tape-stripping of the skin may be used for characterization [2]. Also, ex vivo and in vitro (epi)dermal tissue may be used, but these require mostly special conditions and will not be further considered here in detail. A practical approach is chosen, aiming at increasing the general knowledge and understanding of the relatively inexperienced cosmetic scientist. Various examples are given, which may be used for claim substantiation related, for example, to moisturization, barrier function, microrelief, mechanical properties, scaliness, and the like. Methods to study and quantify sebum surface lipids are also covered because these lipids form an interface between stratum corneum and the external environment and are, although somewhat indirectly, related to the biophysical properties of stratum corneum. Moreover, this method may also be used to quantify the amounts of externally applied cosmetic lipids. Special attention is paid to the relationship between the assessment of the effects of cosmetic lipids and stratum corneum structure and function.

The chapter will be concluded by a brief description of other less frequently used biophysical methodologies to measure stratum corneum parameters such as skin pH, temperature, and color. A list of supplier addresses of most of these biophysical devices is provided at the end of this chapter.

Techniques that are directed more toward getting structural information, such as basic histological techniques or the more sophisticated techniques (e.g., confocal [fluorescence] laser scanning microscopy, magnetic resonance imaging [MRI] and electron spin resonance [ESR]), and other biophysical techniques that are relatively unknown and not (yet) used very often, such as brillanometry/ reflection and frictiometry, are not or only briefly described at the end of this chapter. For more advanced and detailed description of the methodologies mentioned here, the reader is referred to general books such as the ones edited by Marks [3], Serup [4], and Elsner [5]

II. SKIN MOISTURE ASSESSMENT

A. General

Skin moisturization is one of the most relevant parameters characterizing the condition of the stratum corneum [6]. Water acts as plasticizer for the skin and is also an essential part of the highly organized (liquid crystalline or gel phase) lipid barrier of the stratum corneum. As water is evaporating continuously from the surface of the skin—depending among others on the relative humidity and temperature—skin may become dry, brittle, and flaky and even can show fissures. So nature and the cosmetic industry have taken measures to prevent drying out of the skin:

> The human body is an almost infinite source of water, which compensates for this continuous loss of water. This compensation does not result in an equilibrium but in a steady state.

This steady state is governed by the water-binding and water-retarding properties of the stratum corneum.

The stratum corneum gets its water-binding properties from the natural moisturizing factor (NMF) present in the stratum corneum and also from humectants supplied by cosmetics.

The water-retarding properties are governed by barrier properties of the stratum corneum. These properties are strongly related to the structure of the lipid region of the stratum corneum. This structure can be affected by washing (surfactants) but also ameliorated by the lipids and emollients supplied by creams or lotions [see also under Sec. III].

The moisturization can further be improved by (semi-)occlusive layers (sebum, lipids, emollients) at the surface of the skin.

B. Techniques

Accurate assessment of the moisture content of the stratum corneum is difficult for several reasons:

The moisture content of the stratum corneum varies from relatively low (about 15%) in the outer layers to relatively high (about 40%) in the deepest layers [7]. Assessment of the moisture level will result in an average value. In most cases it is not known what the exact contribution is of the outer layers to this average value nor the contribution of the deeper layers.

The moisture content is not evenly distributed in each layer: the moisture content of the corneocytes (horny cells) differs from the moisture content of the lipid regions.

Three types of water are generally recognized in literature: tightly bound water, bound water, and free water [8].

Most assessments are based on an indirect measurement of skin moisture (e.g., the electric properties of the stratum corneum). We cannot be sure that the outcome of these assessments is indeed proportional to the moisture content.

Although the absolute skin moisture is hard to measure, any increase or decrease can easily be detected: currently a couple of devices are on the market that are capable of measuring skin moisture within seconds. These devices are commonly accepted and can be used, for example, for claim substantiation (moisturization, repair of barrier function of the outer layers of the stratum corneum) and for the assessment of skin compatibility of cleansing agents.

Some of the most frequently used commercial devices are the Corneometer® CM820 and the Corneometer® CM825, supplied by Courage + Khazaka.

Others are the DPM® 9003 (NOVA) and the Skicon®-200 (IBS, Hamamatsu). All these instruments measure the electric properties of the skin. They work at different frequencies, using various sizes of probes that are applied with distinct pressure on the skin. All of them give hydration values (expressed in arbitrary units) that are not linearly correlated with the hydration state of the skin [8]. Readings are sometimes influenced by salts and humectants (such as glycerol) present on the skin.

The Corneometer® CM820 and the Corneometer® CM825 (new version) measure capacitance and operate at low frequency (40–75 Hz). They are sensitive to the relative dielectric constant of the measured materials. And as water has a high dielectric constant, the device is very sensitive to water. It estimates the water content in the epidermis to an approximate depth of 60 to 100 µm [9,10]. Yet it is very likely that the (outer) layers of the stratum corneum are overproportionally contributing to the readings because the device is very sensitive to application of moisturizing creams whereas the deeper layers of the skin will have a fairly constant water level. This sensitivity might be due to a direct measurement of the increased moisture content of the stratum corneum or to an indirect effect, namely the deeper penetration of the electric field due to moisturization of the stratum corneum [11].

The DPM® 9003 uses measurements of phase angle. Phase angle is one of the two coordinates of impedance at each particular frequency of stimulation. The device uses frequencies up to 1 MHz. It measures the outer stratum corneum layers. The device is not discriminative for dry skin: if the skin is very dry, the outcome of the skin measurement will be zero. Yet it is very useful for assessing the moisturizing effects of moisturizing creams.

The Skicon®-200 measures the conductance at high frequency (3.5 MHz). Like the DPM 9003, it measures the outer stratum corneum layers.

Other, less frequently used techniques to measure skin hydration are infrared spectrography [12], nuclear magnetic resonance [13], and magnetic resonance imaging [14]. Also, the desquamation and skin elasticity measurements [15] give information about the degree of skin moisturization.

C. Example

As the natural moisturizing factor present in the skin is water-soluble, even washing with water will (partly) remove it, resulting in a slight dehydrating effect. This dehydrating effect can be more pronounced after washing with cleansing products because the surfactants from these cleansing products can remove skin lipids and furthermore are able to disturb the lipid ordering in the stratum corneum. These dehydrating effects are shown in Figure 1 (corneometer readings taken 12–16 hours after last washing). This figure also shows that these dehydrating effects can be (over)compensated by application of a moisturizing product.

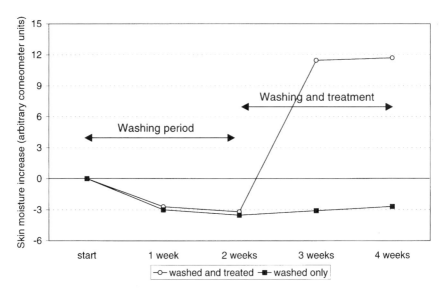

Figure 1 The corneometer CM825 demonstrates the drying-out effect of a shower gel during the first 2 weeks of the test period and the combined restoration and protection effect by a lipid-containing cream during the last 2 weeks (baseline corneometer value 38 arbitrary units).

1. Some Practical Advice

The conditions during the measurements are crucial for a useful assessment of skin moisture [16,17]. For example, perspiration has to be prevented for obvious reasons. Furthermore, the contact between probe and skin should be very close: hairs and product residues (including waxes, oils, and inorganic UV-filters) will influence the readings. So, before any assessment one should clean the skin (e.g., by paper tissues) if there is any doubt about product residues at the skin's surface. However, the assessor should realize that removed product remnants cannot contribute anymore to skin hydration.

In order to substantiate moisturizing claims for moisturizing products, it is advised to apply the product(s) at the volar side of the forearms. As the assessments are not very accurate (yet also not very time-consuming), one should take at least three—but preferably ten—readings from each spot. To prevent occlusive effects caused by the probe, waiting periods of at least 5 seconds should be observed between measurements at identical skin sites. Furthermore, an untreated control site should be included in order to take environmental variables into account.

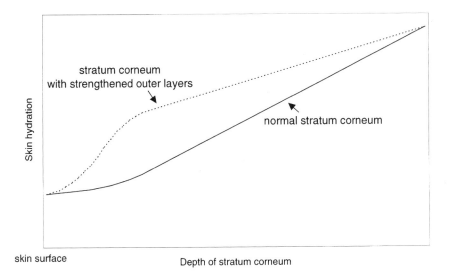

Figure 2 Schematic representation of water distribution in the stratum corneum for normal skin and for skin with improved barrier properties for the outer layers.

For the right interpretation of the data we have to realize that the assessment indicates the average skin moisturization. And as the effect of lipids on moisturization is different from that of humectants, we have to take this into account. Also, differences in stratum corneum thickness may lead to misinterpretations.

2. Theoretical Effect of Skin Strengthening

In general, the deepest layers of the stratum corneum will have the best barrier properties. Treatment of the skin with barrier-improving lipids will have the biggest impact at the more superficial layers, as these layers are easiest to strengthen. Beyond these strengthened layers a new steady state will be established. This is represented in Figure 2, which depicts this relation schematically.

It is clear from this graph that the outer layers have not benefited much by this treatment, although the deeper layers clearly have. Such beneficial effect can easily be assessed by the corneometer, but only with difficulty by devices such as the DPM 9003 and the Skicon-200, because these devices preferentially measure the moisture content of the outer layers.

3. Theoretical Effect of Humectants

In general, use is also made of humectants, which enhance the moisture content of the whole stratum corneum (Fig. 3).

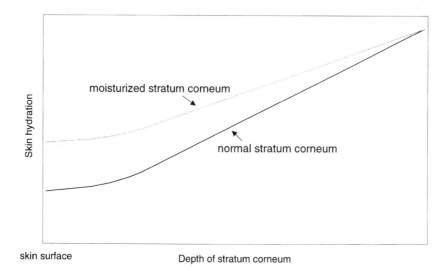

Figure 3 Schematic representation of water distribution in the stratum corneum for normal skin and for skin with applied humectants.

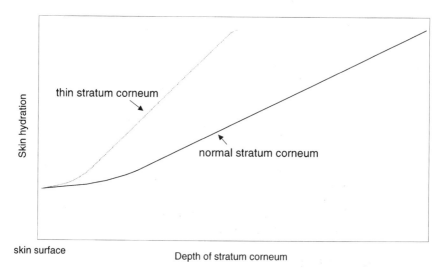

Figure 4 Schematic representation of water distribution in the stratum corneum for normal skin and for thin skin.

Such effects are easily detected by the Corneometer CM825 as well as Skicon-200 and DPM 9003. Note: Application of occlusive products results in similar skin moisture effects; assessment of the skin moisture increase, however, requires the removal of the occlusive layer.

4. Theoretical Effect of Stratum Corneum Thickness

The stratum corneum forms the main barrier for water; thus, the "water level to depth" relation will be steeper for thin skin. This is represented in Figure 4.

Assessment of skin hydration will produce in this case the highest value for thin stratum corneum, although in this example in both cases the hydration of the surface is the same. Even if the surface hydration is lowest for thin stratum corneum, the hydration assessment might result in the highest value. This phenomenon has to be taken into account in comparing young to (relatively thin) old skin, male to (thin) female skin, and also UV-exposed (thick) to nonexposed skin.

III. BARRIER ASSESSMENT

A. General

The barrier function of the skin prevents, on the one hand, the excessive diffusion of water through the epidermis and, on the other hand, the penetration into the skin of external compounds that come in contact with the surface of the skin. This barrier is determined largely by the integrity of the stratum corneum, which is formed by stacked corneocytes embedded in a lipid continuum, mainly consisting of ceramides, cholesterol, and free fatty acids. This lipid continuum is composed of highly ordered lipid bilayers, which form the actual barrier, and is located at the deeper layers of the stratum corneum with thin layers of bound and/or free water in between. A healthy skin in good condition has a proper barrier function, which is characterized by low values of transepidermal water loss (TEWL) [18,19].

In chemically and/or physically damaged skin and in certain pathological skin conditions, the barrier function is impaired, resulting in increased values of TEWL. The barrier function can be assessed by measuring the amount of water that passes through the stratum corneum barrier and leaves the skin as water vapor; it provides information on the integrity of the epidermis and the effects that topical (pharmaceutical or cosmetic) products have on the epidermal barrier [20,21].

Indeed, it is a big challenge and opportunity to develop products that contain cosmetic lipids that will enhance the stratum corneum lipid barrier function of the skin. This will require highly specialized formulation strategies to target cosmetic lipids to become an integral part of the lipid barrier system.

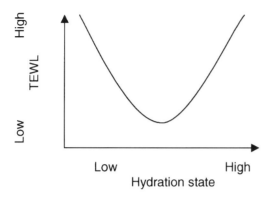

Figure 5 Schematic diagram showing the relation between TEWL and hydration state in skin.

As mentioned in Section II, the strengthening of the barrier is one way of increasing the overall hydration state of the skin, but TEWL and moisturization are not necessarily linked directly to each other. High TEWL and low hydration may coexist, and are in fact a typical phenomenon for dry, xerotic skin [22]. On the other hand, damaging the stratum corneum barrier (e.g., by stripping) results in a skin condition in which high TEWL and high hydration state coexist. Figure 5 shows the relation that in general is expected to exist between TEWL and hydration. The lowest TEWL-value will reflect healthy skin with optimal barrier properties.

B. Techniques

The measurement of TEWL is now generally recognized as one of the most popular and commonly used methods to assess barrier function [23].

The following devices are often used and seem to be the most popular: the Evaporimeter® (Servo Med), TEWA-meter® (Courage + Khazaka) and DermaLab® (Cortex).

The measuring principles of these three devices are basically the same [24,25]. All devices have the advantage that the measurements can be performed non-invasively and relatively fast and accurately. The results obtained are expressed in g/m^2h. The evaporimeter seems to have a somewhat lower sensitivity and a higher susceptibility to air circulation and relative humidity [26].

Under steady state conditions, the human skin surface is surrounded by a water vapor boundary layer, in which an inner-outer gradient exists. This vapor gradient is correlated with the stratum corneum water gradient, which in turn is

Biophysical Methods for SC Characterization

a function of the barrier (quality). The water vapor gradient is measured by a probe that is placed perpendicular on the skin. The probe consists of an open cylinder containing two hygrosensors coupled with two thermistors placed at different distances from the skin surface. At both points the local relative humidity and temperature are measured, and the corresponding vapor pressure is calculated. The difference between the vapor pressure at both points along the gradient is directly related to the rate of the water loss from that particular skin site [27]. Many factors influence proper TEWL measurements with the TEWL devices; the most important ones will be mentioned here and should be given special attention [23,25,28]:

- The temperature of the measuring probe. This is an essential variable. For instance, an increase of the probe temperature from 22°C (~room temperature) to 30°C (~temperature of the skin) may result in an increase of TEWL of more than 100% [26].
- Sweating. Physical, thermal, and emotional sweating need to be controlled because these will have profound influence on TEWL measurements.
- The temperature and relative humidity of the ambient air. Under constant relative humidity, there is an almost linear increase of TEWL as a function of ambient temperature.
- The circulation of the ambient air. Only major changes in air circulation significantly affect the measurements, but care should be taken.
- Condition of the skin. Diseased skin and any damage of the barrier function of chemical and mechanical origin may result in increased TEWL values.

Other techniques used to assess the barrier function of the skin are usually more elaborate and less straightforward. Two of these techniques are only briefly mentioned here:

- A gravimetric method in which the weight increase of hygroscopic salt, attached to the skin, is determined in an unventilated chamber as a measure of water loss. Not so popular anymore [19].
- The measurement of other, indirect biophysical parameters such as skin redness, caused by the penetration of biologically active compounds, such as methyl nicotinate; this is an (indirect) indication of the quality of the stratum corneum barrier [29].

C. Example

Some general examples in which TEWL measurements may be used are as follows:

- The assessment of the damaging effect on the stratum corneum barrier function caused by surfactant-containing rinse-off products.

The assessment of the restoration of a damaged stratum corneum barrier by certain lipid-containing creams.

The assessment of the prevention of damage caused by chemical or physical treatment (e.g., stripping) by certain lipid-containing creams.

The results of a test in which these principles play a role are shown in Figure 6. The volar forearms were washed twice daily with a moderately aggressive shower gel; the surfactants caused a slight barrier damage and consequently TEWL values increased from ~6 g/m^{2}h to ~10 g/m^{2}h. Using this moderately aggressive washing regime, a "damaged steady state" was reached after about 2 weeks, as demonstrated by the leveling off to an (elevated) TEWL plateau. When a lipid-containing cream is applied on the forearm in between the washings with the shower gel, a normalization of the TEWL values can be realized.

In this case, the TEWL normalization is caused by a combination of lipid barrier repair and an increased protection against further barrier deterioration by the shower gel. These findings are in line with the results of the moisturizing test (see Sec. II, Fig. 1), demonstrating that increased TEWL values often result in a drying out effect of the skin (see also Fig. 5).

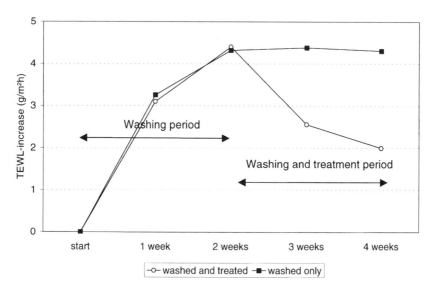

Figure 6 The TEWA-meter TM210 demonstrates the damaging effect of a shower gel during the first 2 weeks of the test period and the combined restoration and protection effect of a lipid-containing cream during the last 2 weeks (baseline TEWL 6.2 g/m^{2}h).

Biophysical Methods for SC Characterization

IV. ASSESSMENT OF THE MICRORELIEF OF THE SKIN

A. General

The skin surface contains wrinkles and lines, which can be classified according to their depth [30]:

Clearly visible wrinkles, grooves, thin folds, and furrows (100 μm to several millimeters)
Primary lines, criss-crossing the skin to form squares, parallelograms, and rectangles (20–100 μm)
Secondary lines, branching off from the primary lines (5–40 μm)
Tertiary lines, forming the borders of the corneocytes (about 0.5 μm)
Quaternary lines, corresponding to the corneocyte relief (about 0.05 μm).

The wrinkles and the primary lines reflect the contours of the dermis; nevertheless, the condition of the stratum corneum (scaling, thickness, and degree of swelling) has a huge impact on the microrelief. For example, swelling (due to increased moisturization and/or uptake of lipids) of the stratum corneum from, say, 20 μm to 25 μm may result in a decrease of the depth of the primary lines, much more than the 5 μm thickness increase of the stratum corneum. This phenomenon is due to squeezing of the lines.

The skin texture reflects the subject's age, the weather, and the environmental conditions and also the effect of topical applications. So, objective measurements of the microrelief and morphological structure of the human skin is important in developing effective cosmetics.

Several parameters are used to describe skin roughness [31]. Commonly accepted are the depth of roughness (R_t) for surfaces dominated by wrinkles and the mean depth of roughness (R_z) for surfaces dominated by primary and secondary lines. These variables explore the variations of the skin profile in the vertical direction.

R_t is the maximum peak to valley depth for a given scan length L. The value for L (usually 5–15 mm) is the length of a chosen line across the (replica) surface. R_z represents the arithmetic average of the different segment depths of roughness calculated from 5 equal sections of L.

Both parameters originate from the metallurgical industry for scanning metallic surfaces and are somewhat outdated but still widely used.

B. Techniques

Two main techniques [32–34] can be distinguished to measure the skin surface:

The skin replica technique: replicas are taken from the skin and analyzed afterward.
The living skin technique: the skin is analyzed directly.

1. Skin Replica Techniques

Skin replicas can be made of polymerizing silicone. Within about 5 minutes, a rather faithful negative reproduction of the skin is obtained. Such replicas can be analyzed by several techniques:

- The oldest technique is the contact stylus method [35]. Replicas are scanned by a mechanical profilometer such as the Hommel Tester® (Hommelwerke): a microprobe consisting of a diamond stylus moves at constant speed along straight lines over the replica. The vertical displacement of the stylus is converted into electrical signals proportional to the deflections. This technique is still used although analyzing the replicas is very time-consuming. It is advised to scan at least ten sweeps, resulting in an analyzing time of about 10 minutes for each replica.
- A similar technique (acoustic profilometry) makes use of a touchless acoustic pickup. Scanning is performed by the Nanoswing®, also supplied by Hommelwerke.
- A modern technique is optical shadowing profilometry [36]. Under grazing illumination of the replica, the shadows behind the crests (i.e., grooves of the skin) are captured by a video camera and used for image analysis. This method has some obvious limitations yet is useful and fast. Furthermore, it can be used to calculate the ratio between the true and the apparent skin surface area, which represents an advantageous profilometric parameter.
- Dynamically focusing laser profilometry is another new method [37] that has resulted in several commercially available devices. The replica is scanned by a laser beam autofocusing at a preselected number of spots. Drawbacks of this method are the high investment required and the time-consuming procedure of scanning (up to hours per replica). This method, however, makes it possible to describe skin roughness with three-dimensional parameters [38].
- Another optical profilometric technique is based on the triangulation method [37,39,40]. In this method, the image of the replica is projected—dot by dot—on a detector. The main advantage is the high scanning speed and the possibility to obtain three-dimensional information of the surface.
- The light transmission technique or transparency profilometry makes use of thin dyed silicone replicas. The replicas are analyzed by the Skin Visiometer® SV500 supplied by Courage + Khazaka. Light is absorbed according to the thickness of the replica according to Lambert-Beer's law. It is a fast method, recently validated [41]. The technique is simple, rather accurate, and not very expensive. A limitation of the method is

that—because of the thickness of the replica itself—grooves with a depth more than about 400 µm cannot be measured accurately.

2. Living Skin Techniques

These touch-free "living skin" techniques are still not fully developed but are promising because time-consuming replica taking is no longer needed. Furthermore, elimination of the replica taking step reduces the chances of obtaining erroneous results. Measurement of living skin, however, makes great demands on the measurement times. Effects of skin movements due to trembling and moving of the subjects, their breathing, and the pulsation of minor arteries have to be excluded.

Two techniques are commercially available:

- In vivo topometry of human skin by projected fringe technique [42–44]. FOITS® (fast optical in vivo topometry of human skin) developed by Breuckmann and PRIMOS® (phaseshift rapid in vivo measurement of skin) developed by GFMesstechnik allow three-dimensional information to be gathered from the surface of the skin in an extremely short time. However, a shortcoming of this technique is that dealing with hairy surfaces seems to be problematic.
- Surface evaluation of living skin by a graphic depiction technique [45,46]. The Visioscan® VC98 device (Courage + Khazaka) is a video camera that monitors the skin surface illuminated under an UV-A light source. Interpretation of the image by the supplied software gives information about skin roughness. A serious drawback of this new technique is that quantitative assessment of the depth of lines and grooves is not yet possible. Also, dark and white spots and hairs seem to influence the outcome of the measurement. A similar device is the Direct Skin Analyzer® [47].

An alternative for the skin replica and the living skin techniques is the visual assessment of the skin relief [48,49]. Especially for assessment of the degree of facial wrinkling, using a photo scale, this technique is valid.

The scanning electron micrography (SEM) technique is used to obtain a high magnification of skin replicas. It can also be used to study ex vivo human skin, skin biopsies, and skin surface strippings [50,51]. This sophisticated but expensive technique is a suitable method for fundamental research on the intra- and extracellular structure of the stratum corneum surface.

C. Examples

The scanning direction is an important factor for the evaluation of (the replica of) the skin because the texture of the skin is highly anisotropic, at least in elderly people. For wrinkles, it is advised to measure perpendicular to the major grooves and to use R_t as a parameter.

Surfaces without wrinkles are better measured in a circular manner because the major direction of the primary lines is not always easily recognized. R_z is in this case the right choice for skin smoothness. However, it has to be realized that a drawback of this choice is that one or more of the five sections for measuring R_z just covers a plateau of the skin and not the primary lines. This is a more serious problem if old subjects are measured than in the case of young subjects because the distance between primary lines increases with age [41]. This effect is illustrated in Figure 7 by the relation between length of L (sum of the five sections) and R_z.

This anisotropy of the skin is gradually developed and is far more marked on the skin of elderly people than on that of children, where the distribution is almost isotropic. Figure 8a shows the forearm pattern of a 3-year-old boy and Figure 8b that of a 76-year-old man. Both pictures are reproduced by the Visiometer SV500, using silicon replicas from the skin.

Some practical advice:

The correct calculation of roughness parameters from the original data is only possible after elimination of superimposed elements of form and waviness

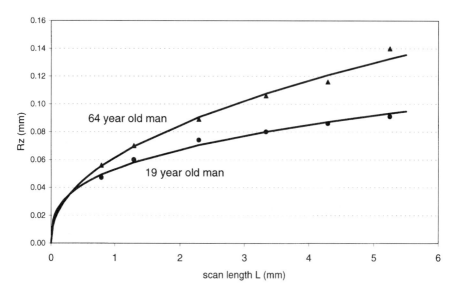

Figure 7 Relation between R_z and scan length L given for a 19-year-old and a 64-year-old man, measured by Visiometer SV500. For each replica analysis, 180 circular arranged lines were used.

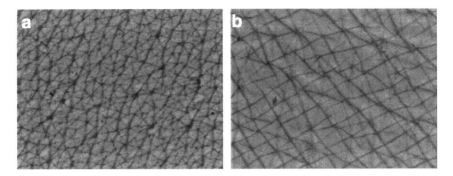

Figure 8 Replicas of the skin reproduced by the Visiometer SV500: (a) replica of the skin of a 3-year-old boy, (b) replica of the skin of a 76-year-old man. For both replicas, the same magnification was used.

of the original skin profiles by mathematical procedures. This elimination (filtering) will in most cases be performed by the software. However, it may be a source of erroneous corrections. Ideally, the software should show what the effect is of these corrections.

Special attention has to be given to the next items:

If the skin surface is covered by a product (e.g., cream), the replica will not yield a faithful image of the skin surface.

Stretching of the skin influences the depth of the lines and wrinkles, so measurements have to be performed under standardized conditions and positions; for example, for measurements of the forearm, the angle between forearm and upper arm has to be fixed (e.g., to 90°).

V. ASSESSMENT OF BIOMECHANICAL PROPERTIES

A. General

Assessment of the biomechanical properties of the skin, and in particular of the stratum corneum, is a difficult task. One of the main reasons is that the skin is a very complex organ and its final biomechanical properties are a result of complex interactions of different layers of the skin (e.g., epidermis, dermis, and hypodermis). This is reflected by the fact that skin has a combination of elastic, viscous, and plastic properties, which as yet cannot be adequately described in an overall mathematical model. Skin mechanics are described mostly by stress (~force)/strain (~deformation) relationships and the course of deformations in time; interpretation is often complicated and factors such as age, skin type, skin thickness, body site, pretreatment, environmental conditions, and so forth should be taken into account and may all have profound effects on these parameters. A more detailed description of the attempts to describe human skin biomechanics is beyond the scope of this chapter, and the reader is referred to Rodrigues [52] and Serup [53].

It is generally accepted that the stratum corneum contributes to the overall biomechanical properties of the skin, albeit less than the viable epidermis, the papillary plus the reticular dermis and the hypodermis.

The main descriptors for skin biomechanics are extensibility (~flexibility) and elasticity, which in turn may be translated into more qualitative benefits such as suppleness, firmness, softness, and resilience. It must be emphasized that interpretation of the relation between extensibility and elasticity is not uniform and definitively not standardized and therefore is often subject to much confusion and discussion. Figure 9 shows a schematic diagram, giving some *subjective* descriptions of skin biomechanical perception in relation to elasticity and extensibility; by no means is it meant to be an accurate account of these perceptions.

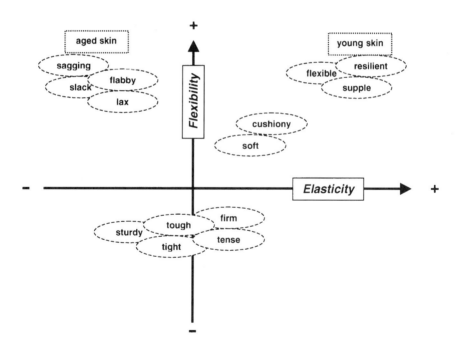

Figure 9 Schematic diagram indicating some qualitative biomechanical perceptions in relation to elasticity and extensibility. (Slightly adapted from Rodrigues [52].)

Biophysical Methods for SC Characterization

The topical application of cosmetic lipids may have profound effects on the biomechanical properties, with special reference to the stratum corneum. Integration of these lipids (e.g., emollients) will generally improve the moisturization and emolliency status of the skin, which in turn will result in improved skin properties such as flexibility and softness.

B. Techniques

There is a wide variety of different techniques that are designed to assess the biomechanical properties of the skin. Mostly, all these techniques are based on variations on the way a force (∼stress) is applied onto the skin and subsequently to look in time if and how the skin retakes its normal position. The forces can be applied either vertically or parallel to the skin surface, which can be further nuanced into different forces such as extension, torsion, indentation, suction, pulling, bouncing, and frictional forces [52,54].

A typical measurement, showing stress-induced skin deformation (usually referred to as strain) and the recovery in time is shown in Figure 10. Various ratios of the deformation parameters are also used to describe the biomechanical status of the skin. For instance, the ratio U_r/U_f (i.e., ratio between immediate

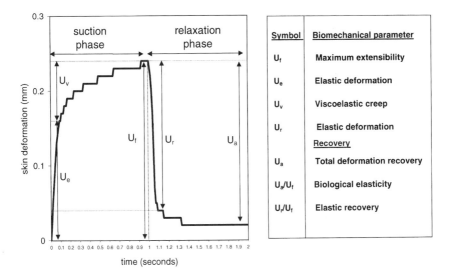

Figure 10 A graphical representation of a typical strain-time curve measured by cutometer SEM575, including the most commonly used parameters.

recovery and maximal extensibility of the skin) is considered a biologically important factor for the characterization of elastic recovery [55].

In this chapter, we describe the three most commonly used devices, each of which is based on different stress/strain methods.

The Cutometer® SEM575 (Courage + Khazaka) is a suction device, whereby a variable vacuum (ranging from 50 to 500 mbar) is applied to the skin surface. Using a hand-held probe, the skin is pulled by this vacuum into the aperture of the probe. The skin adjacent to the opening of the probe is maintained in position by a guard ring. Usually, the strain course under constant vacuum is shown as a function of time and allows the calculation of a series of mechanical parameters related to elasticity, viscosity, and plasticity (Fig. 10).

Special attention should be paid to the following specific variables:

1. The three possible measuring modes—probe diameter aperture, on-set and off-set time, and vacuum strength—will profoundly influence the measurements and should always be indicated. For instance, a smaller aperture will monitor the (visco) elastic properties of more superficial layers of the skin.
2. Pressure application on the skin. Although the probe has a spring system, one should pay extra attention to ensure that constant pressure on the skin surface is ensured, because this will strongly influence the measurements.
3. Repeated suction. Repeated cycles will cause hysteresis and as such may even provide extra information.

The cutometer may measure both the mechanical properties of the superficial skin as well as the deeper layers (e.g., dermis and subdermis) to an unknown extent [56,57].

With the Dermal Torquemeter® (Dia-Stron Ltd), a force parallel to the surface is applied. A disc glued to the skin is set into motion by a torque motor and the torque plus the amount of rotation induced are recorded. The torquemeter will measure the mechanical properties of the more superficial layers (e.g., epidermis). The elongation of the skin can be controlled by a guard ring. It has been claimed that the use of a guard ring and a low torque increases the importance of the stratum corneum contribution of the measured mechanical functions. Usually, like with the cutometer, the strain course under constant torque is shown as a function of time. Reproducibility and accuracy may cause a problem and strict standardization is needed [53]. A variant of this device is called the Twistometer [58].

The GBE, the gas-bearing electrodynamometer (not commercially available) is, contrary to the methods mentioned before, especially designed to measure predominantly the viscoelastic properties of the stratum corneum. As mentioned before, the tissue underneath the stratum corneum has a relative large

Biophysical Methods for SC Characterization 205

impact on the measurement of mechanical skin properties. The GBE device is designed in such a way that the probe (attached to the skin using sticky tape) picks up sensitive deformation of the stratum corneum as a result of a sinusoidal low stress onto the skin. As with the torquemeter, a stress parallel to the skin is applied, and the results of stress and displacement measurements are typically displayed as a hysteresis loop, yielding information about elastic and viscous properties of the skin. For a more detailed description, the reader is referred to Hargens [59]. Recently, a new design, called the linear skin rheometer, has been described, which is said to yield more reproducible results [60].

The GBE is relatively sensitive to changes in, for example, thickness of the stratum corneum; also, positioning adjustment and the imperceptible movements of volunteers are a source of major variability in measurements [61–63].

Various other devices have been described; here, we limit ourselves by mentioning a few of them:

the Dermaflex® [64], like the cutometer, based on suction-force
the Extensometer® [65], based on stretch-force
the Indentometer® [66,67], based on compressing force
the Ballistometer® [68], based on drop impact force

Recently, a new device was introduced by Courage + Khazaka called the Reviscometer® RVM600, based on acoustical shock waves, without the application of external forces.

Certainly, the fact that so many devices exist demonstrates that there is no unique, generally accepted technique that has emerged as a "standard" method, although the cutometer as well as the torquemeter seems to be quite popular at the moment.

C. Example

The relationship between the biomechanical parameters (see Fig. 10) and its physiological significance are often complex and not straightforward. For instance, the extensibility of the skin (U_f) may be increased after treatment with a cosmetic product, but, on the other hand, aging skin shows also a gradual increase in U_f. In studying the aging of the skin, the U_r/U_f ratio (i.e., the elastic recovery) seems to be a particular useful biomechanical parameter, in which typically the more visco-elastic properties of the skin are expressed. Indeed, aging skin shows a continuous decrease in U_r/U_f, which may amount to more than 50% after seven decades (Fig. 11). This considerable decrease is especially manifest on the dorsal forearm, because here the effects are a combination of both intrinsic aging (i.e., biological or chronological aging) as well as extrinsic aging (i.e., photoaging). On the volar forearm the decrease is much less pronounced, indicating that UV-exposure during one's lifetime has a great impact on the elastic properties of

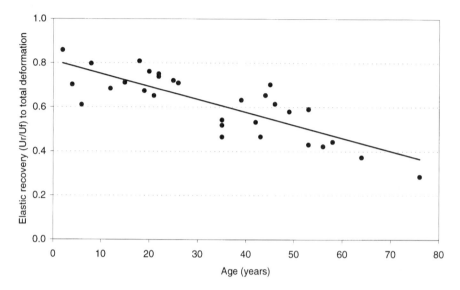

Figure 11 A plot of the elastic recovery (U_r/U_f) of the dorsal forearm versus age for 32 subjects, measured by the Cutometer SEM575. R = 0.82; aperture diameter: 2 mm; pressure: 500 mbar; pressure-on time: 1 sec; pressure-off time: 1 sec repetition: 3; pretime: 0 sec.

the skin. Numerous studies have indeed shown that UV radiation causes photoelastosis, in which, among other factors, massive accumulation of unstructured elastin causes a loss of skin's elasticity [69]. It seems that the cutometer is more appropriate for measuring this phenomenon than the torquemeter, which is in line with the fact that the torquemeter registers the more superficial, rather than the deeper, dermal biomechanical changes of the skin [70].

VI. ASSESSMENT OF DESQUAMATION

A. General

The loss of stratum corneum at the skin surface (desquamation) is an integral and essential part of epidermal physiology, which takes place in a controlled manner by the loss of single corneocytes (fine flakes). The term desquamation is, however, also used to describe the shedding of scales (or squames), defined as flat plates or coarse flakes of stratum corneum, which actually are conglomerates of corneocytes.

Using the first definition, the rate of desquamation is expressed as the number of corneocytes released per square centimeter per hour. This rate is related to the stratum corneum turnover time. Stimulating the epidermal cell renewal can shorten this turnover time and increases the rate of desquamation (and thus gives the skin a "younger" appearance).

Using the second definition, the degree of desquamation is related to the extent of scaliness. And as scaly skin is generally associated with dry skin, a high degree of desquamation is also associated with dry skin. So, depending on the definition used, rather contradictory phenomena can be described, giving rise to confusion.

The mechanical strength of the stratum corneum is based on the coherent lipid layers and on the intercellular junctions (desmosomes). The breakdown of the desmosomes in the outer layers of the stratum corneum is essential for the desquamation process (loss of single corneocytes) and requires the action of proteases such as stratum corneum chymotryptic enzyme (SCCE). Desmosomal degradation must occur to ensure that desquamation proceeds in an orderly fashion. One of the most important factors for these enzymatic reactions is an optimal water and lipid environment, which in turn will lead to optimal desquamation [71,72]. In case of insufficient breakdown, dry, scaly skin will arise. Such skin is characterized by groups of elevated corneocytes and flakes, which give the skin a rough, cracked, gray appearance (xerosis).

Large scales are clearly visible to the naked eye because of the occurrence of air between the scales and the surface of the skin (due to the large differences in refractive indices). Application of a lipid-rich cosmetic product generally will result in replacement of the air by lipid, making the scales less visible. The scales will become clearly visible again after absorption of the lipid or after washing of the skin. Daily application of a well-formulated cosmetic product will result in an improved skin condition and might eventually result in a skin without scales. This latter process will take about one week.

B. Techniques

There are three possible approaches to characterize the process of desquamation: the assessment of the stratum corneum turnover, the measurement of the intracorneal cohesion, and the quantification of scaling.

1. Assessment of the Rate of Desquamation and Stratum Corneum Turnover

Measurement of the rate of desquamation can be direct [73], by counting the number of corneocytes released at the surface with a passive chamber technique (expressed as number of corneocytes released per square centimeter per hour) or with a forced desquamation technique such as the detergent scrub method [74].

An indirect measurement of the rate of desquamation estimates the clearance rate of loss of a stain on the stratum corneum. Staining methods aim to apply a substantive dye, such as the fluorescent dye dansyl chloride, to the stratum corneum. [75]. After penetration, the entire thickness of the stratum corneum should be labeled. When the stained stratum corneum has been shed completely, the stain will have disappeared, and the time taken is the stratum corneum turnover time (usually about 20 days). An alternative for staining is artificial tanning [76] of the skin induced by dihydroxyacetone (DHA). Fading of pigmentation occurs during shedding of the stratum corneum. The evaluation of the intensity of browning of the skin is assessed colorimetrically (see also Sec. VIII).

2. Measurement of the Intracorneal Cohesion

The release of the intracorneal cohesion (ICC) allows desquamation to occur. This binding force can be measured using a cohesograph. This device employs a piston that is stuck to the skin surface with a cyanoacrylate adhesive and the force required to distract a segment of stratum corneum from the surface is measured [77]. One of the drawbacks of this method is its complexity.

3. Quantification of Scaling

Scaly skin can be assessed clinically or instrumentally. The clinical assessment can be performed either by the subject himself or herself or by an expert evaluator [78]. Very helpful for the clinical assessment could be the use of a microscope (in vivo microscopy) or the use of the Visioscan® VC98 device (Courage + Khazaka). This video camera monitors the skin surface illuminated under a UVA light source. The spectrum of the light is optimized in order to exclude reflections from the deeper layers of the skin. This results in sharp images from the scaly skin. These images can be analyzed afterward by a graphic depiction technique [45] or by visual scoring of the degree of scaling.

Visual examination of tape strippings of the stratum corneum can reveal differences in the extent of dryness not noticeable by visual inspection of the skin [79]. New image analysis techniques were developed, following this simple method, to objectively analyze the desquamation of the stratum corneum. Skin surface sampling discs, such as the commercially available D-Squames® (Cu-Derm) and Corneofix® (Courage + Khazaka), can be used to sample loose cells and scales from the superficial stratum corneum. The contact time is usually 5 to 30 seconds. After placement of the disc or tape on a black background, an experienced observer can grade the extent of scaliness. It is a convenient method for evaluating the effectiveness of cosmetic products and also for classifying cleansing products for their irritation/compatibility potential. One option is also to take video images of these discs and to process them afterward with the aid of an image analysis program to calculate the degree of desquamation [80].

When appropriately stained, scales harvested by the adhesive discs develop an intense color. This property is used in the squamometry method to improve the sensitivity [81–83]. The rinsing process, which is needed for the removal of excess of stain, is critical, however: too intense rinsing will remove the scales; mild rinsing will not remove excessive stain completely. Corneosurfametry [84] deals with staining of cyanoacrylate skin surface strippings after contact with surfactant solutions. Both methods (squamometry and corneosurfametry) can be used for comparison of the human skin compatibility of personal care cleansing products.

C. Examples

Figure 12a shows scales sampled from untreated forearm skin by a Corneofix® skin surface sampling disc (image taken by Visiometer SV500). Figure 12b shows the scales sampled from the opposite arm washed with an alkaline soap during one week. From these pictures it is clear that aggressive cleansing products can induce formation of large scales.

Techniques to quantify scaling can be used to select mild cleansing products but can also be used to prove that leave-on products improve the condition of the skin. Figure 13a shows the scaly (untreated) skin of the forearm; Figure 13b the opposite arm after one week of product application. The last picture is taken 18 hours after the last product application.

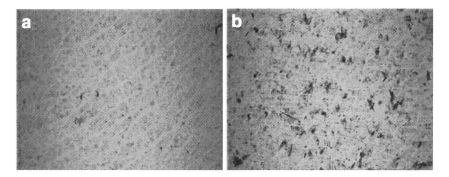

Figure 12 Images taken by Visiometer SV500 from sampled scales: (a) untreated forearm skin, (b) opposite forearm, washed during one week with an alkaline soap.

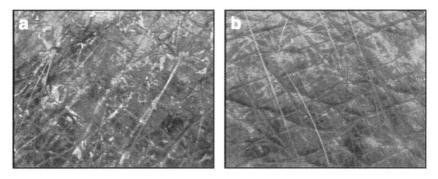

Figure 13 Pictures of forearm skin, taken by the Visioscan VC98: (a) untreated, scaly forearm skin, (b) opposite forearm, after 1-week application of a moisturizing product.

VII. ASSESSMENT OF SURFACE LIPIDS

A. General

The skin is covered by a thin layer of lipids that are derived mainly from sebum excreted from the sebaceous glands via the hair follicle. Although no evident role can be attributed to sebum (e.g., it is not essential for proper moisturization of the skin), interaction with the stratum corneum may be responsible for certain biophysical effects. Indeed, Blanc et al. [85] have demonstrated a substantial capacity of the stratum corneum to absorb sebum. Sebum lipids, which consist mainly of squalene, wax- and cholesterol-esters, and triglycerides, are often called skin surface lipids (SSL). Analytical methods, including contamination problems with epidermal (~stratum corneum) lipids, will not be described in this chapter (see Chapter 6).

Sebum lipids are not an integral part of the stratum corneum epidermal lipids and as such do not play an essential role in the skin's barrier function. Sebum quantities present on the skin surface may be as high as 50 up to 500 $\mu g/cm^2$, much higher than the amount of epidermal lipids, which varies from 5 to 50 $\mu g/cm^2$ [86]. Under normal conditions, sebum is present in an emulsified form with water and/or sweat, constituting the so-called hydrolipidic film [87]. Cosmetic products, especially the lipid fraction (i.e. its emollients and emulsifiers) will interact with this hydrolipidic film [88]; this may result in altered absorption kinetics and biophysical effects of the products, a phenomenon that is often overlooked by cosmetic scientists.

Several skin disorders, such as acne and seborrheic dermatitis, are associated with aberrant sebum secretion levels, and strategies to reduce sebum secretion are an increasing area of interest [89–91].

B. Techniques

Two parameters are of importance in the assessment of surface lipids: the homeostatic, casual level of sebum present on the skin and the sebum excretion rate (SER), which can be determined after proper degreasing of the skin and looking at the refatting kinetics. Alcohol seems to be the best cleansing agent [88]. Amounts of SSL vary greatly depending on number and activity of sebaceous glands, which in turn depend on body site; for example, scalp, face (i.e., the so-called T-zone) and upper side of trunk have a high density of sebaceous glands (ca. 500 per cm^2 vs. <50 per cm^2 on other sites [92,93]).

Various techniques for the collection and quantification of SSL have been described [94]. The two most popular sebumetric methods will be described here.

The Sebutape® (CuDerm Corporation) is a tape containing a polymeric film that can be directly attached to the skin [95]. On places where SSL lipids are absorbed, the color of the film will be changed. Thus, not only a "fingerprint" of the number of follicular openings but also the activity of its sebaceous glands is obtained [96,97]. The technique can be used to determine a variety of parameters such as:

SER
total produced quantity
number of excreting follicles
quantity per follicle

Quantification of these parameters can be realized in different ways: (1) visual assessment, (2) color measurements using reflectance chromametry, (3) densitometric measurement, or (4) image analysis [89]. Moreover, quantitative and qualitative lipid analysis is possible after extraction of the lipids from the tape.

Whatever the parameter under investigation, strictly controlled experimental procedures are required for proper assessment of SSL quantities. Control of temperature, relative humidity, and proper acclimatization period of volunteers are of critical importance.

Some conditions that may influence the SER and SSL quantities are:

History of application of oil/lipid-containing topical products, which may influence SER
Body site: scalp, face, upper chest, and shoulders have highest levels of SSL [92,93]
Age and sex of volunteers: generally, sebum production is low among children and is higher in male adults than in female adults, and a gradual decrease is seen with increasing age. This is mainly due to the fact that sebum secretion is under hormonal (androgenic) control.

The Sebumeter® SM810 (Courage and Khazaka) is a photometric device and is based on the measurement of increased transparency of a frosted plastic film, caused by lipid absorption. The film is pressed on the skin and lipids are transferred from the skin onto the film; its rough surface is smoothed, because the spreading of the lipids fills up the cavities. Incident light is less scattered and more light can pass through the coated surface. A cartridge contains a continuous plastic film that is sufficient for approximately 300 measurements. The cartridge is placed in a measuring device that measures the transmission of light, emitted from a photocell, through the film. The difference in transparency before and after lipid sampling relates to the amount of SSL, and proper calibration converts the data into µg SSL per cm^2. Unlike the Sebutape® method, the sebumeter does not measure the number of excreting follicles and quantity of sebum per follicle.

It should be mentioned here that the sebumetric techniques may also be used to measure the absorption rates of lipophilic fractions of cosmetic products—i.e., the emollients and emulsifiers per se. Of course, proper removal of SSL prior to product application is important.

Various other methods and devices have been described in the literature, such as cigarette paper and bentonite clay methods and the Lipometre (not commercially available) [98]. Another interesting method is the combination of Sebufix® (Courage + Khazaka) and the Visiometer (see Sec. IV), which visualizes the sebaceous secretion in time (not published). An overview of the various methodologies is given by Piérard [99].

With reference to the presence or absence of SSL, four different skin types are in general recognized (see Fig. 14) [100]. Relatively high SSL levels are found in people with oily and mixed skin.

C. Examples

Figure 15 shows a typical course of the reappearance of SSL on the forehead, after it has been removed by a facial cleanser. From these data, a person's typical SER can be calculated. Generally, 3 to 4 hours are necessary to normalize the sebum levels for a normal skin type. For oily skin, this takes much less time.

It may be noted that the starting value in this examples is relatively low (180 vs. 300 µg/cm^2); this is not the subject's casual level of SSL, but it is the result of the fact that the volunteer had washed his face at home in the morning and the test took place quite early in the morning before normalization had occurred. This illustrates once more that the assessor should be alert at all times!

Especially on body sites that have only low endogenous SSL levels (e.g., the volar forearm), the sebumeter can be used to quantify the amount of lipids from a cosmetic product, deposited on the skin and subsequently absorbed.

This is demonstrated in Figure 16, where the absorption of the lipid phase of a cosmetic product in time is shown. The 14% lipid phase consisted of 5%

Biophysical Methods for SC Characterization 213

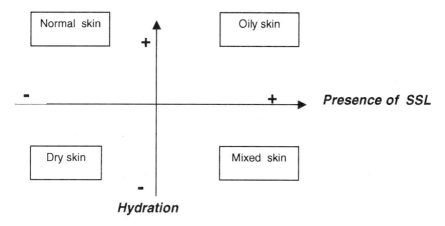

Figure 14 Schematic categorization of different skin types in relation to presence of SSL and hydration state.

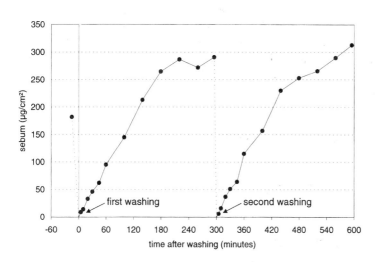

Figure 15 Course of sebum (SSL) levels on the forehead versus time, after two times washing with a facial cleanser; measured by the Sebumeter SM810.

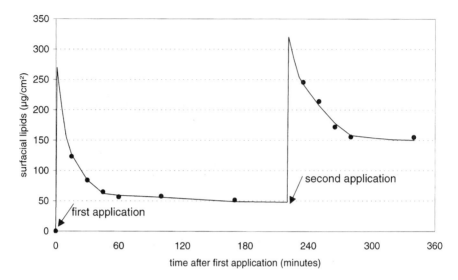

Figure 16 Absorption kinetics on the volar forearm of emollients from a cosmetic product, as measured by the Sebumeter SM810.

cetearyl glucoside/cetearyl alcohol (emulsifier) and 9% of a mixture of isohexadecane, caprylic/capric triglyceride, and ethylhexylpalmitate (emollients). A calculated amount of 270 μg/cm^2 lipids was applied on the *whole* volar forearm. Within 40 minutes approximately 80% is absorbed; Notably, after a second application of the same amount of lipids, the absorption is much lower, i.e. <50% after 40 minutes. This clearly demonstrates that the skin becomes lipid-saturated at high application levels.

VIII. MISCELLANEOUS TECHNIQUES

In this section, some less frequently used in vivo, noninvasive techniques are briefly described. Other techniques are described elsewhere in this book (see Chapters 6 and 8).

A. High-Resolution Ultrasound

High-frequency (20 MHz and 50 MHz) ultrasound measurements can be used to examine the lipid and water content of living epidermis and the intrinsic (biological) and extrinsic (photoaging) of the skin [101,102]. This technique generates two-dimensional pictures representing the skin structure. It can also be used for the in vivo estimation of the thickness of the dermis [103] but is not appropriate to measure the thickness of the stratum corneum.

B. Optical Coherence Tomography

Optical coherence tomography is a new diagnostic in vivo method for tissue characterization. Using this technique, structures of the stratum corneum, the living epidermis and the papillary dermis can be distinguished [104].

C. Magnetic Resonance Imaging

Magnetic resonance imaging is a sophisticated technique allowing deep exploration of the skin, including the hypodermis [105]. It offers possibilities to assess the hydration level of human stratum corneum by measuring proton density and proton relaxation times [14,106].

D. Confocal Laser Scanning Microscopy

A very promising in vivo technique is confocal (laser scanning) microscopy. It produces spatial 3-dimensional information on the skin: epidermal cells can be individually observed [107]. It is also able to measure the thickness of the stratum corneum.

E. Skin Color

Skin color measurements can be performed by scanning reflectance spectrophotometry (e.g., Minolta Chromameter® CR-200), using the L*a*b*-classification system as the preferred system to characterize the color [108]. The parameter L* represents the brightness, a* the green-red, and b* the yellow-blue axis. This means that white skin will show high values for L*, erythemic skin high values for a* [109], and tanned skin high values for b*.

Skin color measurements can also be performed by narrow-band reflectance spectrometry [110]: Erythema/Melanin Meter® (Dia-Stron), Mexameter® MX18 (Courage + Khazaka), and Dermaspectometer® (Cortex). The aim of this technology is to explore the presence of melanin and hemoglobin. The data are expressed as erythema and melanin indices.

It has to be taken into account that skin color is influenced by a variety of factors, including local blood flow, skin surface lipids, and scaling. And as blood flow is also influenced by temperature, pressure exerted by the probe, and position of the measured skin site relative to the heart, these factors might also affect skin color.

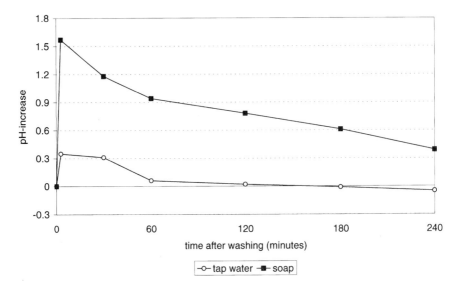

Figure 17 Effect of washing with soap and washing with tap water on skin pH, measured by the Skin-pH-Meter (baseline pH 5.1).

F. pH

Standard pH meters with a flat surface electrode can be used to measure the pH of the skin [111]. Figure 17 shows the effect of washing the skin with an alkaline soap: the pH of the skin increases for several hours. Also washing with tap water increases the pH to some extent, as tap water usually is slightly alkaline.

G. Skin Temperature

Skin temperature may be measured by readily available skin surface temperature measuring devices [112]. To avoid direct contact between skin and probe, infrared thermography is the preferred technique for measuring the temperature of the skin. This technique is also able to visualize the temperature profile of relatively large surfaces such as the face or the forearm [113].

One of the applications of skin temperature assessments is measuring erythema [114].

H. Friction

The frictionmeter measures the power necessary to move an object on the skin surface. The measured parameters are adhesional and glide friction [115]. High friction values are an indication of a rough skin. Nevertheless, most moisturizing creams increase friction after application [116,117]. Interpretation of the obtained data is difficult.

SUPPLIERS
Breuckmann GmbH

Torenstrasse 14, 88709 Meersburg, Germany
Telephone: ++49 7532 1563
Fax: ++49 7532 9377
E-mail: contact@breuckmann.com
Internet site: *http://www.breuckman.com/*

Cortex Technology

Smedevaenget 10, 9560 Hadsund, Denmark
Telephone: ++45 9857 4100
Fax: ++45 9857 2223
E-mail: cortex@cortex.dk
Internet site: *http://www.cortex.dk/*

Courage + Khazaka electronic GmbH

Mathias-Brüggen-Str. 91, 50829 Cologne, Germany
Telephone: ++49-221-956499-0
Fax: ++49-221-956499-1
E-mail: info@courage-khazaka.de
Internet site: *http://www.courage-khazaka.de/*

CuDerm Corporation

17430 Campbell Rd., Suite 106, Dallas, TX 75252, USA
Telephone: ++1-800 690 1933
Fax: ++1-972 248 1094
E-mail: mktgdep2@cuderm.com
Internet site: *http://www.cuderm.com*

Dia-Stron

Unit 9, Focus 303, Business Centre South Way Andover Hampshire, SP10 5NY, United Kingdom

Telephone: ++44 1264-334700
Fax: ++44 1264-334686
E-mail: diastron@compuserve.com

GFMesstechnik GmbH

Warthstrasse 21, D- 14513 Teltow/Berlin, Germany
Telephone: ++49 3328 305185
Fax: ++49 3328 305188
E-mail: gfmess@aol.com
Internet site: http://www.gf-messtechnik.de

Hommelwerke GmbH

Alte Tuttlinger Str. 20, 78056 Villingen-Schwennigen, Germany
Telephone: ++49 77 20 60 20
Fax: ++49 77 20 60 21 23
E-mail: hommelwerke@t-online.de
Internet site: http://www.tipcoeurope.com

MINOLTA

Internet site: http://www.minolta.com

NOVA Technology Corporation

75 Congress Street, Portsmouth, NH 03801-4006, USA
Telephone: ++1-603-422-9595
Fax: ++1-603-446-34066445522-7330
E-mail: info@novatechcorp.com
Internet site: http://www.novatechcorp.com/

Servo Med AB

P. O. Box 47, S-432 21 Varberg, Sweden
Telephone: ++46-340664455
Fax: ++46-34017320
E-mail: info@servomed.se
Internet site: http://www.servomed.se

REFERENCES

1. Rohr M, Schrader K. Climatic influences on cosmetic skin parameters. Curr Probl Dermatol 1998; 26:151–164.
2. Rogiers V, Balls M, Basketter D, Berardesca E, Edwards C, Elsner P, Ennen J, Lévêque JL, Lóden M, Masson P, Parra J, Paye M, Piérard G, Rodrigues L, Schaefer H, Salter D, Zuang V. The potential use of non-invasive methods in the safety assessment of cosmetic products. ATLA 1999; 27:515–537.
3. Marks R, Payne PA. Bioengineering and the skin. Lancaster: MTP Press Ltd, 1981.
4. Serup J, Jemec GBE, eds. Handbook of Non-invasive Methods and the Skin. Boca Raton, FL: CRC Press, 1995.
5. Elsner P, Barel AO, Berardesca E, Gabard B, Serup J. Skin Bioengineering: Techniques and Applications in Dermatology and Cosmetology. Basel: Karger, 1998.
6. Rieger M. Water, water, everywhere. Cosm Toil 1998; 113:75–87.
7. Warner RR, Myers MC, Taylor DA. Electron probe analysis of human skin: determination of the water concentration profile. J Invest Derm 1988; 90:218–224.
8. Berardesca E. EEMCO guidance for the assessment of stratum corneum hydration: electrical methods. Skin Res Technol 1997; 3:126–132.
9. Salter DC. Monitoring human skin hydration in vivo using electrical impedance—a model of skin as a solid ionic conductor. Proceedings IX International Conference on Electrical Bio-Impedance and European Community Concerted Action on Impedance Tomography. Heidelberg, 26–30 September 1995:17–30.
10. Barel AO, Clarys P. In vitro calibration of the capacitance method (Corneometer CM 825) and conductance method (Skicon-200) for the evaluation of the hydration state of the skin. Skin Res Technol 1997; 3:107–113.
11. Martinsen ØG, Grimnes S, Haug E. Measuring depth depends on frequency in electrical skin impedance measurements. Skin Res Technol 1999; 5:179–181.
12. Wichrowski K, Sore G, Khaïat A. Use of infrared spectroscopy for in vivo measurement of the stratum corneum moisturization after application of cosmetic preparations. Int J Cosmet Sci 1995; 17:1–11.
13. Querleux B, Richard S, Bittoun J, Jolivet O, Idy-Peretti I, Bazin R, Lévêque JL. In vivo hydration profile in skin layers by high-resolution magnetic resonance imaging. Skin Pharmacol 1994; 7:210–216.
14. Franconi F, Akoka S, Guesney J, Baret JM, Dersigny D, Breda B, Müller C, Beau P. Measurement of epidermal moisture content by magnetic resonance imaging: assessment of a hydration cream. Br J Dermatol 1995; 132:913–917.
15. Bettinger J, Gloor M, Vollert A, Kleesz P, Fluhr J, Gehring W. Comparison of different non-invasive test methods with respect to the effect of different moisturizers on skin. Skin Res Technol 1999; 5:21–27.
16. Rogiers V, Derde MP, Verleye G, Roseeuw D. Standardized conditions needed for skin surface hydration measurements. Cosm Toil 1990; 105:73–82.
17. Wilhelm K-P. Skin hydration measurements: general considerations and possible pitfalls. SÖFW 1998; 124(4):196–203.
18. Schaeffer H, Redelmeier TE. In: Schaeffer H, Redelmeier TE, eds. Skin barrier; principles of percutaneous absorption. Basel: Karger; 1996:1–87.

19. Wilson DR, Maibach H. TEWL: a review. In: Lévêque JL, ed. Cutaneous Investigation in Health and Disease: Noninvasive Methods and Instrumentation. New York: Marcel Dekker; 1989:113–133.
20. Elias PM, Feingold KR. Lipids and the epidermal water barrier: metabolism, regulation and pathophysiology. Semin Dermatol 1992; 11:176–182.
21. Wertz PW. The nature of the epidermal barrier: biochemical aspects. Adv Drug Delivery Rev 1996; 18:283–294.
22. Takahashi M. Recent progress in skin bioengineering and its application to evaluation of cosmetics. SÖFW 2000; 126:6–18.
23. Rogiers V. EEMCO Guidance for the assessment of the transepidermal water loss in cosmetic sciences. Skin Pharmacol Appl Skin Physiol 2001; 14:117–128.
24. Grove GL, Grove MJ, Zerweck C, Pierce E. Comparative metrology of the evaporimeter and the DermaLab® TEWL probe. Skin Res Technol 1999; 5:1–8.
25. Barel AO, Clarys P. Comparison of methods for measurement of transepidermal water loss. In: Serup J, Jemec GBE, eds. Handbook of Non-invasive Methods and the Skin. Boca Raton, FL: CRC Press, 1995:179–184.
26. Rogiers V. TEWL measurements in patch test assessment: the need for standardisation. Curr Probl Dermatol 1995; 23:152–158.
27. Pinnagoda J, Tupker RA. Measurement of the transepidermal water loss. In: Serup J, Jemec GBE, eds. Handbook of Non-invasive Methods and the Skin. Boca Raton, FL: CRC Press, 1995:173–178.
28. Pinnagoda J, Tupker RA, Agner T, Serup J. Guidelines for transepidermal water loss (TEWL) measurement. Contact Dermatitis 1990; 22:164–178.
29. Issachar N, Gall Y, Borrel MT, Poelman M-C. Correlation between percutaneous penetration of methyl nicotinate and sensitive skin using laser doppler imaging. Contact Dermatitis 1998; 39:182–186.
30. Hashimoto K. New methods for surface ultrastructure: comparative studies of scanning electron microscopy, transmission electron microscopy and replica method. Int J Dermatol 1974; 13:357–381.
31. Cook TH. Profilometry of skin—A useful tool for the substantiation of cosmetic efficacy. J Soc Cosmet Chem 1980; 31:339–359.
32. Lévêque JL. EEMCO guidance for the assessment of skin topography. J Eur Acad Dermatol Vener 1999; 12:103–114.
33. Fischer TW, Wigger-Alberti W, Elsner P. Direct and non-direct measurement techniques for analysis of skin surface topography. Skin Pharmacol Appl Skin Physiol 1999; 12:1–11.
34. Serup J, Jemec GBE. Skin surface contour evaluation. In: Serup J, Jemec GBE, eds. Handbook of Non-invasive Methods and the Skin. Boca Raton, FL: CRC Press, 1995:83–131.
35. Marks R, Pearse AD. Surfometry: a method of evaluating the internal structure of the stratum corneum. Br J Dermatol 1975; 92:651–657.
36. Grove GL, Grove MJ, Leyden JJ. Optical profilometry: an objective method for quantification of facial wrinkles. J Am Acad Dermatol 1989; 21:631–637.
37. Potorac AD, Toma I, Mignot J. In vivo skin relief measurement using a new optical profilometer. Skin Res Technol 1996; 2:64–69.
38. Stout KJ, Sullivan PJ, Dong WP, Mainsah E, Luo N, Mathia T, Zahouani H. The

Biophysical Methods for SC Characterization 221

development of methods for the characterisation of roughness in three dimensions. Commission of the European Communities Report EUR 15178 EN, 1993.
39. Zahidi M, Assoul M, Bellaton B, Mignot J. A fast 2D/3D optical profilometer for wide range topographical measurement. WEAR 1993; 65:197–203.
40. Nita D, Mignot J, Chuard M, Sofa M. 3-D profilometry using a CCD linear image sensor: application to skin surface topography measurement. Skin Res Technol 1998; 4:121–129.
41. De Paepe K, Lagarde JM, Gall Y, Roseeuw D, Rogiers V. Microrelief of the skin using a light transmission method. Arch Dermatol Res 2000; 292:500–510.
42. Rohr M, Schrader K. Fast optical in vivo topometry of human skin (FOITS). SÖFW 1998; 124:3–8.
43. Jaspers S, Hopermann H, Sauermann G, Hoppe U, Lunderstädt R, Ennen J. Rapid in vivo measurement of the topography of human skin by active image triangulation using a digital micromirror device. Skin Res Technol 1999; 5:195–207.
44. Hopermann CHH. Comparison of replica- and in vivo-measurement of the microtopography of human skin. SÖFW 2000; 126:40–46.
45. Tronnier H. Hautschutz. ÄP Dermatologie 1998; 18:346–350.
46. Tronnier H, Wiebusch M, Heinrich U, Stute R. Surface evaluation of living skin. Adv Exp Med Biol 1999; 455:507–516.
47. Takahashi M. Image analysis of skin surface contour. Acta Derm Venereol Suppl (Stockh) 1994; 185:9–14.
48. Griffiths CEM, Wang TS, Hamilton, TA, Voorhees JJ, Ellis CN. A photonumeric scale for the assessment of cutaneous photodamage. Arch Dermatol 1992; 128:347–351.
49. Tsukahara K, Takema Y, Kazama H, Yorimoto Y, Fujimura T, Moriwaki S, Kitahara T, Kawai M, Imokawa G. A photographic scale for the assessment of human facial wrinkles. J Cosmet Sci 2000; 51:127–139.
50. Garber CA, Nightingale CT. Characterizing cosmetic effects and skin morphology by scanning electron microscopy. J Soc Cosmet Chem 1976; 27:509–531.
51. Ryan RL, Hing SAO, Theiler RF. A replica technique for the evaluation of human skin by scanning electron microscopy. J Cutan Pathol 1983; 10:262–276.
52. Rodrigues L. EEMCO Guidance to the in vivo assessment of tensile functional properties of the skin. Part 2: Instrumentation and test modes. Skin Pharmacol Appl Skin Physiol, 2001; 14:52–67.
53. Serup J, Jemec GBE. Mechanical properties of the skin. In: Serup J, Jemec GBE, eds. Handbook of Non-invasive Methods and the Skin. Boca Raton, FL: CRC Press, 1995:319–367.
54. Murray BC, Wickett RR. Correlations between dermal torque meter, cutometer and dermal phase meter measurements of human skin. Skin Res Technol 1997; 3:101–106.
55. Ennen J, Jaspers S, Sauermann G, Hoppe U. Measurement of biomechanical properties of human skin. Cosmet Toil Manuf Worldwide, Aston publishing group 1995: 97–101.
56. Wilhelm KP, Cua AB, Maibach HI. In vivo study on age-related elastic properties of human skin. In Frosch PJ, Kligman AM eds. Non-invasive Methods for the Quantification of Skin Functions; 1993:190–203.

57. Barel AO, Courage W, Clarys P. Suction method for measurement of skin mechanical properties: the Cutometer. In: Serup J, Jemec GBE, eds. Handbook of Non-invasive Methods and the Skin. Boca Raton, FL: CRC Press, 1995:335–340.
58. Agache PG. Twistometry measurement of skin elasticity. In: Serup J, Jemec GBE, eds. Handbook of Non-invasive Methods and the Skin. Boca Raton, FL: CRC Press, 1995:319–328.
59. Hargens CW. The gas-bearing electrodynamometer. In: Serup J, Jemec GBE, eds. Handbook of Non-invasive Methods and the Skin. Boca Raton, FL: CRC Press, 1995:353–359.
60. Matts PJ, Goodyer E. A new instrument to measure the mechanical properties of human stratum corneum in vivo. J Cosmet Sci 1998; 49:321–333.
61. Gillon V, Perie G, Freis O, Pauly M, Pauly G. New active ingredients with cutaneous tightening effect. Cosmetics and Toiletries Manufacture Worldwide, Aston publishing group 1999:22–31.
62. Christensen MS, Hargens CW, Nacht S, Gans EH. Viscoelastic properties of intact human skin: instrumentation, hydration effects and the contribution of the stratum corneum. J Invest Dermatol 1977; 69:282–286.
63. Maes D, Short J, Turek BA, Reinstein JA. In vivo measuring of skin softness using the Gas Bearing Electrodynamometer. Int J Cosmet Sci 1983; 5:189–200.
64. Martelli L, Berardesca E, Martelli M. Topical formulation of a new plant extract complex with refirming properties: Clinical and non-invasive evaluation in a double-blind trial. Int J Cosm Sci 2000; 22:201–206.
65. Thacker JG, Stalnecker MC, Allaire PE, Edgerton MT, Edlich RF. Practical applications of skin biomechanics. Clin Plast Surg 1977; 4:167–171.
66. Manny-Aframian V, Dikstein S. Indentometry. In: Serup J, Jemec GBE, eds. Handbook of Non-invasive Methods and the Skin. Boca Raton, FL: CRC Press, 1995: 349–352.
67. Dikstein S, Hartzshtark A. In vivo measurement of some elastic properties of human skin. In: Marks R, Payne PA, eds. Bioengineering and the skin. Boston: MTP Press, 1981:45–53.
68. Fthenakis CG, Maes DH, Smith WP. In vivo assessment of skin elasticity using ballistometry. J Soc Cosmet Chem 1991; 42:211–222.
69. Gniadecka M, Jemec GBE. Quantitative evaluation of chronological ageing and photoageing in vivo: studies on skin echogenicity and thickness. Br J Dermatol 1998; 139:815–821.
70. de Rigal J, Lévêque J-L. Influence of aging on the mechanical properties of skin. In: Lévêque J-L, Agache PG, eds. Aging Skin. New York: Marcel Dekker, 1993: 15–17.
71. Rawlings A, Harding C, Watkinson A, Banks J, Ackerman C, Sabin R. The effect of glycerol and humidity on desmosome degradation in stratum corneum. Arch Dermatol Res 1995; 287:457–464.
72. Harding CR, Watkinson A, Rawlings AV, Scott IR. Dry skin, moisturization and corneodesmolysis. Int J Cosm Sci 2000; 22:21–52.
73. Roberts D, Marks R. The determination of regional and age variations in the rate of desquamation: a comparison of four techniques. J Invest Dermatol 1980; 74: 13–16.

74. McGinley KJ, Marples RR, Plewig G. A method for visualizing and quantitating the desquamating portion of the human stratum corneum. J Invest Dermatol 1969; 53:107.
75. Jansen LH, Hojko-Tomoko MT, Kligman AM. Improved fluorescence staining technique for estimating turnover of the human stratum corneum. Br J Dermatol 1974; 90:9–12.
76. Piérard GE, Piérard-Franchimont C. Dihydroxyacetone test as a substitute for the dansyl chloride test. Dermatology 1993; 186:133–137.
77. Marks R, Nicholls S, Fitzgeorge D. Measurement of intracorneal cohesion in man using in vivo techniques. J Invest Dermatol 1977; 69:299–302.
78. Serup J. EEMCO guidance for the assessment of dry skin (xerosis) and ichthyosis: clinical scoring systems. Skin Res Technol 1995; 1:109–114.
79. Piérard GE. EEMCO guidance for the assessment of dry skin (xerosis) and ichthyosis: evaluation by stratum corneum strippings. Skin Res Technol 1996; 2:3–11.
80. Lagarde JM, Black D, Gall Y, Del Pozo A. Image analysis of scaly skin using Dsquame® samplers: technical and physiological validation. Int J Cosm Sci 2000; 22:53–65.
81. Piérard GE, Piérard-Franchimont C, Saint Léger D, Kligman AM. Squamometry: the assessment of xerosis by colorimetry of D-Squame adhesive discs. J Soc Cosmet Chem 1992; 47:297–305.
82. Charbonnier V, Morrison BM Jr, Paye M, Maibach HI. Open application assay in investigation of subclinical irritant dermatitis induced by sodium lauryl sulfate (SLS) in man: advantage of squamometry. Skin Res Technol 1998; 4:244–250.
83. Paye M, Cartiaux Y. Squamometry: a tool to move from exaggerated to more and more realistic application conditions for comparing the skin compatibility of surfactant-based products. Int J Cosm Sci 1999; 21:59–68.
84. Piérard GE, Goffin V, Piérard-Franchimont C. Squamometry and corneosurfametry for rating interactions of cleansing products with stratum corneum. J Soc Cosmet Chem 1994; 45:269–277.
85. Blanc D, Saint-Leger D, Brandt J, Constans S, Agache P. An original procedure for quantitation of cutaneous resorption of sebum. Arch Dermatol Res 1989; 281: 346–350.
86. Clarys P, Barel AO. Quantitative evaluation of skin surface lipids. Clin Dermatol 1995; 13:307–321.
87. Schmid MH, Korting HC. The concept of the acid mantle of the skin: its relevance for the choice of skin cleansers. Dermatology 1995; 191:276–280.
88. Rode B, Ivens U, Serup J. Degreasing method for the seborrheic areas with respect to regaining sebum excretion rate to casual level. Skin Res Technol 2000; 6:92–97.
89. Piérard GE, Cauwenbergh G. Modulation of sebum excretion from the follicular reservoir by a dichlorophenyl-imidazoldioxolan. Int J Cosm Sci 1996; 18:219–227.
90. Morganti P. Acne therapy, the cosmetic approach. Soap Perfum Cosmet 1998; 71: 23–26.
91. Hommel L, Geiger J-M, Harms M, Saurat J-H. Sebum excretion rate in subjects treated with oral all-trans-retinoic acid. Dermatology 1996; 193:127–130.

92. Downing DT, Wertz PW, Stewart ME. The role of sebum and epidermal lipids in the cosmetic properties of the skin. Int J Cosmet Sci 1986; 8:115–123.
93. Blume U, Ferracin J, Verschoore M, Czernielewski JM, Schaefer H. Physiology of the vellus hair follicle: hair growth and sebum excretion. Br J Dermatol 1991; 124:21–28.
94. Black D, Lagarde J-M, Auzoux C, Gall Y. An improved method for the measurement of scalp sebum. In: Elsner P, Barel AO, Berardesca E, Gabard B, Serup J, eds. Skin Bioengineering Techniques and Applications in Dermatology and Cosmetology. Curr Probl Dermatol. Basel: Karger, 1998:61–68.
95. Clarys P, Lambrecht R, Barel AO. Does lipid sampling with the Sebutape technique disturb the skin physiology? Skin Res Technol 1997; 3:169–172.
96. Kligman AM, Miller DL, McGinley KJ. Sebutape: a device for visualizing and measuring human sebaceous secretion. J Soc Cosmet Chem 1986; 37:369–374.
97. El-Gammal C, El-Gammal S, Pagnoni A, Kligman AM. Sebum-absorbent tape and image analysis. In: Serup J, Jemec GBE, eds. Handbook of Non-invasive Methods and the Skin. Boca Raton, FL: CRC Press, 1995:517–522.
98. Saint-Leger D, Berrebi C, Duboz C, Agache P. The Lipometre: an easy tool for rapid quantitation of skin surface lipids (SSL) in man. Arch Dermatol Res 1979; 265:79–89.
99. Piérard GE, Piérard-Franchimont C, Marks R, Paye M, Rogiers V. EEMCO guidance for the in vivo assessment of skin greasiness. Skin Pharmacol Appl Skin Physiol 2000; 13:372–389.
100. Kumagai H, Shioya K, Kawasaki I, Horii I, Koyama J, Nakayama Y, Mori W, Ohta S. Development of a scientific method for classification of facial skin types. J Soc Cosmet Chem 1985; 19:1–15.
101. de Rigal J, Escoffier C, Querleux B, Faivre B, Agache P, Lévêque JL. Assessment of aging of the human skin by in vivo ultrasonic imaging. J Invest Dermatol 1989; 93:621–625.
102. Gniadecka M, Gniadecki R, Serup J, Søndergaard J. Ultrasound structure and digital image analysis of the subepidermal low echogenic band in aged human skin: diurnal changes and interindividual variability. J Invest Dermatol 1994; 102:362–365.
103. Serup J, Keiding J, Fullerton A, Gniadecka M, Gniadecki R. High-frequency ultrasound examination of skin: introduction and guide. In: Serup J, Jemec GBE, eds. Handbook of Non-invasive Methods and the Skin. Boca Raton, FL: CRC Press, 1995:239–256.
104. Wetzel J, Lankenau E, Birngruber R, Engelhardt R. Optical coherence tomography of the human skin. J Am Acad Dermatol 1997; 37:958–963.
105. Querleux B. Nuclear magnetic resonance (NMR) examination of the epidermis in vivo. In: Serup J, Jemec GBE, eds. Handbook of Non-invasive Methods and the Skin. Boca Raton, FL: CRC Press, 1995:133–139.
106. Gilard V, Martino R, Malet-Martino M, Riviere M, Gournay A, Navarro R. Measurement of total water and bound water contents in human stratum corneum by in vitro proton nuclear magnetic resonance spectroscopy. Int J Cosm Sci 1998; 20:117–125.
107. Corcuff P, Gonnord G, Piérard GL, Lévêque JL. In vivo confocal microscopy of

human skin: a new design for cosmetology and dermatology. Scanning 1996; 18: 351–355.
108. Piérard GE. EEMCO guidance for the assessment of skin colour. J Eur Acad Dermatol Vener 1998; 10:1–11.
109. Wilhelm KP, Maibach HI. Skin color reflectance measurements for objective quantification of erythema in human beings. J Am Acad Dermatol 1989; 21:1306–1308.
110. Clarys P, Alewaeters K, Lambrecht R, Barel AO. Skin color measurements: comparison between three instruments: the Chromameter®, the DermaSpectrometer® and the Mexameter®. Skin Res Technol 2000; 6:230–238.
111. Wickett RR, Trobaugh CM. Personal care products: effects on skin surface pH. Cosm Toil 1990; 105:41–46.
112. Agner T, Serup J. Contact thermography for assessment of skin damage due to experimental irritants. Acta Derm Venereol 1988; 68:192–195.
113. Artmann C, Röding J, Stanzl K, Zastrow L. Thermography and skin temperature measurement: an in vivo method for the characterization of cold protection creams. Euro Cosm 1994; 1:30–33.
114. Lock-Andersen J, Gniadecka M, de Fine Olivarius F, Dahlstrøm K, Wulf HC. Skin temperature of UV-induced erythema correlated to laser Doppler flowmetry and skin reflectance measured redness. Skin Res Technol 1998; 4:41–48.
115. Highly DR, Coomey M, DenBeste M, Wolfram LJ. Frictional properties of skin. J Invest Dermatol 1977; 69:303–305.
116. Lodén M, Olsson H, Skare L, Axéll T. Instrumental and sensory evaluation of the frictional response of the skin following a single application of five moisturizing creams. J Soc Cosmet Chem 1992; 43:13–20.
117. Koudine AA, Barquins M, Anthoine P, Aubert L, Lévêque JL. Frictional properties of skin: proposal of a new approach. Int J Cosm Sci 2000; 22:11–20.

8
Detection of Cosmetic Changes in Skin Surface Lipids by Infrared and Raman Spectroscopy

Thomas Prasch and Thomas Förster
Henkel KGaA, Düsseldorf, Germany

I. INTRODUCTION

Human skin consists essentially of three layers: the deep subcutaneous fatty layer, the dermis, and the epidermis. The outermost layer of the epidermis is the horny layer, or stratum corneum (SC). The barrier function of the skin resides almost entirely in the outermost portion in the SC [1]. It presents a system that prevents dehydration of the underlying tissues and excludes the entry of noxious substances from the environment [2,3]. The stratum corneum consists of 10–15 layers of flattened, anucleate, keratinized cells embedded in a lipidic matrix. Typically, the dry SC is approximately 10 μm thick.

The barrier function is mainly the result of the unique structure of proteins and lipids in the skin. The lipids in the stratum corneum are arranged into lamellar structures that are covalently linked to proteins of cornified envelopes. In that way, a water-impermeable layer of lipids is formed in the SC. The number of disulfide cross-links in the SC is significantly lower than in nail or in hair. In the SC, a noteworthy amount of sidechain mobility is found. This is due to the loose packing of keratin filaments in the SC.

The main constituents of the SC are protein, 75–80%; lipids, 5–15%; and unidentified materials, 5–10% on a dry weight basis [4,5]. The protein fraction within the cells consists mainly of α-keratin (approx. 70%) with a smaller amount of β-keratin (approx. 10%) and cell wall envelopes (approx. 5%). In the SC, several types of lipids are found: neutral lipids, sphingolipids (ceramides), polar lipids, and cholesterol sulfates.

The stratum corneum is not a static but instead a dynamic barrier that is continually being formed by the dermis by differentiation of keratinocytes to corneocytes while on average one layer of corneocytes is lost from the upper surface per day. Small-angle x-ray scattering (SAXS) experiments have shown that human skin lipid layers are arranged in two types of parallel layers, which have interlayer distances of 6.4 nm and 13.4 nm [6,7]. In contrast to normal cell membranes, these lipid multilayers are not liquid-crystalline at skin temperature but exhibit a complex polymorphism of mainly solid phases in which the lipid alkyl chains are tightly packed and immobile [8,9].

Infrared [10,11] and Raman [12] spectroscopy are two techniques that have proved especially suited for the investigation of the stratum corneum and the interaction of cosmetic products with the stratum corneum [13]. By using the IR-spectra of the facial skin, an investigation panel is classified into different skin types according to the sebum content. A comparison with the position of the CH_2 stretching vibrations of synthetic lipid mixtures with a known degree of order allows quantification of the degree of order of the skin lipids in vivo. The effect of a mild facial cleansing is investigated using a cleansing milk as an example and compared with the use of a surfactant-based shower gel. Oils and lipids, the chief constituents of skincare cosmetic creams, show completely different interactions with the skin lipids; this can be utilized positively as the basis for the development of biomimetic lamellar creams [13].

In vitro skin cultures are another research area that has contributed tremendously to a deeper understanding of skin function in the past years. Today, not only are epidermal skin cultures commercially available and used routinely but also the much more complex full-skin models have been successfully cultured in some laboratories [14]. Due to their similarity to normal human skin on one hand and their inherently low biological variability on the other hand, in vitro skin cultures are especially useful for basic cosmetic research. The analysis of IR and Raman spectra helps to obtain a fundamental understanding of the similarities and differences in the stratum corneum barrier between normal human skin and in vitro skin models.

II. TEST METHODS

A. Biophysical Techniques Applied to the Study of SC and Skin Lipids

Several biophysical techniques such as differential scanning calorimetry (DSC) [15,16], x-ray diffraction [7], nuclear magnetic resonance (NMR) [17], and electron spin resonance (ESR) [18] have been applied to study and characterize the

SC. One important approach is the use of vibrational spectroscopy to probe the molecular nature of the skin and especially the SC. The early work concentrated on the use of Fourier transform infrared (FT-IR)-spectroscopy [19]. The water concentration in the tissue has been determined by means of IR-spectroscopy [20]. It has been used successfully to probe the tissue in vivo using attenuated total reflection methods [21,22]. For the exact characterization of the lipid layer structure in the stratum corneum, IR spectroscopy has proved itself in practice. The exact spectral position of the symmetrical CH_2 stretching vibrations in the IR spectrum can provide information about the average degree of order of the lipid alkyl chain conformation [23]. An alternative vibrational spectroscopic technique, which has gained considerable interest, is Raman spectroscopy.

B. Description of the Basics of Vibrational Spectroscopy

Molecules consist of atoms that have a certain mass and that are connected by elastic bonds: As a result, they can perform periodic motions. They have vibrational degrees of freedom. Nonlinear polyatomic molecules with n atoms possess $3n-6$ normal vibrations that define their vibrational spectra. Linear molecules possess only $3n-5$ vibrational degrees of freedom.

Molecular aggregates such as complexes or biomembranes behave like "super-molecules" in which the vibrations of the individual components are coupled. The two most important methods of vibrational spectroscopy are IR and Raman spectroscopy. They provide complementary images of molecular vibrations, because in these spectroscopic techniques the mechanisms of the interaction of light quanta with molecules are quite different. Further theory and background of vibrational spectroscopy can be found in the literature [24–26].

C. Description of IR Spectroscopy

Infrared spectra are usually recorded by measuring the transmittance of light quanta with a continuous distribution of the sample. The frequencies of the absorption bands are proportional to the energy difference between the vibrational ground and excited states. The absorption bands due to the vibrational transitions are found in the wavelength region of $\lambda = 2.5$ μm up to 1000 μm, which corresponds to a wavenumber range of $\tilde{v} = 4000 \text{ cm}^{-1} - 10 \text{ cm}^{-1}$.

Interaction of infrared radiation with a vibrating molecule is possible only if the electric vector of the radiation field oscillates with the same frequency as the molecular dipole moment does. The consequence is that vibration is only infrared active if the molecular dipole moment is modulated by the normal vibration. The following condition needs to be fulfilled to observe an IR-band

$$\frac{\partial \mu}{\partial q} \neq 0 \tag{1}$$

In this expression, μ is the molecular dipole and q stands for the normal coordinate describing the motion of the atoms during a normal vibration.

D. Description of the Attenuated Total Reflectance Technique

One special IR technique, that is important for the investigation of the skin barrier is the attenuated total reflectance technique. It belongs to the class of reflectance techniques that are often applied when normal transmittance methods cannot be successfully applied. The sample is brought into optical contact with the surface of a special crystal. The IR-beam, propagating within the crystal by means of several total reflections, senses the material on the surface by penetrating into the layer over a distance, which is calculated below. The measuring beam is directed through the crystal to be reflected at the interface to the sample; for this reason the technique is also referred to as internal reflection. For detailed information on the ATR technique, the books of Harrick [27,28] and Mirabella [29] are recommended.

1. Theory of the ATR Technique

An explanation of the basic theory of the ATR method follows: Radiation is totally reflected at the boundary between a medium with a refractive index n_1 and a medium with lower refractive index n_2 if it hits this boundary with an incident angle greater than the critical angle $\theta_c = \arcsin(n_2/n_1)$. The reflected radiation energy penetrates the boundary as a so-called evanescent wave. The penetration depth d_p is the thickness in which the intensity decreases to 1/e of the intensity at the boundary. The penetration depth d_p depends on the refractive indices n_1 and n_2, the incident angle θ, and the wavelength λ (in vacuum):

$$d_p = \frac{\lambda}{2\pi n_1 \sqrt{\sin^2\theta - n_{21}^2}} \text{ with } n_{21} = \frac{n_2}{n_1} \tag{2}$$

The penetration depth for skin is usually on the order of 0.3–3 μm, depending mainly on the wavelength of interest, the material of the ATR crystal, and the hydration level of the sample.

The sample absorbs the evanescent field so that the totally reflected wave is attenuated accordingly. The consequence is that the reflectance resembles closely the corresponding transmission spectrum. The resultant attenuated radiation is measured as a function of wavelength by the spectrometer and gives rise to the absorption spectral characteristics of the sample.

Some sensitivity is lost with angles of reflection well above the critical one. But it can be regained by multiple internal reflections. The signal/noise ratio

IR-Radiation

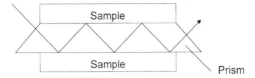

Figure 1 Schematic representation of the ATR-technique.

of the spectrum obtained by means of the ATR-technique can be adjusted by the number of reflections used. It depends on the size of the sample-coated area of the slab: whether both surfaces of the crystal, only one, or only a fraction of it is exploited for the measurement.

2. Practical Aspects of the ATR Technique

The crystals (also called internal reflection elements, IRE) used in ATR cells are made from materials that have a proper refractive index to allow internal reflection. Moreover, they should offer a low solubility in water, and the material must show transparent regions in the infrared. Commonly used materials are zinc selenide (ZnSe, refractive index 2.4), germanium (Ge, refractive index 4.0), and thalliumiodide/thalliumbromide (KRS-5, refractive index 4.0). The shapes of the commonly used crystals are trapezoid or parallelepiped.

The basic experimental setup of an ATR experiment is illustrated in Figure 1.

E. Description of Raman Spectroscopy

1. Theoretical Background

The second important type of vibrational spectroscopy is Raman spectroscopy. If a sample is irradiated with an intense beam of monochromatic radiation (e.g., from a laser) of a frequency v, most of the radiation is transmitted by the sample. Only a very small fraction of the exciting radiation (approx. 1 photon in 10^4) is elastically scattered. This fraction of the radiation is called Rayleigh scattering. The wavenumber of the Rayleigh scattering is equal to that of the incident radiation. An even smaller portion (approx. 1 photon of 10^8) scatters inelastically with wavenumbers different from the incident radiation ($v \pm v_i$, where v_i represents a vibrational frequency of the molecule). This inelastically scattered radiation is designated as Raman scattering.

The Raman scattering derives from the excitation of vibrational and rotational levels of the electronic ground state of the molecules. This causes an energy loss of the incident radiation and a band shift to longer wavelengths as compared to the Rayleigh line, which is known as Stokes shift. Anti-Stokes lines can be observed when the molecules under consideration are in excited vibrational levels before the interaction with the laser source. At room temperature, these anti-Stokes lines are weaker than the Stokes lines.

When a molecule is exposed to an electric field, electrons and nuclei are forced to move in opposite directions. A dipole moment is induced which is proportional to the electric field strength and to the molecular polarizability α. A molecular vibration can be observed in the Raman spectrum only if there is a modulation of the molecular polarizability by the vibration. The condition

$$\left(\frac{\partial \alpha}{\partial q}\right) \neq 0 \tag{3}$$

needs to be fulfilled for a vibration to be observed in the Raman spectrum.

2. Complementarity of IR and Raman Spectroscopy

Raman and infrared spectroscopy are complementary techniques that differ in their interactions with the incident radiation. For a vibrational mode to be infrared active, the dipole moment of a bond must change, whereas for a mode to be Raman active the polarizability of a bond must alter. The differing selection rules for activity in these two techniques are the reasons for the complementary information content of IR and Raman spectra.

For a C=O moiety with an existing dipole, a change in the dipole moment can easily be induced and consequently the mode is strongly infrared active. But such a dipole-distorted bond cannot easily have its polarizability altered and so only a weak Raman effect results. Homopolar bonds such as C=C moieties can easily have their electron clouds distorted and so are polarizable and Raman active, whereas they possess no dipole moments and hence are infrared inactive.

The basic principles of IR and Raman spectroscopy are compared in Figure 2.

3. Advantages and Disadvantages of Raman versus IR Spectroscopy

One practical consequence of the different selection rules for Raman and IR-spectroscopy is that water is a very weak Raman scatterer but strongly absorbs infrared radiation. This is a clear advantage for Raman spectroscopy as a tool in the analysis of the molecular structure of biological materials. Moreover, the wavenumber range that is accessible to Raman spectroscopy is greater than that of infrared. But Raman spectroscopy is rather insensitive because it examines

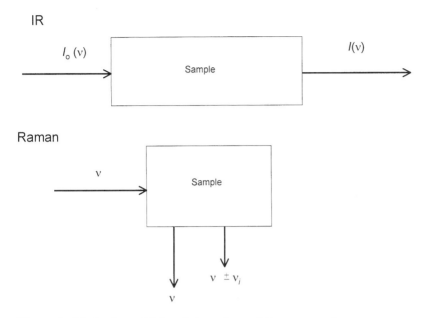

Figure 2 Mechanisms of infrared absorption and Raman scattering.

very weak scattered radiation. Therefore, Raman spectroscopy is not such a versatile tool in the analysis of minor compounds.

4. The Main Components of IR and Raman Spectrometers

Today there are two main types of IR and Raman spectrometers. They use either the dispersive or the Fourier transform technique. Both comprise four main instrumental components: a radiation source, optics to illuminate the sample and to collect and focus the radiation to be analyzed, a monochromator (for dispersive spectrometers) or an interferometer (for FT-instruments), and a detector.

Radiation Sources Used in IR Spectroscopy. For the generation of the infrared radiation in middle infrared (MIR), usually a thermal radiation source is used. Today the most popular radiation source for MIR spectrometers is the so-called globar (glow bar). It is a rod or a tube with a diameter of 6–8 mm consisting of SiC. It is heated by a high current at low voltage. Its emission is about 75% of that of a black body at about 1400 K.

Radiation Sources for Raman Spectroscopy. The excitation of Raman scattering is (at least theoretically) not restricted to a specific region of the electromagnetic spectrum. Usually an ultraviolet, visible, infrared, or near-infrared excitation source is used. Raman scattering is much weaker than Rayleigh scattering. Therefore, a very strong excitation source is required. The advent of lasers has revolutionized Raman spectroscopy. Lasers provide strong, coherent monochromatic light in a wide range of wavelengths. The selection of the most appropriate laser radiation source depends on sample color, stability, and molecular features that may result in fluorescence. Fluorescence is often much more intense than the Raman effect. Especially, biological samples such as skin tend to have a pronounced fluorescence effect. Fluorescence usually gets weaker as the laser moves toward the infrared region of the electromagnetic spectrum. For this reason, manufacturers have developed near-infrared Fourier transform Raman spectrometers. The Nd:YAG (neodymium-doped yttrium aluminum garnet) with a wavelength of 1.064 µm is the most widespread excitation source for NIR-Raman-spectrometers, but Raman spectra can also be obtained by means of laser excitation in the ultraviolet, visible or infrared region.

Dispersive and Fourier Transform Spectrometers. Today there are two main types of IR and Raman spectrometers. They utilize either the dispersive or the Fourier transform technique. In dispersive spectrometers, a grating monochromator is used. The essential problem with the dispersive spectrometer lies in its monochromator. This contains narrow slits at the entrance and exit that limit the wavenumber range of the radiation reaching the detector to one resolution width. These limitations can be overcome through use of Fourier transform infrared spectrometers. In Fourier transform spectrometers, the basic optical component is a Michelson interferometer. The essential experiment to obtain an FTIR spectrum is to produce interferograms, both with and without sample in the beam, and then transform these interferograms into spectra by means of the mathematical operation of a Fourier transformation, which transfers frequency domain data into time domain data and vice versa. The ratio of the source with sample absorptions and without them corresponds to the IR-spectrum of the sample. The FT can be calculated rapidly on-line with modern microcomputers connected to the spectrometer. FT instruments simultaneously measure the signal of multiple data points by analysis of a single signal from a detector providing the full spectral region from only one scan. The data accumulation is superimposable, so that FT spectrometers allow accurate coaddition of scans.

FTIR spectrometers offer many significant advantages over dispersive spectrometers. The Fellgett (or multiplex) advantage is the result of an improvement in the S/N ratio per unit time, which is proportional to the square root of the number of resolution elements being monitored. It results from the large number of resolution elements being monitored simultaneously.

The Jacquinot (or throughput) advantage derives from the fact that no restricting device such as a slit is necessary in FTIR spectroscopy. Particularly at low resolution, the total radiation source output can be passed through the sample continuously. This results in a substantial gain in energy at the detector, which leads to an improved S/N ratio.

Moreover, there is a speed advantage. It is possible to obtain FTIR spectra on a millisecond time scale. FTIR spectroscopy also allows one to determine the exact position of an IR band with extremely high precision, is much better than that achieved with a dispersive spectrometer.

Detectors. For standard applications often a pyroelectric detector such as deuterium triglycin sulfate (DTGS) is used. Liquid-nitrogen-cooled semiconductor quantum detectors, such as mercury cadmium telluride (MCT) detectors, are frequently used for more demanding applications in IR and Raman spectrometers. They offer a very high sensitivity and an advantageous low noise equivalent power (NEP). For detection in the visible and ultraviolet region, photographic plates, photomultipliers, and charge-coupled diode (CCD) array detectors are commonly applied.

F. Comparison of IR and Raman Spectra of SC

The assignment of the bands of proteins and lipids in the stratum corneum has been done for IR and for Raman spectra [5]. The number of visible modes in the Raman is nearly twice as high as in the IR spectrum. By means of lipid extraction of samples of human stratum corneum, the origin of bands could be determined.

Several comparisons of human and animal tissues have been published [30,31]. The molecular structure of snake skin is markedly different from that of human stratum corneum. The ceramides, characteristic of the lipids of human SC, are replaced in snake skin by phospholipids. Moreover, snake skin possesses an outer β-keratin sheet layer. SC from pigs resembles more closely that of the human SC. By means of a Fourier self-deconvolution technique, the main assignments and peak positions in Table 1 were confirmed [32].

In the IR spectrum, there is a strong water feature around 3300 cm^{-1} that is largely absent in the Raman spectrum. The positions of the CH stretching modes around 3100 cm^{-1} to 2700 cm^{-1} indicate the conformation of the intercellular lipid chains. The fingerprint region of the spectra between approx. 1800 cm^{-1} to 1000 cm^{-1} show some characteristic differences. The Amide I band (mainly C=O stretching vibration) is strong in the infrared, but it is comparatively weak in the Raman. However, the CH$_2$-scissoring modes of the lipid chains are stronger in the Raman. However, most of the bands appear in both techniques.

The IR spectrum below 1000 cm^{-1} tends to lose some definition, and peaks are difficult to observe because of the strong IR-absorbance of water. Characteristic for the Raman spectrum is a series of weak but important bands between 1200 cm^{-1} and 1000 cm^{-1}. They can be assigned to the C—C stretching modes of

Table 1 FTIR and FT Raman Band Assignments Consistent with the Vibrational Modes from Human Stratum Corneum

Raman frequency (cm^{-1})	IR frequency (cm^{-1})	Assignment
424		δ(CCC) skeletal backbone
526		ν(SS)
600		ρ(CH) wagging
623		ν(CS)
644		ν(CS); amide IV
746		ρ(CH$_2$) in-phase
827		δ(CCH) aliphatic
850		δ(CCH) aromatic
883		ρ(CH$_2$)
931		ρ(CH$_3$) terminal; ν(CC) α-helix
956		ρ(CH$_3$); δ(CCH) olefinic
1002		ν(CC) aromatic ring
1031		ν(CC) skeletal *cis* conformation
1062	1076	ν(CC) skeletal *trans* conformation
1082		ν(CC) skeletal random conformation
1126		ν(CC) skeletal trans conformation
1155		ν(CC); δ(COH)
1172		ν(CC)
1207		not assigned
1244	1247	δ(CH$_2$) wagging; ν(CN) Amide III disordered
1274		ν(CN) and δ(NH) Amide III α-helix
1296	1298	δ(CH$_2$)
1336		not assigned
	1366	δ[C(CH$_3$)$_2$] symmetric
1385	1389	δ(CH$_3$) symmetric
	1401	δ[C(CH$_3$)]
1421		δ(CH$_3$)
1438	1440	δ(CH$_2$) scissoring
	1451	δ(CH$_3$) antisymmetric
	1460	δ(CH$_2$)
	1515	not assigned
1552	1548	δ(NH) and ν(CN) amide II
1585		ν(C=C) olefinic
1602		not assigned
1652	1650	ν(C=O) amide I α-helix
	1656	ν(C=O) amide I disordered
1743	1743	ν(C=O) amide I lipid
1768		ν(COO)
2723		ν(C−CH$_3$)

Table 1 Continued

Raman frequency (cm^{-1})	IR frequency (cm^{-1})	Assignment
2852	2851	$\nu(CH_2)$ symmetric
	2873	$\nu(CH_3)$ symmetric
2883		$\nu(CH_3)$ symmetric
	2919	$\nu(CH_2)$ antisymmetric
2931		$\nu(CH_2)$ antisymmetric
2958	2957	$\nu(CH_3)$ antisymmetric
3000		not assigned
3060		$\nu(CH)$ olefinic
	3070	1st overtone, amide II at 1548 cm-1
	3287	$\nu(OH)$ of H_2O

Abbreviations: δ = deformation; ν = stretch; ρ = rock.
Source: Ref. 5.

the lipid chains. The conformation of the C—C bonds is correlated with the position and intensity of the Raman-bands in that region. The Raman spectrum shows further spectral features such as the stretching modes of S—S bonds from proteins down to around 400 cm^{-1}.

1. Determination of SC Moisture from the Amide Band Ratio

In the infrared spectrum, the ratio of the band heights for amide I (1580 cm^{-1}–1720 cm^{-1}) and amide II (1475 cm^{-1}–1580 cm^{-1}) is a measure of the skin moisture. The amide I band is underlaid by water so that the band height of this absorption increases as the skin moisture increases. The larger it is, the greater the skin moisture.

In the Raman spectrum, the amount of bound water is determined from the ratio of the CH band height around 2940 cm^{-1} and the OH stretching band height around 3250 cm^{-1} [33]. This band ratio increases with the skin moisture.

2. Determination of the Greasiness

The greasiness is determined from the IR spectrum by using the band height of the ester absorption and the fatty acid absorption (1690 cm^{-1}–1790 cm^{-1}). This value is placed in relationship to the height of the amide I band.

3. Determination of the Fatty Acid/Ester Ratio

The determination of the fatty acid/ester ratio is similar to that for the determination of the amide band ratio. In this case, the band heights of the fatty acid absorption (1690 cm^{-1}–1720 cm^{-1}) and the ester absorption (1720 cm^{-1}–1790 cm^{-1}) are determined and placed in relationship to each other.

4. Determination of Lipid Alkyl Chain Order

ATR-FTIR has been extensively used to study the phase behavior of lipid membranes [34,35]. The disorder of the SC lipids can be studied by means of that technique because it was shown that the degree of the alkyl chain disorder results in a shift of the C—H stretching absorbances to higher wavenumber [36]. This can be correlated to the introduction of gauche conformers in the alkyl chain [37].

In Raman spectroscopy, the degree of alkyl chain order is given by the lateral interaction parameter S_{lat} [38]. From the ratio of the band heights for the C—H stretching modes at 2890 cm^{-1} and 2850 cm^{-1}, the order parameter for the lateral interaction S_{lat} is calculated:

$$S_{lat} = (I_{2890}/I_{2850} - 0.7)/1.5 \qquad (4)$$

S_{lat} indicates the packing order of the lipidic alkyl chains. It changes from 0 for the liquid state to 1 for the crystalline state [38].

III. VIBRATIONAL SPECTROSCOPY OF THE STRATUM CORNEUM

Some examples from the literature shall illustrate potential application areas of IR and Raman spectroscopy for the investigation of the SC and for SC lipids. They are characteristic but by no means comprehensive examples of scientific work in that field and give an overview of the potential of vibrational spectroscopy.

A. Studies on Model Skin Lipid Mixtures

FTIR was employed to investigate the thermotropic phase behavior of stratum corneum lipid multilamellae [39]. It has been demonstrated that the lipids of the stratum corneum provide the primary electrical and transport resistance in the skin. These lipids are unusual in their composition, structure, and localization; they contain only cholesterol, fatty acids, and ceramides and they form broad, multilamellar sheets, which are located extracellulary. The FTIR results from both the symmetric CH$_2$ stretching and the CH$_2$ scissoring vibrations suggest that the SC lipids exhibit polymorphic phase behavior below the main phase transition temperature. The multiple phases are most likely crystalline mixtures of different alkylchain packings, along with solid-liquid phases. Similarities between the

FTIR results reported here for SC lipids and those obtained for cholesterol-containing gel phase phospholipids suggest that the nonuniform distribution of cholesterol occurs in each system.

The domain structure of the SC lipid multilamellae has been characterized by FTIR spectroscopy in a three-component SC lipid model consisting of ceramide III, cholesterol, and perdeuterated palmitic acid [40]. At physiological temperature, the CD_2 scissoring mode of the palmitic acid methylenes, and the CH_2 rocking mode of the ceramide methylenes, are each split into two components. This indicates that both components exist in separate, conformationally ordered phases, probably with orthorhombic perpendicular subcells. The magnitude of the splitting indicates that the domains are at least 100 chains in size. The thermotropic behavior of the CD_2 stretching vibrations demonstrates that conformational disordering of the palmitic acid commences at 42°C with a transition midpoint of 50°C. The CH_2 stretching frequency indicates the ceramide chains remain ordered until 50°C, then they disorder with a midpoint of 67°C. The results provide a molecular characterization for the complex low temperature (10–40°C) dynamic behavior suggested also by recent ^2H NMR experiments [9].

The position of the CH_2 stretching vibrations alters with great sensitivity with the fraction of *trans* and gauche conformers in the alkyl chains of the lipids. Solid-liquid phase boundaries of well-defined model lipid mixtures with a simple composition can therefore provide information about the maximum possible alteration of the band position of the CH_2 stretching vibrations of the alkyl chains of the stratum corneum lipids. If the temperature is increased from 25 to 60°C, the solid gel phase of the investigated lipid/water mixtures melts completely to form a liquid-crystalline phase that can be recognized in the DSC measurements as melting transitions (Table 2).

Table 2 Melting Transitions of Model Lipid Mixtures

	Lipid mixture A	Lipid mixture B
Stearic acid	9.78	4.77
Palmitic acid	33.43	16.40
Oleic acid	30.75	16.11
Cholesterol	0.00	36.96
Water	26.04	25.76
Melting transition/°C	40–50 (peak at 48)	40–50 (peak at 47)
Melting enthalpy/J/g	63	28
CH_{2sym}/cm^{-1} at 25°C	2848.0	2848.3
CH_{2sym}/cm^{-1} at 37°C	2848.1	2848.4
CH_{2sym}/cm^{-1} at 60°C	2852.9	2850.9

Source: Ref. 13.

The addition of cholesterol to the fatty acid mixture (Lipid mixture B) does not broaden the melting range but it clearly reduces the melting enthalpy. The symmetrical CH_2 stretching vibration shifts during the melting process from 2848.0 cm^{-1} to up to 2852.9 cm^{-1} for the fatty acid/water mixture and from 2848.3 cm^{-1} to up to 2850.9 cm^{-1} for the fatty acid/cholesterol/water mixture. This means that the addition of cholesterol has different effects on the different phases: at 25 and 37°C, the solid gel phase is present and cholesterol leads to a positive shift of the wavenumber of the CH_2 stretching vibration, which indicates a lower degree of order of the lipid alkyl chains. In contrast, at 60°C in the liquid-crystalline phase, cholesterol increases the degree of order of the lipids in model mixture B. This adjustment of the degree of order of the alkyl chains in the lamellar gel and liquid-crystalline phases manifests itself in the smaller melting enthalpy of model mixture B in the melting transition.

B. Analysis of Porcine Stratum Corneum Ex Vivo

The most informative lipid absorbances are those originating from the hydrophobic alkyl chain in the analysis of SC vibrational spectra. The antisymmetric (approx. 2850 cm^{-1}) and symmetric (approx. 2920 cm^{-1}) carbon-hydrogen stretching mode are sensitive to the trans/gauche ratio of the alkyl chain. These two bands are used in the analysis of the conformation of the hydrocarbon chains. The environment within the interior of the bilayer can be analyzed by means of the antisymmetric (approx. 2956 cm^{-1}) and symmetric (approx. 2870 cm^{-1}) stretching modes of the terminal methyl groups. The lateral packing within the lamellae can be described by the methylene scissoring (1462–1474 cm^{-1}) and rocking (720–730 cm^{-1}) vibrations [36]. Melting of hydrocarbon chains and phase transitions are monitored by the analysis of the methylene stretching vibrations. The temperature dependence of the symmetric methylene stretching frequency of porcine SC has been studied. One obtains a sigmoidal curve with an abrupt increase in frequency between 60 and 80°C, which is typical for SC. From the absolute frequencies together with the magnitude of the thermal shift, the degree of hydrocarbon chain order can be deduced. This can be interpreted as the transition from gel to liquid-crystalline phase. This transition is characterized by an increased amount of gauche rotational conformers along the alkyl chains. A higher energy is required to excite the corresponding vibrations for these gauche conformers and thus they are occurring at higher IR frequencies. For stratum corneum lipids obtained from pigs a shift in the position of the symmetrical CH_2 stretching vibrations of the SC lipids from 2849.4 to 2853.6 cm^{-1} ($-10°$–$100°C$) has been reported [39]. In addition to this increase in vibrational energy, an increase in bandwidth is observed on heating. The latter finding correlates with the increase in the number of conformational states and in motional freedom of the alkyl chains.

The temperature dependence of the terminal methyl group conformation has also been scrutinized for porcine SC [39]. In contrast to the methylene stretching frequency, the methyl stretching frequency increased progressively with increasing temperature and no distinct phase transition was found. This can be interpreted as a consequence of the less constrained environment toward the center of the bilayer, compared to the methylene groups closer to the polar headgroups.

C. Study of Human Stratum Corneum Lipids

According to the literature, human stratum corneum lipids obtained from skin biopsies alter the spectral position of their symmetrical CH_2 stretching vibrations in the temperature range from 25 to 60°C only from 2849.7 to 2850.8 cm^{-1} [23]. In an in vivo study of humans, the position of the symmetrical CH_2 stretching vibrations on the cheek was found between a minimum of 2849.4 cm^{-1} for dry and a maximum of 2850.8 cm^{-1} for greasy, sebum-rich skin. On the lower arm, the sebum content is lower and the symmetrical CH_2 stretching vibrations had an average of 2849.1 cm^{-1}. The degree of lipid order in vivo is therefore less than in the model lipid mixtures like A or B in Table 2 [13].

In Raman spectra, the positions of the CH stretching modes around 3100 cm^{-1} to 2700 cm^{-1} indicate the conformation of the intercellular lipid chains, and they have been used to investigate the phase transitions in model lipid systems [38].

The separation between the CH_2 antisymmetric and symmetric stretching mode was used as an indicator of lipid order. In this Raman spectroscopic study, human SC was heated from 25°C to 100°C and three lipid transitions were observed. This is in good accordance with the results from thermal analysis. The analysis of the scissoring mode which is reflected in a Raman band at 1440 cm^{-1} shows the three aforementioned lipid transitions at 37–40°C, 70–72°C, and 80°C. Additionally a further transition can be observed at 55°C, which suggests that a minor reorganization of the methylene groups of the lipid alkyl chains takes place. It is also found that bands at 1126 cm^{-1} and 1065 cm^{-1} which are characteristic for C—C bonds in the *trans* configuration decrease in intensity on heating. It can be concluded that higher temperature gradually increases the amount of free rotation around the C—C bond.

D. Penetration Enhancer Studies Using Raman Spectroscopy

Signal overlap is a severe problem encountered in the study of molecular interactions between penetration enhancers and lipids in the SC by means of vibrational spectroscopy. There is significant interference of some vibrational modes of the enhancer with those of the tissue. This is found especially for the C—H stretching modes. Therefore, deuterated enhancers have been introduced, which shift the C-D modes to lower wavenumbers [41]. Raman spectroscopy allows the linear subtraction across the whole wavenumber range without the need for complex calibration curves. Therefore, spectra can be generated with an enhancer present and subsequently the signal of the enhancer can be removed by normalization on a unique band. After this procedure, the spectrum of the enhancer modified SC can be analyzed. This method has been used for tissue treated with dimethyl sulphoxid (DMSO) and the terpene derivative 1,8-cineole, respectively [42,43].

The C—C stretching modes were disturbed by DMSO treatment [42]. The observed effect is indicative of a transition of a *trans* gel to a trans-gauche liquid-crystalline phase. The α-helical keratin in the SC was altered to a β-sheet conformation. The effect could be semiquantified by curve fitting and intensity measurements. The data from 1,8-cineole indicated a direct modification of the lipid domain [43].

E. Comparison of in Vitro Skin Models with Normal Human Skin

The infrared spectra of human skin in vivo and in vitro (epidermis model and full-skin model) resemble each other in principle (example in Figs. 3 and 4).

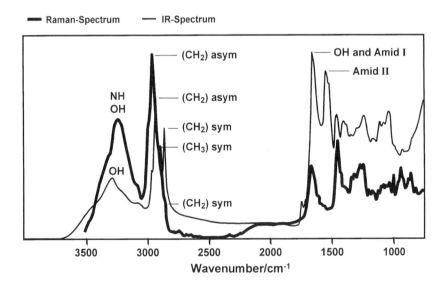

Figure 3 Infrared and Raman spectra of normal human skin.

Detection of Changes in Skin Lipids by IR and Raman Spectroscopy 243

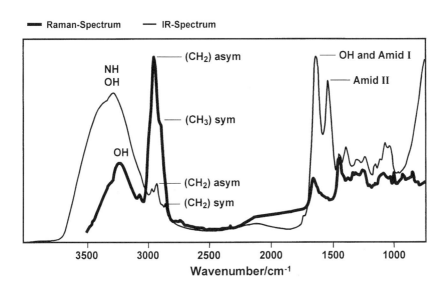

Figure 4 Infrared and Raman spectra of a full-skin model.

There are slight differences in the absolute wavenumbers of the amide vibrations (Table 3). However, the calculation of moisture content from the ratio of amide I to amide II band leads to nearly the same moisture content for the full-skin model as for human dry skin, but to a higher value for the epidermis model. The C—H stretching bands are much smaller for both skin models compared to human skin in vivo and therefore not detectable in the underlying noise of the spectrum. As a consequence, an exact positioning of the C—H-stretching vibrations is no longer possible.

Table 3 Infrared Data on Human Skin in vivo and in vitro

Skin sample	Human dry skin	Epidermis model	Full-skin model
ν (C=O)/cm^{-1} Amide I	1645.8 ± 0.3	1640.4 ± 0.5	1636.5
ν (C—N)/cm^{-1} Amide II	1539.6 ± 0.3	1536.2 ± 0.5	1542.8
Moisture/arb. units	1.66 ± 0.05	1.28 ± 0.02	1.61
ν (CH$_2$)$_{sym}$/cm^{-1} Alkyl chain order	2849.7 ± 0.16	2849.3 ± 0.4	not determined
ν (CH$_2$)$_{asym}$/cm^{-1} Alkyl chain order	2915.9 ± 0.13	2915.9 ± 0.3	not determined

Mean values; n = 4, if SEM indicated, otherwise n = 2.
Source: Ref. 13.

The Raman spectra (examples in Figs. 3 and 4) provide complementary information regarding the skin status. The absolute wavenumbers for the epidermis as well as the full-skin model differ slightly from the ones of normal human skin ex vivo (Table 4). It is interesting to see that moisture content—as determined from the Raman spectra—differs considerably from moisture content as determined from the amide I to II ratio from the infrared spectra. According to the Raman results, moisture content is the same for the full-skin model and normal human skin ex vivo. However, the epidermis model seems to be less hydrated. A possible reason is faster drying out under the comparably long near-infrared laser irradiation during the Raman measurement.

Information on the lipid packing order is easily obtained from the Raman spectra, even in the case of the skin models. The evaluation of the lateral order parameter S_{lat} reveals a distinct reduction in lipid packing order for the epidermis and the full-skin model compared to normal human skin. The intercellular lipids are obviously less organized in the skin models, as has been found also in an earlier ATR-FTIR study [2].

F. Determination of Skin Type from Infrared Spectra

IR-spectroscopy was used as a tool to classify an investigation panel of 35 female test subjects into different skin types. Skin-type characterization by IR spectroscopy is carried out on the basis of the spectra recorded 2 hours after the skin had been cleansed. By clustering into four skin type subgroups (dry, normal, greasy, mixed) based on the sebum content (greasiness), the standard deviation of 0.80 obtained for the whole group from the band heights of the triglyceride absorption measured on the forehead, which is a measure of the sebum content of the skin, could be considerably reduced (Table 5). After this clustering, the following standard deviations for the individual groups were obtained: 0.39 for dry skin, 0.63 for normal skin, 0.79 for greasy skin, and 0.36 for mixed skin.

If the corresponding standard deviations are now investigated when the skin types are assigned according to the visual skin typing, then the following values are obtained: 0.66 for dry skin, 0.78 for normal skin, 1.10 for greasy skin, and 0.60 for mixed skin. In this case the standard deviations are clearly larger than those obtained by clustering based on FTIR data. Assignment according to IR spectroscopy and according to classical-visual skin typing therefore differ from each other. Each sorts the skin types according to different criteria. Depending on the application range, one or other of these two systems is to be preferred. For example, effects on the size of skin pores are best described by visual skin typing. The IR method offers the advantage of better reproducibility and quantification in the investigations of the effects of cosmetic products on the stratum corneum.

Table 4 Raman Data on Human Skin ex vivo and in vitro

Skin sample	Human skin		Epidermis model	Full-skin model
	36y	55y		
ν (C=O)/cm^{-1} Amide I	1666.5 ± 0.3	1666.3 ± 0.6	1653.8 ± 0.5	1654.6 ± 0.7
δ (NH) and ν(C—N)/cm^{-1} Amide III	1271.5 ± 0.1	1271.3 ± 0.3	1271.1 ± 0.7	1268.0
ν(C—C) in proteins/cm^{-1}	938.8 ± 0.1	938.8 ± 0.1	936.7 ± 0.5	938.3 ± 1.2
δ (OH)$_{sym}$ of water/cm^{-1}	3223.7 ± 1.4	3225.2 ± 2.5	3235.5 ± 10.1	3230.5 ± 1.5
δ (CH$_2$)$_{asym}$ in proteins/cm^{-1}	2942.5 ± 0.2	2942.2 ± 0.1	2932.6 ± 0.3	2936.0 ± 0.1
Moisture/arb. units	0.5 ± 0.0	0.6 ± 0.0	0.05 ± 0.005	1.16 ± 0.03
ν(CH$_3$)$_{sym}$ in lipids/cm^{-1}	2881.6 ± 0.1	2881.6 ± 0.3	2878.9 ± 1.5	2882.0 ± 0.9
ν(CH$_2$)$_{sym}$ in lipids/cm^{-1}	2848.4 ± 0.2	2849.3 ± 0.2	2848.5 ± 0.1	2848.8 ± 0.2
Alkyl chain order parameter S_{lat}	1.02 ± 0.03	0.93 ± 0.02	0.68 ± 0.02	0.75 ± 0.03
δ (CH$_2$)(CH$_3$) in proteins and lipids/cm^{-1}	1451.2 ± 0.2	1451.4 ± 0.1	1448.7 ± 0.5	1450.6 ± 0.3

Mean values; SEM for n = 4.

Table 5 Skin type classification by FTIR on the cheek; SEM for n = 32

Skin type	Dry skin	Mixed skin	Normal skin	Greasy skin
Moisture	1.66 ± 0.05	1.58 ± 0.04	1.6 ± 0.03	1.52 ± 0.04
Greasiness	0.44 ± 0.11	0.59 ± 0.09	0.69 ± 0.05	1.0 ± 0.14
Free fatty acids/esters	0.09 ± 0.02	0.16 ± 0.07	0.09 ± 0.02	0.23 ± 0.09
Position of the symmetrical CH_2 stretching vibrations/cm^{-1}	2849.7 ± 0.16	2849.8 ± 0.25	2849.8 ± 0.16	2850.8 ± 0.23
Position of the antisymmetrical CH_2 stretching vibrations/cm^{-1}	2915.9 ± 0.13	2915.7 ± 0.11	2916.1 ± 0.12	2916.3 ± 0.16

Source: Ref. 13.

The other skin parameters that can be derived from the IR spectrum are skin moisture, fatty acid to ester ratio, and position of the CH_2 stretching vibrations, which gives the degree of order of the lipid alkyl chains. The different skin types differ slightly in their moisture. However, for classification the "moisture" parameter appears to be unsuitable; in the case of "dry" skin, it is even misleading. The degree of order of the lipid alkyl chains in the stratum corneum, given by the position of the antisymmetrical CH_2 stretching vibration, increases as the sebum content decreases. The skin barrier properties differ significantly between the greasy skin and dry skin subgroups.

IV. EFFECTS OF COSMETIC PRODUCTS ON THE STRATUM CORNEUM

A. Introduction

Apart from other environmental influences, daily cleansing with soap or surfactants can lead to an impairment of the skin barrier. The cleansing surfactants do not just wash off dirt and excess sebum as required, but can also remove part of the intercellular lipid mixture [44–47]. Even when the amount of lipid removed is only small [48] the selective removal of free fatty acids and cholesterol esters

Detection of Changes in Skin Lipids by IR and Raman Spectroscopy

can lead to a significant alteration in the composition of the intercellular lipids [49,50]. In addition, surfactants penetrate the stratum corneum, adsorb on the keratin of the corneocytes [51], and mix themselves with the intercellular lipids [52]. The result is an impairment of the skin barrier.

In the cosmetics industry, great efforts are made to avoid these undesirable effects. Mild surfactants attack the intercellular lipids to a considerably lesser extent than more aggressive surfactants with a similar cleansing performance. The lipid film, or at least a part of it, can be restored by refatting oil additives or creams applied after washing [53–55]. In this case, differentiation should be made between physical effects on the stratum corneum and biological effects on the metabolism of the living epidermis. On skin whose barrier has been damaged, paraffin oil causes an immediate partial restoration of the skin barrier properties of the stratum corneum, and physiological lipid mixtures slowly penetrate into the epidermis and rebuild the barrier in a natural way by accelerated metabolization in the lamellar bodies, playing a part in the recovery of squamous skin following surfactant damage.

B. Effects of Rinse-Off Products

1. Cleansing with a Shower Gel

FTIR spectroscopy is an elegant method of following the skin cleansing process quantitatively with respect to the sebum content (Fig. 5). The cleansing effect has been quantified here on the cheek and lower arm by evaluation of the band heights of the esters standardized against the amide I-absorption [13]. During facial cleansing with a commercially available surfactant-based shower gel, the fatty acid ester content of the cheek is reduced by 53% (from 0.84 units to 0.28 units). Within 2 hours, sebum secretion has increased it again to 84% of the initial value.

On the forearm, the sebum content is lower by almost a factor of 10 and the cleansing effect is correspondingly weaker. The ester content is reduced from 0.09 to 0.03 units.

2. Facial Cleansing with a Cleansing Milk

An important reason for the different preferences and expectations of consumers is provided by individual skin differences, which are particularly evident in the facial area. In addition, facial skin reacts particularly sensitively to external stimulation, so that facial cleansing was selected for a differentiation between various cleansing products.

During a mild facial cleansing process with a commercially available cleansing milk, the sebum content sinks as expected (Fig. 6) [13]. In comparison to a surfactant-based shower gel, the defatting effect of the cleansing emulsion

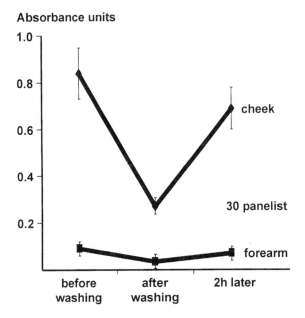

Figure 5 Defatting of the facial skin by a surfactant-based shower gel and endogenous refatting, measured as the averaged band heights of the esters on the cheek and forearms of 30 volunteers. (From Ref. 13.)

is somewhat less (-0.50 units instead of -0.56 units on the cheek for the whole group).

A comparison of the cleansing performance for the different skin types is interesting. The defatting effect is considerably more noticeable on greasy skin (-1.1 units) than on dry skin (-0.2 units). Alterations to the skin moisture content or the ester/fatty acid ratio are not observed. In the dry skin, the initial fat content has already been reestablished 2 hours after cleansing by sebum secretion. Endogenous refatting can also be demonstrated for greasy skin, but after 2 hours this has not reached the original level. The cleansing milk is therefore ideally suitable for removing dirt and sebum from facial skin in a manner suited to the skin type without unwanted side effects.

C. Effects of a Leave-On Product

1. Strengthening the Skin Lipid Film by Creams

An important requirement for skin care creams is that the skin barrier is strengthened. Conventional creams, particularly of the water-in-oil (w/o) type, reduce

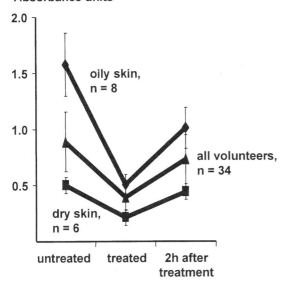

Figure 6 Defatting of the facial skin by a cleansing milk and endogenous refatting, measured as the band height of the esters on the cheeks of 34 volunteers. (From Ref. 13.)

skin permeability by the so-called occlusive effect: the oil components of the cream spread out on the skin surface and form a continuous film that inhibits transepidermal water loss and thus causes an increase of moisture in the skin [56]. The occlusive effect happens quickly but fades away after a short time when the oil slowly penetrates the lipid layers of the stratum corneum and mixes with the skin lipids.

A long-term strengthening of the skin barrier is to be expected when cream components interact directly with the skin lipids and in this way increase the degree of order and thus also the packing density of the lamellar lipid film in the stratum corneum [44,57]. A cream with this effect would ideally contain lipids that themselves build up lamellar gel structures and whose structure therefore resembles the morphology of the lipid film in the stratum corneum.

The influence of a water-in-oil cream and a lamellar cream on the degree of order of the skin lipid film and on skin moisture was investigated by FTIR measurements on the lower arm before and after cream treatment [13]. Six hours after cream application, the cream components had penetrated the stratum corneum; this could be demonstrated by the lack of characteristic IR-bands of the applied cream in the skin print. Whereas the W/O cream reduced the degree of order of the lipids, the lamellar cream increased the degree of order (Fig. 7). This

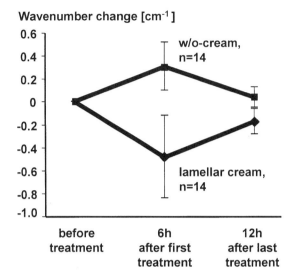

lipid packing density from CH-stretching mode; ± SEM

Figure 7 Degree of order of the lipid alkyl chains after use of a w/o cream and a lamellar cream on the forearm, measured as the band position of the antisymmetric CH_2 stretching vibrations on 13 volunteers for the w/o cream and 14 volunteers for the lamellar cream. (From Ref. 13.)

effect could be observed throughout the whole course of the daily cream treatment over a period of 2 weeks and, in the positive case of the lamellar cream, could still be recognized even after 12 hours with a statistical significance of 75%. This means that the lipid-rich lamellar cream increases the conformational degree of order of the alkyl chains of the lamellar lipid film in the stratum corneum, whereas the liquid oil of the w/o cream actually reduces the degree of order. The lamellar cream therefore strengthens the skin lipid film in a biomimetic manner.

The parallel IR spectroscopy measurement of the skin moisture via the amide band ratio shows that both creams, which have the same moisture factor, also increase the skin moisture in a similar way (Fig. 8). Comparative measurements with the corneometer could not determine any significant effect on the moisture after 6 or even 12 hours.

This shows how sensitive ATR-FTIR measurements can detect cosmetic alterations to the skin moisture and morphology of the skin lipid film quantitatively in vivo.

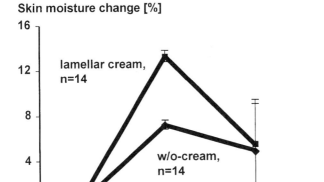

Figure 8 Skin moisture after use of a w/o cream and a lamellar cream, measured by the amide I/amide II band ratio on 16 and 17 volunteers, respectively. (From Ref. 13.)

V. CONCLUSIONS

FTIR and Raman spectroscopy provide valuable insights into the structure and dynamic of the SC. Both methods are versatile tools to unravel the structural heterogeneity of the SC. The central role of the intercellular lipid domains in SC barrier function has been elucidated by means of FTIR and Raman spectroscopy. The understanding of this complex membrane mechanisms and functions has improved during the past years. The effects of cosmetic products on the skin surface lipids can now be studied successfully and verified by vibrational spectroscopy.

REFERENCES

1. Blank HI. Cutaneous barriers. J Invest Dermatol 1965; 45:249–256.
2. Pouliot R, Germain L, Auger FA, Tremblay N, Juhasz J. Physical characterization of the stratum corneum of an in vitro human skin equivalent produced by tissue

engineering and its comparison with normal human skin by ATR-FTIR spectroscopy and thermal analysis (DSC). Biochim Biophys Acta 1999; 1439:341–352.
3. Scheuplein RJ. Mechanism of percutaneous adsorption. I. Routes of penetration and the influence of solubility. J Invest Dermatol 1965; 45:334–346.
4. Wilkes GL, Brown IA, Wildnauer RH. The biomechanical properties of skin. CRC Crit Rev Bioeng 1973; 1:453–495.
5. Barry BW, Edwards HGM, Williams AC. Fourier transform Raman and Infrared vibrational study of human skin: assignment of spectral bands. J Raman Spectrosc 1992; 23:641–645.
6. Friberg SE, Osborne DW. Small angle X-ray diffraction patterns of stratum corneum and a model lipid structure of its lipids. J Disp Sci Technol 1985, 6:485–495.
7. Bouwstra JA, Gooris GS, van der Spek JA, Bras W. Structural investigation of human stratum corneum by small angle X-ray scattering. J Invest Dermatol 1991, 97:1005–1012.
8. Schaefer H, Redelmeier TE. Skin Barrier: Principles of percutaneous Absorption, Basel: S. Karger AG, 1996.
9. Kitson N, Thewalt J, Lafleur M, Bloom M. A model membrane approach to the epidermal permeability barrier. Biochemistry 1994, 33:6707–6715.
10. Potts RO, Francoeur ML. Infrared spectroscopy of stratum corneum lipids. Drugs Pharm Sci 1993; 59:269–291.
11. Naik A, Guy RH. Infrared spectroscopic and differential scanning calorimetric investigations of the stratum corneum barrier function. Drugs Pharm Sci 1997, 83:87–162.
12. Williams AC, Barry BW. Raman spectroscopy. Drugs Pharm Sci 1999, 97:499–514.
13. Prasch T, Knübel G, Schmidt-Fonk K, Ortanderl S, Nieveler S, Förster T. Infrared spectroscopy of the skin: influencing the stratum corneum with cosmetic products. Int J Cosmetic Sci 2000; 22:371–383.
14. Collombel C, Damour O, Gagnieu C, Marichy C, Poinsignon F. Biomaterials with a base of collagen, chitosane and glycosaminoglycans, process for preparing them and their application in human medicine, French Patent 8708252. 1987. European Patent 884101948, 8-2112/Q, 1988. US Patent PCT/FR/8800303 (1989).
15. Van Duzee BF. Thermal analysis of human stratum corneum. J Invest Dermatol 1975; 65:404–408.
16. Goodman M, Barry BW. Differential scanning calorimetry of human stratum corneum: effect of penetration enhancers Azone and dimethyl sulphoxide. Anal Proc 1986; 23:397–398.
17. Fenske DB, Thewalt JL, Bloom M, Kitson N. Models of stratum corneum intercellular membranes: 2H NMR of macroscopically oriented multilayers. Biophys J 1994; 67:1562–1573.
18. Rehfeld SJ, Plachy WZ, Hou SYE, Elias PM. Localization of lipid microdomains and thermal phenomena in murine stratum corneum and isolated membrane complexes: an electron spin resonance study. J Invest Dermatol 1990; 95:217–223.
19. Golden GM, Guzek DB, Harris RR, McKie JE, Potts RO. Lipid thermotropic transitions in human stratum corneum. J Invest Dermatol 1986; 86:255–259.

20. Potts RO, Guzek DB, Harris RR, McKie JE. A noninvasive, in vivo technique to quantitatively measure water concentration of the stratum corneum using attenuated total-reflectance infrared spectroscopy. Arch Dermatol Res 1985; 277:489–495.
21. Mak VHW, Potts RO, Guy RH. Percutaneous penetration enhancement in vivo measured by attenuated total reflectance infrared spectroscopy. Pharm Res 1990; 7:835–841.
22. Bommannan D, Potts RO, Guy RH. Examination of stratum corneum barrier function in vivo by infrared spectroscopy. J Invest Dermatol 1990; 95:403–408.
23. Gay L, Guy RH, Golden GM, Mak VHW, Francoeur ML. Characterization of low-temperature lipid transitions in human stratum corneum. J Invest Dermatol 1994; 103:233–239.
24. Schrader B. Infrared and Raman spectroscopy. New York: VCH-Wiley, 1995.
25. Stuart B, George WO, McIntyre PS. Modern infrared spectroscopy, John Wiley & Sons Ltd. Chichester, England, 1998.
26. Griffiths PW. Chemical infrared fourier transform spectroscopy. Wiley. Canada, 1986.
27. Harrick NJ. Internal reflection spectroscopy. New York: Wiley-Interscience; 1967.
28. Harrick NJ. Internal reflection spectroscopy. 2nd Ed. New York: Harrick Scientific Corporation; 1979.
29. Mirabella FM, Harrick NJ, eds. Internal reflection spectroscopy: review and supplement. New York: Harrick Scientific Corporation; 1985.
30. Williams AC, Barry BW, Edwards HGM. Comparison of Fourier transform Raman spectra of mammalian and reptilian skin. Analyst 1994; 119:563–566.
31. Edwards HGM, Farwell DW, Williams AC, Barry BW. Raman spectroscopic studies of the skin of the Sahara sand viper, the carpet python and the American black rat snake. Spectrochim Acta 1993; 49A:913–919.
32. Edwards HGM, Farwell DW, Williams AC, Barry BW, Rull, F. Novel spectroscopic deconvolution procedure for complex biological systems: vibrational components in the FT-Raman spectra of Ice-man and contemporary skin. J Chem Soc Faraday Trans 1995; 91:3883–3887.
33. Gniadecka M, Faurskov Nielsen O, Christensen DH, Wulf HC. J Invest Dermatol 1998; 110:393–398.
34. Melchior DL, Steim JM. Thermotropic transitions in biomembranes. Annu Rev Biophys Bioeng 1976; 5:205–238.
35. Melchior DL, Steim JM. Lipid-associated thermal events in biomembranes. Prog Surf Membr Sci 1979; 13:211–296.
36. Casal HL, Mantsch HH. Polymorphic phase behaviour of phospholipid membranes studied by infrared spectroscopy. Biochim Biophys Acta 1984; 779:381–401.
37. Asher IM, Levin IW. Effects of temperature and molecular interactions on the vibrational infrared spectra of phospholipid vesicles. Biochim Biophys Acta 1977; 468:63–72.
38. Gaber BP, Peticolas WL. On the quantitative interpretation of biomembrane structure by Raman spectroscopy. Biochim Biophys Acta 1977; 465:260–274.
39. Ongpipattanakul B, Francoeur ML, Potts RO. Polymorphism in stratum corneum lipids. Biochim Biophys Acta 1994; 1190:115–122.
40. Moore DJ, Rerek ME, Mendelsohn R. Lipid domains and orthorhombic phases in

model stratum corneum: evidence from Fourier transform infrared spectroscopy studies. Biochem Biophys Res Commun 1997; 231:797–801.
41. Ongpipattanakul B, Burnette RR, Potts RO, Francoeur ML. Evidence that oleic acid exists in a separate phase within stratum corneum lipids. Pharm Res 1991; 8:350–354.
42. Anigbogu ANC, Williams AC, Barry BW, Edwards HGM. Fourier Transform Raman spectroscopy of interactions between the penetration enhancer dimethyl-sulfoxide and human stratum corneum. Int J Pharm 1995; 125:265–282.
43. Anigbogu ANC, Williams AC, Barry BW, Edwards HGM. Fourier transform Raman spectroscopy in the study of interactions between terpene penetration enhancers and human skin. In: Brain KR, James V, Walters KA, eds. Prediction of Percutaneous Penetration: Methods, Measurements, Modelling. Cardiff: STS Publishing, 1993: 27–36.
44. Denda M, Koyama J, Namba R, Horii I. Stratum corneum lipid morphology and transepidermal water loss in normal skin and surfactant-induced scaly skin. Arch Dermatol Res 1994, 286:41–46.
45. Gloor M, Munsch K Friederich HC. Über die Beeinflussung der Hautoberflächenlipide durch Körperreinigungsmittel I. Derm Mschr 1972; 158:576–581.
46. Gloor M, Tretow CW, Friederich HC. Über die Beeinflussung der Hautoberflächenlipide durch Körperreinigungsmittel II. Derm Mschr 1974; 160:291–296.
47. Gloor M. Influence of surface active agents, emollients, and cosmetic applications on skin and hair lipids. Cosmetics Toiletries 1977; 92:54–62.
48. Froebe CL, Simion FA, Rhein LD, Cagan RH, Kligman A. Stratum corneum lipid removal by surfactants: relation to in vivo irritation. Dermatologica 1990; 181:277–283.
49. Fulmer AW, Kramer GJ. Stratum corneum lipid abnormalities in surfactant-induced dry scaly skin. J Invest Dermatol 1986; 86:598–602.
50. Dykes P. Surfactants and the skin. Int J Cosmetic Sci 1998; 20:53–61.
51. Rhein LD, Robbins CR, Fernee K, Cantore R. Surfactant structure effects on swelling of isolated human stratum corneum. J Soc Cosmet Chem 1986; 37:125–139.
52. Friberg SE, Goldsmith L, Suhaimi H, Rhein LD. Surfactants and the stratum corneum lipids. Colloids Surfaces 1988; 30:1–12.
53. Imokawa G, Akasaki S, Minematsu Y, Kawai M. Importance of intercellular lipids in water-retention properties of the stratum corneum: induction and recovery study of surfactant dry skin. Arch Dermatol Res 1989, 281:45–51.
54. Mao-Qiang M, Brown BE, Wu-Pong S, Feingold KR, Elias PM. Exogenous nonphysiologic vs physiologic lipids. Arch Dermatol 1995; 131:809–816.
55. Förster T, Guckenbiehl B, Ansmann A, Hensen H. Neuartige Körperpflegemittel auf Basis von Mikroemulsionen mit Alkylpolyglykosiden. SÖFW 1996, 122:746–753.
56. Potts RO. Stratum corneum hydration: experimental techniques and interpretations of results. J Soc Cosmet Chem 1986; 37:9–33.
57. Potts RO, Francoeur ML. Lipid biophysics of water loss through the skin. Proc Natl Acad Sci 1990; 87:3871–3873.

9
Lipidic Ingredients in Skin Care Formulations

Ghita Lanzendörfer
Beiersdorf AG, Hamburg, Germany

I. INTRODUCTION

The history of the use of lipids (waxes and oils) is closely connected with the cultural development of mankind. Lipids served first as food, but their use was extended to the preservation of other foodstuffs—for example, in the sealing of amphorae with beeswax. Lipids were basic ingredients in pigment dispersions and were used for wall or body painting, for greasing the skin, and for impregnating leatherware and wood. With the development of better isolation and refining methods for lipids, purer fractions could be used for the extraction of fragrances from plant leaves and their preservation or for medicinal treatment. In medieval times, so-called witches were accused of using this kind of extraction method. It was believed that they were boiling toads in fat and using the so-obtained extract as a body salve to be able to fly.

Lipids also served as fuel for illuminating the nights with a steady burning flame in oil lamps and candles. In different ancient cultures, various lipids were used for those purposes depending on the kinds available (tallow, beeswax, olive oil, etc.). In those days, no distinction was made between medicine, cosmetics or body care, and religious purposes. Several recipes from ancient Egypt, India, and Greece show that body hygiene and cosmetics were on comparably high levels. But as good as ''cosmetic formulations'' were at those times, they were based entirely either on lipids or on solutions in water—stable emulsions were not known then. Ancient cosmetics consisted almost entirely of lipids.

Galen's formula for cold cream is thought to be the first example of an emulsion written down in the first *London Pharmacopoeia* in 1618. However,

Table 1 Ceratum Refrigerans

Cera alba:	130 g
Oleum amygdalarum:	535 g
Aquae rosae:	330 g
Borax:	5 g

This formula shows an example of a cold cream, Unguentum Leniens.
Source: Ref. 1.

the original formula did not contain borax, which was introduced in 1890 for the manufacture of cold creams, as borax was discovered in the California desert in 1859.

As this cream mainly consists of a wax and a liquid triglyceride, the lipid components still make up the major part of the formula. But with water contents higher than 45%, it is possible to obtain O/W emulsions. For centuries, cold cream remained "state of the art" when it came to incorporating water into oil.

In addition to beeswax, wool fat was known to be of cosmetic value since ancient times (mentioned by Dioscorides about A.D. 60, *de materia medica*). Wool fat can absorb up to 300% water, but its use was limited because of its distinct smell and the odor of the products made from it. This changed substantially after 1882, when Braun and Liebrich succeeded in the production of the first purified wool lipids and named it lanolin [2]. With the development of the industrial processes of washing and deodorizing wool lipids, an acceptable raw material for the development of ointments was available. With refined wool lipids, almost odorless products can be obtained.

Another lipid very much used in centuries past for application on skin was sperm oil. In the golden age of whale hunting, abundant quantities of this material were isolated. Sperm oil and wax were extracted from the hollows of the skulls of sperm whale, where the oil was accumulated. Oil and wax were separated by crystallization. Later, when whale hunting was banned, sperm oil was replaced by jojoba oil of plant origin and the wax by cetylpalmitate of synthetic origin.

A result of the industrialization was the discovery of petrolatum in 1859 by Robert August Chesebrough. He discovered a paraffin-like residue on the oil fields of Pennsylvania that was used by the oil workers to soothe cuts, burns, and wounds. He perfected the extraction method as well as the refining process to sell odorless, tasteless petroleum jelly (Brochure, Vaseline Worldwide Research, Cairns & Associates, 1994). Paraffin oil, also a product of industrial oil production and distillation, became a substitute for triglycerides. The use of paraffin resulted in whiter and more stable preparations (not prone to oxidation).

Lipids in Skin Care Formulations

Also discovered in the 19th century, and industrially isolated from 1920 on, is lecithin, which was at first isolated from eggs and is now mainly obtained from soya beans. It was soon found to be essential for the structural integrity and function of living cells. Walker stated that lecithin also has marked action on skin respiration [3]. Lecithin had its breakthrough in cosmetics applications in the late 1980s; it was used to form liposomes.

Together with the evolution of lipid extraction, purification, and synthesis and also with the change of attitude toward medicine and religion, the knowledge of "cosmetics" did not remain in monasteries but became the know-how of pharmacists. And in the past century, it also became industrial and scientific knowledge.

Because the synthesis of new raw materials—emulsifiers, lipids, hydrophilic and lipophilic thickeners, antioxidants, antimicrobials—and the optimized refining processes for natural products in the past century, development of cosmetic products grew exponentially. This progress supported the development of a huge variety of cosmetics and today's topical preparations used for daily skin care. Nowadays, emulsions typically contain 60–80% water. Furthermore, even emulsions containing up to 99% water have been made, although at the moment they are more of academic interest.

High water contents improve the cosmetic properties (convenience and efficacy) of emulsions, make them easier to apply, and improve skin feel during and after application. Today, lipids are no longer the only ingredient of cosmetics—they even can be left out—but still contribute considerably to their characteristics.

The following chapters will focus on the chemical characteristics and the application of lipids for skin care formulations, concentrating on lipids that are typically used in the oil phases of emulsions. The recent investigations and findings covering the interaction of lipids with the skin are reviewed.

The characteristics of amphiphilic lipids are described thoroughly by Small [4], Israelachvili [5], Lipowski and Sackmann [6], and Seddon and Templer [7].

II. CHARACTERISTICS OF LIPIDS

Lipids are one of the four major classes of biomolecules which are essential for life. In contrast to the other three classes (proteins, carbohydrates, nucleic acids), lipids cannot be defined by a common chemistry like for instance proteins (characteristic building blocks are amino acids and the bondings are of peptide type). The shared characteristic of lipids is their hydrophobicity, which is a result of the presence of long chain hydrocarbon residues, which can be linear, branched or even cyclic.

Lipids are found in all living organisms and are a major source of cellular energy. They are also essential for maintaining the integrity of all living matter, because the amphiphilic molecules form a barrier separating the living cell from the outside world.

Additionally, they have various roles in the microscopic and the macroscopic world and, because of their diversity, can function in many specific ways: as antigens, for example, and as receptors, sensors, electrical insulators, and biological detergents. Many hormones are lipids—for example, steroid hormones (cortisol, estrogens, progesterones, androgenes), prostaglandins, and leukotrienes.

Transportation of lipids to and from animal cells takes place in the form of small aggregates called lipoproteins.

Lipids are essential for the cells of plants and animals: they are used as building blocks for membranes during growth and repair. They also fulfill specific functions—for example, some waxes in plants secreted on the leaves form films on the surface, thus protecting the leaves from drying out or preventing them from the attack of certain harmful insects. In seeds, triglycerides are accumulated to protect the seeds, serve as energy deposit in growth, and solubilize antioxidants.

Lipids are stored in large quantities in certain aquatic animals; they are used as insulators and for buoyancy. Waxes coat the feathers of aquatic birds to increase the feather-water surface tension and thus prevent immersion.

Lipids also play an important role in industry and modern technology. Petroleum, for example, is still the world's major source of energy. The food industry uses lipids not only for their nutritional value but also as emulsifiers, stabilizers, and moisturizers, which are so important in processed food technology.

Lipids can be spread on water for wave damping in harbors and are used to prevent excessive water evaporation from reservoirs. In modern technology, lipidic components are used in liquid crystalline displays (LCD), where due to their molecular arrangement, optical effects can be produced.

All these functions are related to the chemical structure of lipids and the resulting possible amphiphilic interactions between lipid molecules or with other media such as water.

Table 2 depicts a classification of lipids based on their polarity, solubility, and saponifiability. All saponifiable lipids are either acids or have ester bonds that can be subjected to alkaline saponification. In the group of the saponifiable lipids, neutral and polar lipids are discerned; neutral lipids dissolve in acetone and polar lipids do not. Complex lipids also contain ester groups, but the esterified phosphate-groups, amines, or sugar residues render them more polar than lipids that are esterified with aliphatic alcohols.

It is obvious that the physical behavior of such chemically divergent molecules will be quite different. Indeed, one of the most interesting characteristics of lipids is their tremendously varied behavior in aqueous systems, ranging from almost total insolubility (paraffin oil, sterol esters) to nearly complete solubility (soaps, detergents, bile salts, and gangliosides). These characteristics also can be used to group lipids (Table 3).

A very similar grouping also can be achieved by analyzing the molecular shape of lipids [5]. For that purpose, the relative sizes of the polar and the nonpolar portion of the molecule are taken into account and a packing parameter is defined. Large hydrophilic groups and relatively small lipophilic tails, for instance, favor an assembly in the form of a spherical micelle. In molecules in which the molecular dimensions of the polar and nonpolar part are relatively similar, less curved arrangements are favored (i.e., lamellar layers).

The use of lipids in cosmetic formulations is not limited to the naturally occurring compounds. Modern synthetic methods enable selection from a wide variety of synthetic esters, polymers, silicones, or chemically modified natural compounds. Often those compounds are preferred to the natural compounds because of their availability, stability, or price. The main chemical groups used for cosmetic formulations are depicted in Table 4.

A. Waxes

In Table 4, wax esters are mentioned. These are not the only compounds that count as waxes: the term "wax" does not describe a certain class of lipids but classifies lipids according to their physical characteristics. These are kneadability, melting behavior, and crystallinity. The following criteria have to be fulfilled for a substance to be classified as a wax:

Kneadable at 20°C
Solid to brittle hard
Coarse to fine-crystalline
Transparent to opaque, but not glasslike
Melting above 40°C, without decomposition
Above melting point relatively low viscous
Strong temperature-dependent consistency and solubility
Polishable under low pressure

Taking these characteristics into account, substances of many different chemical groups can be waxes: paraffins, high or low density polyethylenes (mol weight from 2000 to 9000), esters of long-chain alcohols and acids, long-chain fatty acids and their soaps. Also, stearic acid and cetyl alcohol belong to the group of waxes because of their physical behavior.

Table 2 Classification Scheme for Lipids

Lipid class	Structure (example)	
Free (not esterified) fatty acids	∿∿∿∿COOH	Neutral lipid Acetone soluble Simple lipids
Waxes (Waxesters, paraffins, fatty alcohols, fatty aldehydes)	R^1—CO—O—R^2	
Glycerides (fats)	CH_2—O—CO—R^1 | CH—O—CO—R^2 | CH_2—O—CO—R^3	

Phosphatides
Glycerolphosphatides
	H_2C—O—CO—R^1
	|
	HC—O—CO—R^3
	| O
	| ‖
	H_2C—O—P—O—Choline
	        |
	       O^-
		Polar lipids
		Acetone insoluble
		Complex lipids
		Saponifiable

Sphingophosphatides	O
	                     ‖
	R^1—CH—CH_2—O—P—O—Choline Sphingolipids
	     | |
	   NH O^-
	     |
	   CO—R^2	

Glycolipids Sphingoglycolipids	R^1—CH—CH_2—O—Galactose | NH | CO—R^2	Sphingolipids
Glycerolglycolipids	CH_2—O—CO—R^1 | CH—O—CO—R^2 | CH_2—O—Galactose	
Aminolipids	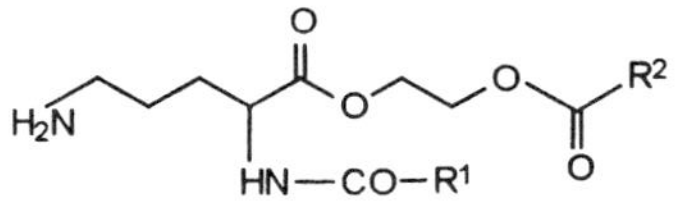	
Isoprenoids Terpenes	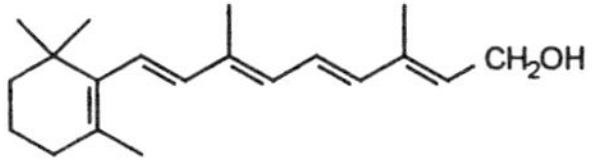	Neutral lipids Acetone soluble Unsaponifiable
Steroids	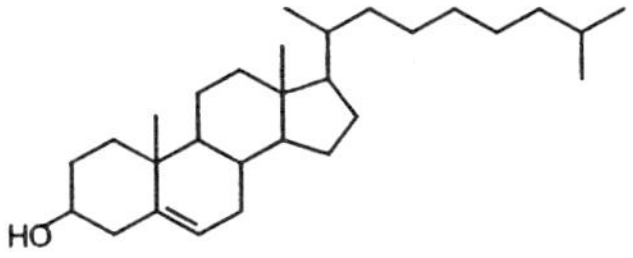	

Source: Ref. 8.

Table 3 Interaction of Lipids with Water

Class	Interaction with water		Examples
	in water	on the surface	
Nonpolar lipids*	crystals or oil	do not spread	Squalene, paraffins
Amphiphatic lipids*			
Insoluble, nonswelling*	crystals or oil	form stable film	fatty acids, wax esters, cholesterol, fatty alcohols, vitamin A, D, E and K
Insoluble, swellable	liquid crystals	form stable film	monoglycerides, phosphatides, lecithins
Soluble lipids			
Micelle-forming lipids with lyotropic mesomorphic behavior	micelles	unstable films	lysophosphatides, anionic soaps, cationic detergents, betaines
Micelle-forming lipids without lyotropic mesomorphic behavior	micelles	unstable films	bile acids and their salts

* Groups of lipids which are commonly used in skin care preparations as lipids.
Source: Ref. 9.

B. Polymers

Also defined by physical properties are the substances termed polymers. Polymeric compounds are macromolecules derived from the polymerization of monomers having average molecular weights of above 10,000 Daltons. They are mainly found in the classes of the hydrocarbons or silicones, but triglycerides also can be polymerized.

C. Some Lipids in Detail

At this point, some important cosmetic lipids shall be discussed in more detail.

1. Vaseline

Vaseline is not the name of a chemically homogeneous substance. Vaseline is a mixture of mostly saturated n-, iso-, and cyclo-paraffins of different molecular weight, thus rendering a crystalline and a fluid phase. The relative amounts of these compounds differ between natural vaselines and is determined (a) by the

Table 4 Overview of Lipids Important for Cosmetic Formulations

Chemical class		Examples
Hydrocarbons, Paraffins	Liquid	Natural, distilled paraffins: mineral oil, synthetic isoparaffins Squalane (natural, modified)
	Semisolid	Petrolatum, polyisobutenes
	Solid	Microcrystalline wax, paraffin wax, ceresin, ozokerite
Silicones		Dimethicone, cyclomethicone, silicone copolymers, perfluorinated silicones
Triglycerides		Synthetic triglycerides, natural oils, hydrogenated tryglycerides
Esters	Derived from branched or unbranched long chain fatty acids	Isopropyl palmitate, cetyl palmitate
	Derived from acids with chain lengths < C12	C12–15 Alkyl benzoate, myristyl lactate
	Derived from branched alcohols or polyhydric alcohols	Neopentyl dicaprate
	Wax esters	Jojoba oil, beeswax, carnauba wax
	Lecithins	Phospholipids (Phosphatidylcoline, -ethanolamine, -inositol)
	UV-Filters	Octylmethoxycinnamate
Alcohols	Aliphatic alcohols	Fatty alcohols unbranched and branched C-chains, Guerbet alcohols
	Sterols	Cholesterol, phytosterols, lanolin alcohols
	Tocopherols	Tocopherol
Ethers		Dicaprylyl ether, PPG-ethers, ethyleneoxide ethers, steareth-20
Other Lipids	Ceramides	Ceramides, sphingolipids
	Fatty acids	Stearic acid

Table 5 Composition of Vaseline*

Composition	c[%]	C-chain distribution	main C-chain	C-chain average
n-Alkanes:	16–19	C17–C51	C28	C29
iso-Alkanes:	11–15	C20–C48	C30	C32
"Oilpeak":	66–73	C21–C37	C27	C27
Full vaseline		C17–C51	C27–28	C28

* According to Häusler and Führer, who determined the qualitative and quantitative composition of vaseline using gaschromatography.
Source: Ref. 10.

composition of the crude oil from which vaseline is isolated and (b) by the isolation method, which accounts for the relations of n-, iso-, and cyclo-paraffins.

Vaseline is described as a colloidal system, in which the iso-paraffins form a solid amorphous region that accounts for the oil-binding capacity of vaseline. They build up "Fransenmicelles," a structure that immobilizes the fluid paraffins.

2. Beeswax

Beeswax is derived from honeybees, which belong to the genus *Apis*. The bees "synthesize" the wax by enzymes from carbohydrates from their food and excrete it from the wax glands. One thousand bees produce about 9 g of wax in only 9 to 10 days of their life span, which lasts about 35 to 45 days.

Beeswax is a mixture of wax esters of long-chain fatty acids, free fatty acids, and alcohols.

Table 6 Composition of Beeswax

Composition	c[%]
Waxesters:	
Myricyl palmitate, myricyl ceroate, myricyl hypogaeate, cetyl hydroxypalmitate, esters of dicarbonic acids, esters of polyhydric alcohols, cholesterolester	71
Free Acids:	
Lignoceric acid, cerotinic acid, montan acid, melissinic acid, physillinic acid ($C_{33}H_{66}O_2$)	13–15
Free alcohols	1
Hydrocarbons (C25–C31):	10–13
Pentacosan	

Source: Ref. 11.

Like other esters, the esters of beeswax can be hydrolyzed or modified, resulting in lipidic derivatives with different characteristics (i.e., higher polarity), facilitating the incorporation in emulsions [12]. Beeswax is a long-known ingredient for cosmetic formulations and is used, for instance, in lipsticks and cold creams.

3. Lanolin

Lanolin is the unctuous secretion of the sebaceous glands of sheep; it is deposited onto the wool fibers and is also referred to as wool wax. It serves to soften the fleece and protect it from the elements. It is a complex mixture of esters, di-esters, and hydroxy-esters, methyl-esters of high molecular weight lanolin alcohols (69 aliphatic alcohols, chain lengths from C12 to C36 and 6 sterols) and high-molecular-weight lanolin acids (approximately 138, chain lengths from C7 to C41) [13].

The distribution of the hydrocarbon chain lengths of the saponifiable fraction of lanolin, the lanolin alcohols, ranges from C12 to C36, including ω-1 and ω-2 methyl-substituted alcohols, and alkane-diols. Also, sterols (cholesterol, lanosterol, agnosterole, and others) are known as lanolin alcohols.

The lanolin acids are composed of unbranched, ω-1 and ω-2 methyl-substituted acids, α-hydroxy and ω-hydroxy acids, and combinations of methyl and hydroxy substitution.

The composition of lanolin alcohols and lanolin acids is thoroughly reviewed by Motiuk [14,15].

Lanolin is an effective emollient, which—by subjective evaluation—effects a softening and improvement of rough skin caused by lack of sufficient natural moisture retention. Idson reported that lanolin causes the water in the skin to build up to its normal level of 10–30% by retarding without completely inhibiting transepidermal waterloss (TEWL) [16]. Lanolin also accounts for the enhancement of the adhesion of the formula to skin, thus rendering a long-lasting water-impermeable film.

In past years, the allergy and sensitization potential of lanolin was discussed thoroughly, resulting in different views of the matter. For some time, especially the iso- and anteiso compounds of hydrogenated lanolin were described as sensitizers or allergens [13], whereas today it is understood that many of the formerly found allergies resulted from impurities in the lanolin [17–19]. The data published on lanolin allergy, especially lanolin alcohols, were reviewed thoroughly by Kligman, who showed that almost all of the so-called allergic responses to lanolin and lanolin alcohols were false positives due to testing on persons with diseased or damaged skin [20,21]

Lanolin can easily be fractionated or chemically modified, resulting in a variety of compounds for different applications (i.e., emulsifiers, oils, or waxes).

Lanolin oil, a fraction of lanolin, is a preferred ingredient in lipsticks due to its good pigment-dispersing properties [22] and its skin care efficacy, measured as reduction of the dryness of the lips [23].

4. Jojoba Oil

Jojoba oil is derived from the fruits of the jojoba bush, which grows in the semi-desert areas of the southwestern United States and in the northwest of Mexico. The oil, which is a fluid wax ester, is composed of the esters of the unsaturated C20–C24 alcohols and unsaturated C20–C22 acids. The chain length distribution is as follows:

C16:0: 1.1%
C18:1: 9.0%
C20:1: 71.5%
C22:1: 14.2%
C24:1: 1.6%

The main components are the esters of the C20 and C22 acids and alcohols. The double bond in both molecule moieties renders the substance fluid. Its big success in cosmetic formulations began when a substitute for sperm oil was looked for. Jojoba oil can be partially or completely hydrogenated, giving solid products with much more waxlike appearance.

5. Esters

Derived from the reaction of alcohols with acids, a tremendous diversity of esters with very different characteristics can be obtained, making them the biggest group of lipids. With variation of the chain length and grade of branching on either side of the molecule, the physical properties (melting points, polarity, viscosity) can be modulated (see Table 7). For example, esters are also used to wet pigments, to dissolve chemical sunscreens, and to promote spreading of sunscreens on skin.

From the reaction of saturated linear fatty acids and alcohols, solid products such as cetyl palmitate can be obtained. These compounds are used as bodying agents for emulsions because they increase viscosity and impart a dry skin feel. Introducing an alkyl branch in either the alcohol portion or the acid portion, the esters (compared to the unbranched compounds of similar molecular weight) are liquid at room temperature and have good spreading properties.

Some examples of the diversity of the acids used are C12–C15 alkyl benzoate when an aromatic acid is used or alkyl lactates by the use of the short chain acid lactic acid. Using polyols as the alcoholic component, one can obtain compounds such as triglycerides with glycerol. One prominent example is caprylic/capric triglyceride, which certainly is the most important synthetic triglyceride. Other polyoles, such as pentaerythritol, can be used, leading, for instance, to

Table 7 Examples of Esters and Their Physical Properties

Alcoholic part	Acidic part	Melting point	Spreading	Polarity [mN/m] interfacial tension against water
Butyl	Stearate	22°C	—	—
Cetearyl	Stearate	55°C	—	—
Myristyl	Myristate	41°C	—	—
Isocetyl	Stearate	−2°C	Medium	17
Ethylhexyl	Stearate	10°C	Good	32
Isopropyl	Palmitate	14°C	Good	29
Isopropyl	Stearate	19°C	Good	22
Diisopropyl	Adipate	−6°C	Very good	14
Lauryl	Lactate	−3°C	N.D.	N.D.
Myristyl	Lactate	13°C	N.D.	N.D.
Ethylhexyl	Hydroxystearate	15°C	Poor	N.D.
Pentaerythrityl	Tetraisostearate	< −10°C	Poor	10
Isodecyl	Neopentanoate	< −30°C	Medium–good	30
Butylene Glycol	Caprylate/ Caprate	−10°C	Good	21.5

N.D. = not determined.

pentaerythrityl tetraisostearate. Also, use of polyhydric acids is possible, leading to products such as Tri C12–13 alkyl citrate.

By partial esterification of polyhydric alcohols or acids, the hydrophobic characteristics are reduced, leading to amphiphilic compounds that typically are used as emulsifiers (e.g., glyceryl mono stearate).

6. Natural Triglycerides

Triglycerides form the group of lipids longest known for their use as food and for skin care. This group of compounds also is referred to as fats. Fats that are liquid at room temperature are commonly defined as oils. Chemically, they are glycerol esterified with three fatty acids, which can vary in chain length and saturation.

Triglycerides derived from animal fatty tissue normally are solid at room temperature due to the high amounts of bound stearic acid. Fish oils are different from other animal-derived triglycerides because they are rich in unsaturated long-chain fatty acids with up to 22 C-atoms. Additionally, these acids contain— calculated as a C20 acid—on average more than five double bonds. This unsaturation renders them liquid but also makes them susceptible to oxidation, which is responsible for the distinct fish-like smell.

Vegetable triglycerides also contain considerable amounts of unsaturated fatty acids, thus making them liquid and more susceptible to oxidation than saturated animal-derived triglycerides. For the effective inhibition of oxidation, the oils contain tocopherols. However, tocopherol quantities are not high enough to preserve the lipids in cosmetics. Synthetic antioxidants have to be added.

Triglycerides are extractable from all kind of seeds or fruits, and modern extraction methods make clear, almost odorless oils available. The natural triglycerides differ in fatty acid composition, which is characteristic for the species the oil is obtained from. Also, the positions of fatty acids in the glyceride in natural oils are not random. The more unsaturated acids favor the 2-position and the more saturated ones favor the 1- and 3-position of the glyceride [24].

Triglycerides with specific properties isolated by fractionated crystallization are used as additives in the food industry—for example, to achieve a specific mouthfeel in chocolates. Isolated by transesterfication, specific triglycerides are also used in the medical field for the formulation of parenteral nutritionals [25]. Those fractions are of less importance for skin care formulations.

Vegetable oils are used in skin care formulations because of their specific fatty acid profile. Olive oil contains high amounts of oleic acid, sunflower seed oil contains linoleic acid, and evening primrose oil γ-linolenic acid. Linoleic and γ-linolenic acids are so called essential fatty acids [26]. These acids are ω-6 unsaturated fatty acids, which are deficient in certain pathological skin conditions such as atopic dermatitis, in which the enzyme Δ-6 desaturase is impaired [27,28].

Essential fatty acids achieved their popularity after it was shown that skin condition could be improved after topical application of products containing these acids [29,30]. Recently, seabuckthorn oil [31] and hemp oil [32] gained interest as vegetable oils that also contain high amounts of α-linolenic acid. Meadowfoam seed oil is interesting also because of its fatty acid profile, containing considerable amounts of C20:1 and eicosanoic acid, which are rare in vegetable oils (Fig. 1) [33].

7. Silicones

Silicone is a generic name for many classes of organo-silicone polymers with repeating siloxane (Si-O) units. Silicone fluids are also called silicone oils because they are hydrophobic. Many families of silicones exist with a wide range of possible molecular weights, including linear, cyclic, and cross-linked varieties. The introduction of various functional groups, such as amino, phenyl, alkyl, or polyether, changes the hydro- and lipophobic behavior of silicones. It is also possible to introduce amphiphilicity along the chain allowing silicones to act as emulsifiers.

Silicone oils are used because of the following characteristics:

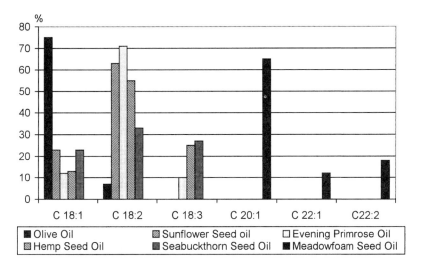

Figure 1 Fatty acid profile of some vegetable oils with different acid profiles. Olive oil contains high amounts of oleic acid. Sunflower seed oil and evening primrose oil both contain high amounts of linoleic acid, but only evening primrose oil contains γ-linoleic acid. Hemp oil and seabuckthorn oil contain considerable amounts of α-linoleic acid; in seabuckthorn oil the ratio of oleic to linoleic to α-linoleic acid is almost 1:1:1. Meadowfoam seed oil mainly contains unsaturated eicanoic acid.

1. They exhibit a very good resistance to high temperature for extended periods of time and are resistant to oxidation.
2. Their viscosity changes less over a wider temperature range than non-silicone fluid.
3. They are chemically inert and usually have high flash points. Due to their low surface tension, they are easily spreadable and have high surface activity. They are resistant to breakdown by mechanical shearing and are noncorrosive [34].

These physicochemical properties of silicones lead to an impressive array of properties and benefits in skin care formulations, including detackification (reduction of the greasy/oily feel of other ingredients); permeability ("breathable films"); lubrication of the skin and characteristic smooth feel (aesthetics); barrier effects (protection); substantivity (long-lasting); anti-adhesive effects; and easy spreading of other materials such as pigments or actives (efficacy and advantages in wetting surfaces).

A large variety of alkylmethylsiloxanes has been synthesized with different physical properties (Fig. 2).

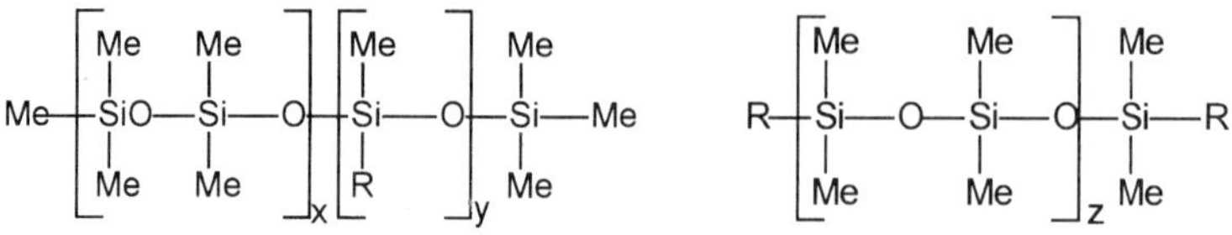

$$\text{Me}\underset{\text{Me}}{\overset{\text{Me}}{-}\text{SiO}-\underset{\text{Me}}{\overset{\text{Me}}{\text{Si}}}-}\text{O}\left[\underset{\text{R}}{\overset{\text{Me}}{\text{Si}}}-\text{O}\right]_x\left[\underset{\text{Me}}{\overset{\text{Me}}{\text{Si}}}-\text{O}\right]_y\underset{\text{Me}}{\overset{\text{Me}}{\text{Si}}}-\text{Me} \qquad \text{R}\underset{\text{Me}}{\overset{\text{Me}}{-}\text{Si}}-\text{O}-\underset{\text{Me}}{\overset{\text{Me}}{\text{Si}}}-\text{O}\left[\underset{\text{Me}}{\overset{\text{Me}}{\text{Si}}}-\right]_z\text{R}$$

Figure 2 Chemical structure and structural elements of polyorganosiloxanes. R = H to C45H91, x = 0 to 1000, y = 0 to 1000, z = 0 to 1000.

1. Volatile fluids are typically materials with molecular weights in the range of 300 to 600 and with hydrocarbon-to-silicone ratios ranging from 0.25 to 1. Alkyl chain length is low.
2. Nonvolatile fluids may have molecular weights up to 100,000; however, the R-substituent is typically not longer than C18.
3. Silicone waxes may range in molecular weight from 500 to 100,000, and the R substituent is typically C16 or longer. The softening points of the waxes range from 25°C to 70°C.

8. Ceramides and Pseudoceramides

Because ceramides constitute a major portion of the skin lipids, extraction from natural sources (animal and plant origin) as well as synthesis of natural and so-called pseudoceramides for cosmetic purposes has been a field of high interest. The structural element of ceramides is a sphingosine backbone amide-linked to a long-chain fatty acid residue. Thus, ceramides are formed with a polar head to which two aliphatic chains are attached (Figs. 3 and 4). The naturally occurring ceramides are optically active and have a D-erythro configuration.

Chain length, saturation and hydroxylation vary in either the fatty acid residue as well as in the sphingosine or phytosphingosine residue of natural ceramides (see also Chapter 1). Synthesis of ceramides with D-erythro configuration can only be obtained by classical chemistry in multi-step synthesis [35]. Alternatively, biosynthetically derived ceramides are available, but their availability is restricted to the phytosphingosine derivatives so far.

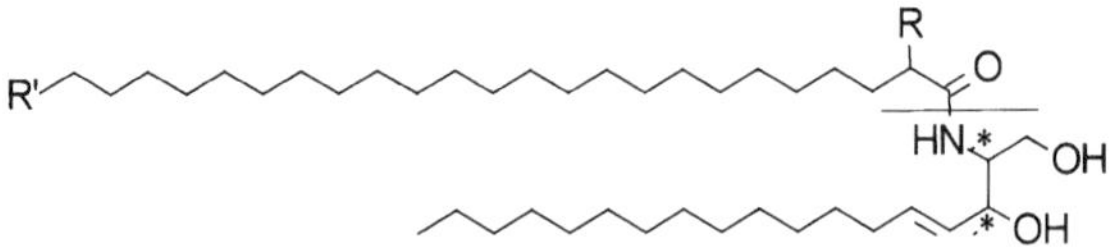

Figure 3 General structure of a ceramide, R = H or OH, R′ = H or ester linked fatty acid; the asterisks indicate chiral positions; the line marks the sphingosine- and fatty acid residue. When R = OH, another chiral C-atom is present.

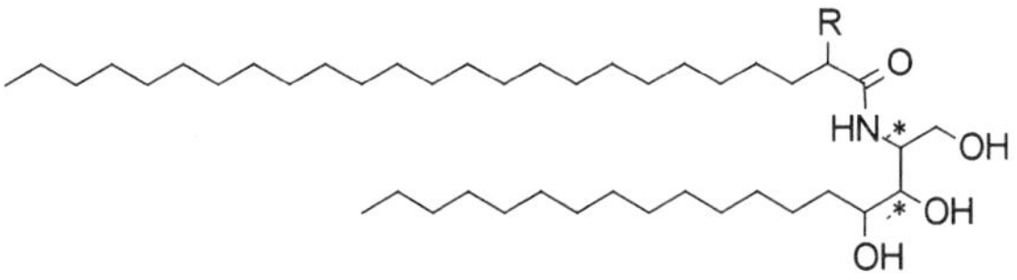

Figure 4 General structure of ceramides on basis of a phytosphingosine, R = H or OH.

The structural elements of pseudoceramides are similar to those of ceramides, but they lack chirality and often possess a different polar head group arrangement. Examples are trihydroxypalmitamido-hydroxypropyl myristyl ether (Fig. 5) or myristyl-propylene glycol hydroxyethyldecanamide.

Ceramides and many pseudoceramides are hard-to-brittle substances with high melting points; they are insoluble in water and only partly soluble in cosmetic lipids, thus making them difficult to formulate in high amounts into cosmetic formulations.

III. FUNCTION OF LIPIDS IN COSMETICS

Of the variety of skin care formulations (Fig. 6), emulsions are by far the biggest and most important group of formulas applied, followed by lipsticks and oils. Therefore, in this chapter mainly the effect of lipids in emulsions will be discussed.

A. General Aspects

Cosmetic emulsions normally contain about 5–30% (w/w) of lipids, rendering them the second major component in cosmetic emulsions after water. The choice of lipids and of the emulsifying system determines the characteristics of the product.

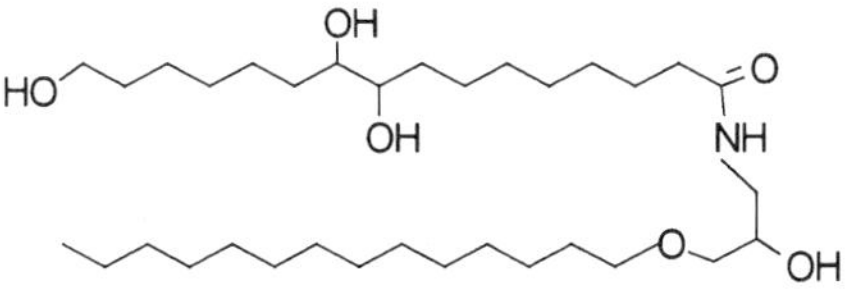

Figure 5 Structure of trihydroxypalmitamido-hydroxypropyl myristyl ether.

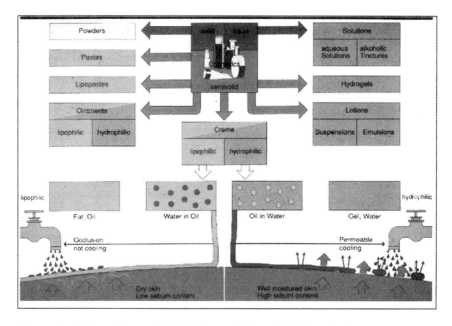

Figure 6 Product forms (with friendly permission of Dr. Sven Gohla).

The influence lipids exert covers stability (important for water-in-oil [W/O] emulsions), viscosity, skin feel, wettability of pigments and solubility of active compounds, and skin care efficacy of the final formulation.

Lipids also account for the film-forming properties of the formulation on the skin, thus rendering it waterproof, for example for sunscreen products, or transfer-resistant for decorative products. Additionally, the film-forming properties are connected with skin care, as the film prevents the skin from drying out [36].

In anhydrous systems such as lipsticks or ointments, lipids are important. Additionally, waxes are included that build the matrix in which the more liquid lipids are incorporated. The characteristics they exhibit depend on the system in which they are incorporated.

B. Influence of Lipids on Emulsion Stability and Viscosity

The factors influencing the rheology of emulsions are volume fraction of the disperse phase (inner phase), viscosity of the disperse droplets, droplet size distribution and the viscosity of the outer phase, and the interfacial rheology of the emulsifier film [38].

The most important factor influencing viscosity and stability of W/O emulsions is the fraction of the disperse phase (water), also called phase volume ratio. For better understanding, a critical phase volume ratio is defined that corresponds to close-packed spheres (for monodisperse distributed droplets, $\Phi c = 0.74$). At this ratio, the water droplets get into contact with each other, and the rheological behavior of the emulsion changes from a liquid particle-solution to that of a close-packed solid network of water droplets [36]. For all practical purposes, that means that the more oil incorporated in a W/O emulsion system, the lower the viscosity becomes.

The viscosity of W/O emulsions also is regulated by the viscosity of the oil phase (viscosity of the outer phase), meaning that the viscosity of the final product is proportional to the viscosity of the oil phase [37,38].

Stable W/O emulsions are preferably formed with nonpolar lipids such as paraffins. W/O emulsions formed with triglycerides exhibit higher viscosities than those with mineral oil [39]. Nevertheless, the choice of emulsifier and the amount of energy used for the dispersion of the water droplets significantly affect the factor of stability and viscosity.

For oil-in-water (O/W) emulsions, there is no preferred class of lipids for achieving stability because of the gel network established by the emulsifier system, which acts independently of the lipids [40].

In O/W emulsions, viscosity normally is not regulated by lipid content but by the choice of emulsifying system, the co-emulsifier (fatty alcohol or monoglyceride), incorporation of solid lipids, and hydrophilic thickener.

C. Influence of Lipids on the Sensory Properties of Skin Care Products

Lipids in emulsions directly influence viscosity and skin feel of the formulation. These are two main characteristics that are also perceived by consumers and influence their choice of product. Therefore, these parameters should be adjusted thoroughly (see also Chapter 11). For that reason, lipids have been studied thoroughly for the sensory characteristics they cause. Many efforts have been made to find a sensory characterization matrix that covers all sensory patterns of lipids. As parameters influencing the skin feel, spreading rate, surface tension, skin absorption (viscosity, solidification temperature), and the "quality" of the afterfeel can be evaluated and compared.

Therefore, not only were the lipids alone tested, but their characteristics in emulsions can be evaluated as well [41,42]. However, in all studies conducted so far, no general relationship between the sensory properties and the physical constants of the lipids could be found. Nevertheless, some correlations are valid.

As a general rule, spreadability values decrease with increasing viscosity of the oil. The spreadability of an oil itself correlates with the subjectively experienced absorption capability into the horny layer of the epidermis and the time course of the fatty character [43]. And the higher the viscosity of the oil, the higher the subjectively experienced "fatty feel" [44].

The correlation between viscosity and "skin absorption" also is valid for triglycerides. In that group of lipids, synthetic medium-chain triglycerides show a balanced pattern of spreading, absorption, and after-feel, whereas most of the natural triglycerides have a pronounced fatty after-feel, which additionally can be correlated with low spreading and absorption.

Synthetic esters normally show good spreading and absorption characteristics and account for a dry after-feel.

By comparing lipids of the same viscosity, Zeidler additionally found some rules that are valid for different lipid classes: Spreadability increases with decreasing surface tension; C12–15 alkyl benzoates (relatively high surface tension, very poor spreadability) → decyl oleate → octyl stearate (relatively low surface tension, medium spreadability).

Table 8 Examples of Lipids with Different Viscosities, Surface Tensions, and Spreadability Values

International Nomenclature of Cosmetic Ingredients	Viscosity [mPas]	Surface tension [mN/m]	Spreadability [class]
Octyl octanoate	3.3	26.8	Good
Dicaprylyl ether	3.3	27.4	Good
Cyclopentasiloxane	3.8	18.3	Very good
Isopropyl myristate	4.6	28.4	Good
Hexyl laurate	5.1	28.7	Medium
Isopropyl palmitate	6.1	28.9	Good
Octyl laurate	6.3	28.8	Medium
Isopropyl stearate	7	29.2	Medium
Octyl palmitate	10.7	29.9	Medium
C12–15 Alkyl benzoate	11.8	31.6	Very poor
Octyl stearate	12.2	30.2	Medium
Dimethicone	13	19.8	Medium
Caproic triglyceride	13.2	29.2	Medium
Decyl oleate	13.5	31	Poor
Octyldodecyl neopentanoate	14.6	28.1	Medium
Cetearyl isononanoate	15.5	29.5	Medium
Isocetyl palmitate	22	30.1	Very poor
Caprylic/Capric triglycerides	23.7	29.4	Very poor
Paraffinum liquidum	25.4	29.7	Poor
Octyldodecanol	44.7	28.7	Very poor
Avocado oil	64.3	32.1	Very poor

Source: Ref. 36.

When the surface tension remains nearly constant, the spreadability decreases with increasing viscosity; isopropyl palmitate (low viscosity) → caproic triglycerides → octyldodecanol (high viscosity).

If the viscosity remains unchanged, alcohols spread more quickly than hydrocarbons, which spread faster than esters. In all cases, it is found that some branched-chain compounds spread better than unbranched ones [44].

Tackiness of a formulation normally is an unwanted sensory characteristic. Lipids are responsible for this sensation, and it occurs when substances having high surface tension and high viscosity are incorporated.

In that respect, silicone oils are favored ingredients for modification of the sensory characteristics of emulsions because they promote spreadability of the incorporated oils due to their low surface tension [36].

IV. FUNCTION OF LIPIDS ON SKIN

A. Behavior of Emulsions after Application to Skin

The dynamics of the structural change of emulsions due to water loss after application to the skin are suspected to have a major impact on product performance. Therefore, this behavior has been the subject of investigation by several groups, and attempts were made to visualize emulsion behavior on skin [45]. Our own results showed that the evaporation of water from O/W emulsions occurred within the first 10 minutes after application and is complete after that time.

A valuable method is the microscopic observation of the structural changes of emulsions applied to glass. Figure 7 shows microscopic pictures (magnification 625-fold) made under phase-contrast illumination [46]. The thickness of the emulsion film applied is 30 µm.

Surprisingly, discrete droplets stay intact for several minutes, and very abruptly (from minute 7 to 8) the surface flattens and the surroundings of the droplets appear smeared. What is left looks like a continuous oil film.

The structural changes of an emulsion can also be studied under polarized light, where the change of birefringence can be measured. By image analysis the brightness of the discrete pictures can be calculated and plotted as a function of time.

Wepf investigated the penetration behavior of the water and lipid phases of emulsions applied to porcine skin with confocal laser microscopy in vitro. These results suggested that the water contained in the emulsion does not penetrate the skin as a whole [47]. For this investigation, the water and lipid phase of the emulsion were colored with different fluorescent dyes (Fig. 8). For the water phase, a green color (Rhodol-green) was chosen, and for the lipidic phase, the red color DID. Confocal laser microscopy was carried out 30 minutes after application of the emulsion. Images of the skin surface and depth profiles of the epidermis were obtained. On the surface, the red and green color are still persis-

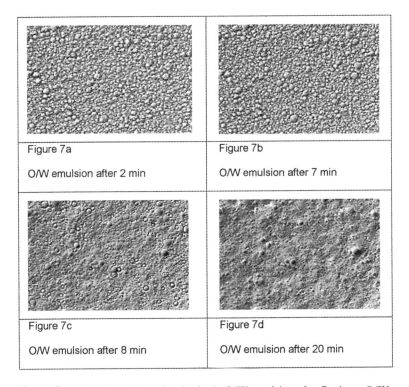

Figure 7 a: O/W emulsion after 2 min. b: O/W emulsion after 7 min. c: O/W emulsion after 8 min. d: O/W emulsion after 20 min.

tent. However, the red color is also seen in the upper layers of the stratum corneum, which suggests that lipids or lipophilic components may penetrate.

The ability of lipids to penetrate the upper layers of the stratum corneum found in this study is in good agreement with results found by Ghadially [48] and Brown [49]. Ghadially examined the fate of topically applied petrolatum, which appears to reach all levels of the stratum corneum in both acetone-treated and untreated stratum corneum. Brown studied the penetration of hydrocarbons into intact skin and found a similar behavior for hydrocarbons.

A good example of an emulsion being able to penetrate the uppermost layers of the SC was provided by Pfeiffer, who could detect residues of a W/O emulsion by Cryo raster electron microscopy in the upper layers of the SC [50].

All these investigations show that an emulsion converts into a kind of a lipid-emulsifier-gel after topical application, which itself accounts for the coverage of the skin, film formation, or occlusivity, as well as for the homogeneous

Lipids in Skin Care Formulations

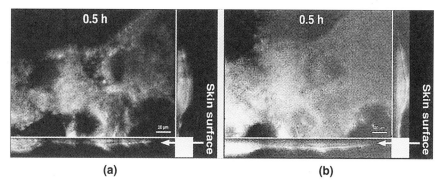

Figure 8 Green channel (a); red channel (b). Confocal fluorescence image of a W/O cream applied to porcine skin in vitro. The water and lipid phases are colored individually with fluorescent colors, green and red respectively. Figure 8a shows the signal of the green channel (representing the water phase), which is restricted to the surface of the skin. Figure 8b shows the signal of the red channel (representing the lipid phase), which clearly is found in deeper layers of the SC than the green signal.

distribution of UV filters or the penetration of active ingredients. Lipids are the major part of that "residue" and influence its characteristics.

B. Film Formation—Occlusivity

Film formation contributes to the efficacy of emollients and lipids in general. A good film on the skin surface (homogeneous and of certain thickness), leading to an optimal occlusivity, should promote endogenous moisturization due to reduction of transepidermal water loss (TEWL) of the skin.

Film formation on skin can only be measured indirectly. Therefore, some of these indirect methods shall be discussed. Mostly, water vapor transmission experiments are conducted in vitro and in vivo. Furthermore, water tightness or SDS challenges of pretreated skin are valuable tools for estimating the protective properties of such films on skin surface.

Wepierre et al. [51] and Choudhury et al. [52] discussed the optimal combination of lipid and emulsifier to obtain films of good quality. For that purpose, they formed mixtures of water, oil, and emulsifier at different HLB values and determined the area of the isotropic oil phase in the phase diagram. They could establish a correlation between the occlusivity of the O/W emulsion and the area of the isotropic phase. Furthermore, they noticed that the occlusivity of an O/W emulsion with a given emulsifier system is directly influenced by the choice of the lipid.

Occlusivity can be estimated in vivo by water permeability measurements through a membrane either gravimetrically or by TEWL measurement. In vivo,

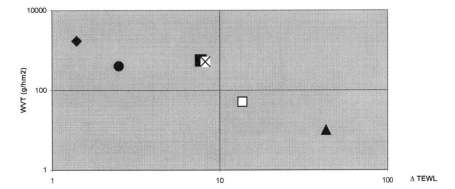

Figure 9 Correlation of in vivo TEWL with in vitro water vapor transmission ◆ = Dibutyl sebacate, ● = Cetyloctanoate, ■ = Caprylic/capric triglyceride, ⊗ = Dimethicone, □ = Mineral oil, ▲ = Petrolatum.

the measurement of occlusivity can be estimated by measuring the reduction of the TEWL after application of an emollient.

Frömder [53] systematically studied the occlusivity of certain lipids in vivo and in vitro and established a correlation between the TEWL reduction in vivo and the water permeation rate in vitro after application of single lipids (Fig. 9).

In general, however, occlusive properties measured with different methods are not directly comparable because they are influenced by the membrane and application techniques chosen [54–59].

Nevertheless, some trends can be seen. In all studies, it is found that hydrocarbons and especially petrolatum are very occlusive components, synthetic branched-chain esters are less effective [59], linear and cyclic silicone oils are not occlusive in a wide range of viscosity and molar weight, and even silicones with high molecular weights do not show substantial occlusivity [55,60].

Only silicones with long-chain alkyl residues show occlusive properties [61]. Beeswax also has good occlusive properties [58].

C. The Role of Skin Barrier Lipids in Skin Condition

The stratum corneum (SC) constitutes the main barrier to the diffusion of substances into the skin. It consists of corneocytes and three intercellular lipid classes, ceramides, sterols, and free fatty acids. Its integrity requires the organization of the lipids into intercellular membranes subsequent to the secretion of epidermal lamellar body contents at the stratum granulosum–SC interface [62].

Recently, Norlén [63] proposed a different barrier lipid composition by including cholesterol esters that constitute a large portion of barrier lipids.

Lipids in Skin Care Formulations

Table 9 Proposed Stratum Corneum Lipid Composition

	Cholesteryl esters	Free fatty acids	Cholesterol	Ceramides
Weight percentage	18	11	24	47
Molar percentage	15	16	32	37

Source: From Ref. 63.

It is interesting that he also discovered that inter-individual differences in relative amounts of free fatty acids, cholesterol, and ceramides are rather high (greater than 100%), but the variance in the ratios of cholesterol/ceramides is only 10.1%.

Several pathological skin conditions with perturbed barrier function have been demonstrated to have abnormal lipid composition, such as X-linked ichthyosis [64] or lamellar ichthyosis [65]. A disrupted lipid organization has also been demonstrated to result in a decreased barrier function in psoriasis and atopic dermatitis [66]. In all these skin conditions, also a paucity of ceramide 1 occurred [67–69]. The deficiency of ceramide 1 was compensated by a surplus of ceramide 5 [70].

Also, aging seems to have an effect on the composition of the SC lipids. Sauermann and Schreiner showed that the sterol content of xerotic skin of elderly people is decreased [71]. In another study, Schreiner found an apparent increase in free fatty acids and a compensatory decrease in ceramides in healthy aged skin [72]. Nevertheless, he could not find any clues on morphological alterations of the intercellular membranes, in contrast to Ghadially [73], who studied volunteers of above 80 years of age.

1. Influences on Skin Condition

The interrelationship of season, individual parameters, and lipid content of the skin on skin condition has been the subject of several investigations. It was found that season is a major factor for the generation of dry skin condition, which itself can also be characterized by dryness score, skin roughness, or increased TEWL [74–77]. According to Yoshikawa, skin adapts to dry environmental conditions by an increase in ceramide content of the barrier. This is not the case in skin sites that typically show symptoms of dryness, such as the calf [74].

In contrast, prolonged exposure of hairless mice to a dry environment (relative humidity [RH] below 10% for 2 weeks) resulted in an increased SC thickness, increased amount of SC lipids, and accelerated barrier recovery compared with mice exposed to a humid environment (RH > 80%). Exposure to low humidity produced histologic evidence of inflammation, epidermal hyperplasia, and other

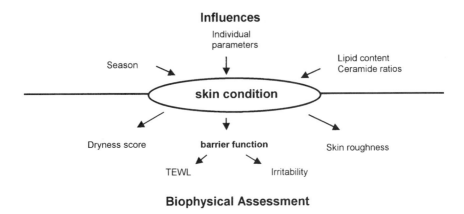

Figure 10 Influences on skin condition with respect to barrier function.

changes in skin. These could be prevented by an occlusive plastic membrane, petrolatum, or glycerol (10%) [78,79].

Low ceramide content can be related to irritability of skin and is influenced by season as well as by atopic disposition. Reduced ceramide content of the SC lipids account for higher irritability of the skin [80,81].

Besides individual differences in irritability of the skin [82] there are inter- and intraindividual differences in skin thickness and TEWL [83]. These findings support the assumption that dry skin and barrier disorders are related.

Not only do lipid contents of the stratum corneum and the irritability of the skin depend on seasonal changes but so does the biophysical appearance of skin (e.g., dryness or roughness) (Figs. 10 and 11) [84,85].

2. Different Roles of Ceramides in Barrier Function of the SC

Ceramides have been shown to play a vital role in maintaining the barrier properties and the functionality of the stratum corneum. However, not much is known about the importance of the ceramide subspecies 1–7 for barrier function.

Ceramide 1 has been shown to be essential for barrier integrity: as humans who lack ceramide 1 have an elevated TEWL and dry skin [75]. The structural impact of ceramide 1 on the stratum corneum lipid lamellar phase formation was discovered by studying the lipid ordering in artificial lipids mixtures by x-ray diffraction measurements [86]. It was detected that ceramide 1 is important for the formation of the long periodicity phase. The long periodicity phase is also formed by the stratum corneum lipids in situ (see also Chapter 2).

By investigating the relationship of SC lipid composition and membrane organization in vivo, Schreiner could show that a lack of the long periodicity

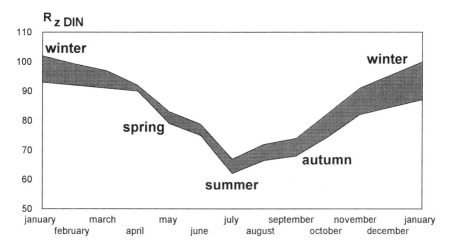

Figure 11 Seasonal changes of skin microtopography as measured by changes of mean R_{zDin} of untreated skin of 120 healthy human volunteers (standard deviation given by width of the curve). (From Ref. 85.)

phase coincides not only with a paucity of ceramide 1 but also with a reduced ceramide 4/ total ceramide ratio and elevated ceramide 2 and ceramide 5 ratios [72]. However, a diminished ceramide 1 content occurred more often in dry skin types than in normal skin. This clearly demonstrates that the changes in barrier function can be related to changes in ceramide ratios, which themselves might be caused by alterations of the ceramide biosynthesis or glycosylceramide processing.

D. Penetration Enhancement by "Normal" Cosmetic Lipids

Because the skin, especially the stratum corneum, is an effective barrier against penetration of substances through the skin, the lipid composition of the SC is thought to be of major importance. The integrity of the lipid barrier often is assessed by the ability of certain substances to penetrate the skin. On the other hand, certain classes of lipids are suspected to be absorbed by the SC lipids and either to be incorporated into the structure or form separate domains, thus altering the barrier properties of the SC or altering the penetration ability of other substances.

For the investigation of the effects of lipids on the stratum corneum lipids and on penetration enhancement, several different approaches are used. Most often, in vitro penetration studies, either with porcine or mouse skin, are conducted. Those skins, although not exactly comparable with human skin, nevertheless are valid in vitro systems for the investigation of the interaction with stratum corneum lipids [87–89].

For in vivo investigations, TEWL is regarded as an indicator of the enhancement capacity of skin permeation enhancers using water vapor as a "model drug"; also it is used to monitor intrinsic skin barrier properties. It is used in in vivo studies in addition to Laser Doppler Flowmetry or visual scoring, which are used for assessing the irritation achieved with a compound.

1. Fatty Acids

It is well known that unsaturated fatty acids can act as effective penetration enhancers for certain substances. The underlying mechanism has been the subject of investigation by several groups. The effect of oleic acid on stratum corneum lipids was investigated by Differentially Scanning Calorimetry (DSC) [90,91] and Fourier Transform Infrared Spectroscopy [92] in studies of the thermal phase behavior of the SC lipids as well as the penetration enhancement of selected drugs.

The mechanism attributed to the penetration-enhancing action is the disturbance of lamellar packing order of the SC lipids by the unsaturated (and therefore bent) alkyl chain of the unsaturated fatty acid, leading to a fluidization of the lipid mixture, which can be seen as a drop in the phase transition temperature. The other possible mechanism is the formation of a separate phase, which was found by Ongpipattanakul in 1991 to account for the penetration-enhancing action of oleic acid [92]. These findings were confirmed using deuterated oleic acid in in vivo experiments, although stratum corneum lipid fluidization could not be ruled out completely [93].

Chain Length, Degree of Unsaturation, and Number of Double Bonds. After oleic acid became known as a penetration enhancer, several fatty acids were investigated in order to elucidate their structure-activity relationship. Therefore, investigations with hydrophilic [94,95] as well as lipophilic model drugs [96–98] in vitro and in vivo were conducted.

Shifting the position of the double bond toward the hydrophilic end obviously enhances the penetration-enhancing activity. Cis-omega-12-octadecenoic acid (petroselinic acid) was found to be a better penetration enhancer than oleic acid (cis-omega-9-octadecenoic acid) [94], and for gondonic acid (cis-11-eicosenoic acid), maximum enhancement of skin permeation of indomethacin was found [95].

On the other hand, Tanojo found monounsaturated *cis*-6-, 9-, 11-, 13-octadecenoic acids to enhance the permeation of p-amino benzoic acids to the same extent. The polyunsaturated fatty acids—linoleic, α-linolenic, and arachidonic acids enhance PABA permeation more strongly than monounsaturated acids, but additional double bonds did not further increase the degree of enhancement [97]. For the saturated fatty acids with 6 to 12 carbons, using PABA as penetrant, he found a parabolic correlation between enhancement effect and chain length, with a maximum at nonanoic and decanoic acids.

Barrier-Perturbing Effects of Fatty Acids. By comparing in vitro penetration-enhancing effects with in vivo TEWL measurements, Tanojo found that barrier disruption (measured as TEWL) does not necessarily correlate with penetration enhancement [98]. From the in vitro studies, the saturated fatty acids with C-chains of 9 to 12 were revealed as excellent permeation enhancers with an equal or even higher capability compared to unsaturated fatty acids. In the in vivo study, however, unsaturated fatty acids increased both TEWL and irritation effects as assessed by measurement of epicutaneous blood flow with laser doppler velocimetry (LDV) to a higher extent than the saturated fatty acids. Thus, the enhancement capability of saturated fatty acids as shown for some compounds might not have a connection with the perturbation of skin lipids in vivo.

The mechanism of action for unsaturated fatty acids has been shown to be exclusively the perturbation of intercellular lipid domains in the skin (SC), because they enhance TEWL but not the permeation of hexyl nicotinate. The lack of correlation between the average TEWL values and the irritation potential (LDV, visual scoring) after the application of unsaturated fatty acids suggests that the degree of irritation does not necessarily correlate with the degree of barrier modulation of fatty acids.

2. Other Compounds

Schmalfuss et al. [99,100] investigated the penetration enhancement of a hydrophilic drug from a microemulsion system. They found that 10-methyl palmitic acid caused a decrease in drug penetration, and cholesterol caused an increase. They did not find the penetration-enhancing action of oleic acid in that system higher than that of the investigated microemulsion alone.

Walters [101] could show different penetration-enhancing capacities of fatty alcohol ethoxylates in hairless mouse skin for methyl nicotinate. Laureth-10 was found to be the most effective enhancer for increasing permeation rate.

Also the subject of investigation were alkanols with chain lengths C2 to C12; a maximum penetration enhancement was found for alkanols with a chain length of C6 (in vitro measurements on hairless mouse skin). Further increase in the chain length of the alkanol led to a decrease in penetration of the nicotinamide despite the higher skin uptake of alkanol. From these investigations it is concluded that alkanols also have effects on the stratum corneum lipids [102].

Leopold and Maibach investigated the action of lipids on the stratum corneum lipids of human skin using DSC measurements and an in vivo method using methyl nicotinate as the model drug, which was dissolved in the investigated lipid. It was found that isopropyl myristate and light mineral oil acted as penetration enhancers but the medium chain triglycerides and dimethicone did not [103]. Leopold and Lippold investigated the mechanism in in vitro studies with human stratum corneum by observation of the change in thermal phase transitions (enthalpies) of the SC lipids after pretreatment with the lipids by DSC measurements. They showed that branched materials such as light mineral oil and isopropyl myristate induced a reduction of the phase transition temperatures and enthalpies [104].

All these studies show that cosmetic lipids can affect SC lipids when applied in the right medium. But penetration-enhancing experiments have to be interpreted with care because many factors besides the choice of penetration enhancer influence the permeation of substances. The solubility of the substance of choice in the medium has to be regarded as well as the release pattern of the "formulation." In the studies cited above, therefore, the investigations are restricted to comparably simple systems such as dispersions in propylene glycol. Cosmetic formulations are more complex systems because the emulsifier and the lipids act differently on the solution of a "permeant," so that emulsions often have "bad" release characteristics.

E. Efficacy of Lipids Applied to Skin

For cosmetic purposes, lipids normally are not applied as pure compounds on the skin but in the form of an emulsion, which is a far more convenient way of application. The required cosmetic or dermatological results are (1) improvement of skin barrier properties, (2) skin protection, and (3) skin care expressed as reduction of skin roughness or scaliness and improvement of skin moisturization.

Because emulsions are complex systems, the efficacy of a product is not solely the result of the lipid, but of the whole system as such. Comparison of the efficacy of different lipids, therefore, often is difficult, because efficacy of the emulsion itself can be high and the additional effect of lipids comparably small. In addition, the test strategies applied for efficacy measurements vary considerably. Only in the area of skin roughness and moisturization are test methods somewhat standardized. For the assessment of barrier integrity and skin protection, a huge variety of test strategies exist, which makes an overall comparison rather difficult. Here, some examples shall be discussed.

1. Recovery of Skin Barrier Function

The most obvious possibility for recovering the skin barrier function is by the replacement of the intercellular lipids. The skin of hairless mice damaged with acetone has provided a useful in vivo model to study the restoration of barrier function and has been thoroughly investigated by the group of Elias [105–107].

These studies have demonstrated that after extraction of skin lipids with acetone, the barrier function can be partially restored by the application of a mixture of skin-type lipids. TEWL is used as an indicator of barrier integrity and barrier recovery is measured as percentage recovery as a function of time compared to untreated. It was found that for optimal barrier recovery, the lipid mixtures consisted of free cholesterol, ceramide (3 and 4), and fatty acids. The effect was optimized at a cholesterol: ceramide: palmitic acid: linoleic acid molar ratio of 4:2:1:1 [105].

Inspired by these results, the application of ceramides or pseudoceramides alone or as a component of an emulsion has been in the focus for skin care as well. Imokawa investigated the improvement of water-retention properties of the skin [108]. He induced dryness by using a mixture of acetone/ether and measured the improvement in skin condition following the application of pseudoceramides and ceramides. Pseudoceramides are less effective in repairing the barrier than the natural ceramides. However, these synthetic lipids were superior to the untreated control.

Möller et al. used an in vitro model system using porcine skin and measured the TEWL to rank pseudoceramides; they found alkyl-succinic acid derivatives to be most effective [109]. Lintner studied the effect of a synthetic ceramide (N-stearoyldihydrosphingosine) after barrier perturbation with SDS in vivo [110]. Incorporated in concentrations of 0.5 and 1% in an emulsion, the synthetic ceramide effectively reduced TEWL, whereas the emulsion itself showed no effect.

Despite the efficacy shown in the above-mentioned studies, ceramides and pseudoceramides until today were not deployed in cosmetic formulations. This may be explained in part by their poor solubility in common cosmetic ingredients and their high melting point and the resultant difficulties in formulating stable emulsions containing ceramides and pseudoceramides.

The restoration of the epicutaneous barrier is of major importance in pathological skin conditions such as atopic dermatitis. Atopic skin, even if non-lesional, shows elevated TEWL, signs of skin dryness, and a deficiency in barrier lipids [67]. The efficacy of emulsions containing evening primrose oil on the barrier restoration in atopic skin were demonstrated in several studies [111–113]. Jánossy demonstrated that the application of an evening primrose oil–containing cream on atopic dry skin for 3 weeks significantly reduced the TEWL on the treated skin sites [111]. Maas-Irslinger additionally found a stronger reduction of skin roughness for an evening primrose oil–containing cream compared to placebo [112]. Gehring et al. discovered that the effect on skin barrier function might even depend on the vehicle applied. The TEWL-reducing effect of evening primrose oil was significantly stronger if it was applied in an W/O emulsion instead of an O/W emulsion [113]. Therefore, they concluded that the vehicle is of crucial importance for the efficacy of the active ingredient.

In dry skin conditions induced by prolonged exposure towards water or water-soluble irritants (i.e., detergents), the epidermal barrier is perturbed, which can result in severe irritation and skin problems. Therefore, the skin irritation model with SDS often is used to investigate effects on barrier recovery after irritation.

Using such a test design, Lodén showed the effectiveness of canola oil and the sterol-enriched fraction of canola oil in skin barrier recovery [114]. Phytosteroles are thought to exert beneficial effects to the skin barrier because of their structural relationship to cholesterol [115,116].

2. Skin Protection

Due to their hydrophobicity, lipids also act as a barrier towards water and water-soluble irritants on the skin. For maximum protection, the lipids should completely cover the skin site and form a homogeneous and durable film. Unfortunately, they do not form a barrier against lipophilic ingredients and often favor their solubility and the persistence of those substances on the skin.

In order to measure the efficacy of lipids or barrier creams against the irritating action of water soluble-substances, three main test strategies are applied:

Single application and single irritation (occlusive)
Repeated applications and repeated irritation (occlusive)
Application and open irritation, single or repeated ("wasching test")

Sodium dodecyl sulfate (SDS), sodium hydroxide, and lactic acid are recommended as model irritants. However, SDS is the irritant that is most frequently used.

Petrolatum is thoroughly tested in many different approaches, and because of its good performance it is known as the prototype skin protectant [117]. It also has been tested in in vivo skin irritation models [118] as well as in in vitro investigations [119]. Petrolatum obviously combines all the properties of a good barrier lipid. It is nonpolar and semisolid and although it is known to penetrate the SC, it forms a separate nonlamellar phase in the corneocyte interstices [73]. Accordingly, petrolatum exerts an immediate protective action on the skin surface and a prolonged effect due to its incorporation in the SC.

Other lipidic ingredients also have been shown to be beneficial in terms of skin protection. These are waxes such as beeswax or paraffinic waxes [120] as well as silicones [60], perfluorhydrocarbons [121], paraffin, or natural oils [122].

3. Skin Care

The most important cosmetic property of lipids is their skin care efficacy. Skin care comprises moisturization and smoothing of the skin as well as a visual reduction of the signs of dryness. The term emolliency often is used in order to express the efficacy of lipids to influence the skin's softness. The improvement of the

Lipids in Skin Care Formulations

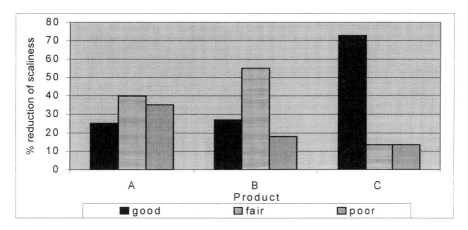

Figure 12 Reduction of scaliness after application of the lipstick for 14 days.

skin condition is also connected with the ability of the lipid to form an occlusive film on the skin surface, where the lipids reduce TEWL and therefore account for a better moisturization of the skin [123]. Petrolatum or lanolin oil are examples of occlusive lipids (see also previous chapters). Additionally, not only the kind but also the amount of lipid incorporated in an emulsion accounts for the occluding properties of the final product [123]. Furthermore, the ability of lipids to form liquid crystalline lamellar phases is thought to be an important factor for skin care efficacy [115,124].

The use of lanolin oil in lipstick preparations has been shown to have beneficial effects on the scaling of lips [23]. Three lipstick formulations were compared (Fig. 12).

Table 10 Composition of Lipsticks

Formula code	A	B	C
Paraffin	30%	30%	30%
Petrolatum	30%	20%	30%
Liquid petrolatum (350–360 cps)	30%	30%	20%
Polyethyleneglycol 400 distearate	10%	10%	10%
Liquid lanolin	—	—	10%
Wool fat, anhydrous	—	10%	—

Source: Ref. 23.

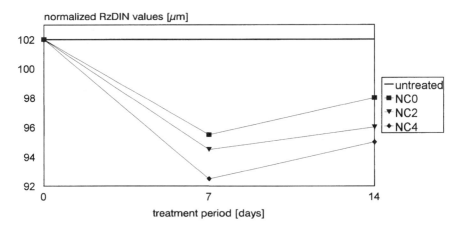

Figure 13 Decrease of skin roughness measured microtopographically after application of a W/O emulsion containing 0, 2, and 4% of Eucerit (containing lanosterol and cholesterol). n = 24 healthy volunteers; area tested is volar forearms; application twice daily; treatment was stopped about 14 hr before measurement. (From Ref. 71.)

By comparing the dryness of the lips before and after treatment and grouping the results in three classes, Silverman found the formula containing liquid lanolin to be the most effective in reduction of scaliness.

Sterols are one of the three lipid classes in human stratum corneum. Sauermann and Schreiner published the effect of sterols on skin roughness when applied in an emulsion system [71] (Fig. 13). They found a dose-response curve for the effect of sterols on skin roughness reduction. This finding is in agreement with the good skin care properties proposed for sterols and lanolin derivatives.

Fatty acids are the second class of barrier lipids. Consequently, topical application of saturated fatty acids should also exert skin care benefits. An augmentation in the content of cholesterol, free fatty acids, and tryglycerides in the SC could be detected after application of an emulsion containing the same. The enrichment of the lipids correlates with increased skin moisturization [72] (Fig. 14).

F. Discussion

The actions lipids can exhibit on skin are very diverse. They render a characteristic skin feel, give a certain occlusivity to the product, and can penetrate into the SC interstices to a certain extent.

The activity of emulsions is even more complex as emulsifiers and water are combined with lipids. Furthermore, lipid and water-soluble active ingredients

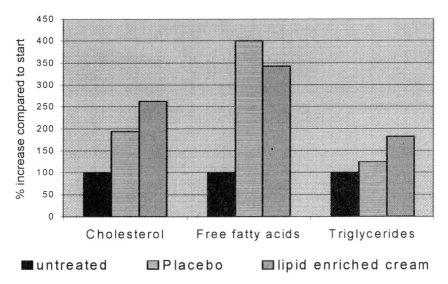

Figure 14 Significant incorporation of skin-related lipids in the stratum corneum after 14 days of application of a lipid-enriched cream. (From Ref. 125.)

can be incorporated. Using such complex systems, the investigation of effects caused by a single ingredient is very difficult. Measurement of the in vivo effects such as skin moisturization or reduction of skin roughness are valid parameters for the assessment of skin care products but do not give any information about the mechanism underlying these effects.

In that respect, a quote from Peter Flesch is remarkable; he published an article on the hydration of the skin surface by fats and emulsions and voiced some thoughts about the effects of lipids and emulsions on the skin which in general are still valid today [126]:

> The question was raised in the beginning concerning the respective merits of fats and oils versus emulsions in promoting hydration of the skin surface. In the absence of adequate tests and satisfactory quantitative data, no final statement can be made. The following represents a personal educated guess rather than definite recommendations. Oils and fats can have only a single effect on hydration. This could be occlusion of the surface or added hygroscopicity. Much too little is known about absorption to be considered here.
>
> On the other hand, emulsions having several types of components may have a variety of effects: they can occlude the surface, can add water directly to the horny layer and may increase its hygroscopicity if they contain some suitable hygroscopic compound.

Some of these questions can be answered today, such as occlusive effects or film forming characteristics. However, new questions arose and certainly will arise. In the future, there will be systems available to investigate effects of lipids on the structural arrangements of SC lipids selectively, and as a consequence they will enable the cosmetic chemist to specifically choose substances that further enhance barrier function of the skin.

On the other hand, knowledge of the interaction of different ingredients in emulsions applied to skin will rise, thus making it possible to choose an even better combination of ingredients for maximum activity.

Also, in terms of subjectively perceived skin care, progress will be made, because the consumer acceptance still is and will continue to be the yardstick for the quality of a skin care product. Consumer choice is influenced mainly by sensorial factors such as the smell of the product and the feeling it leaves on the skin. Furthermore, the subjectively experienced care effect of a product is not necessarily identical with the objectively measured skin care efficiency of the product. Because lipids are the main carriers for sensory effects (see also Chapter 11), the combination of lipids that have proven skin care benefits with those transporting the sensory effect will garner more interest.

I think in the future we can expect many developments in that direction.

ACKNOWLEDGMENTS

I would like to express my thanks to my colleagues Volker Schreiner, Andreas Bleckmann, Roger Wepf, and Rainer Kröpke, who supported this work with their knowledge and many helpful suggestions. I am also very thankful to Ralph Schimpf, Volker Skibbe, and especially Günther Schneider, who judiciously read this chapter and with their comments helped to make it complete. Last but not least, I am grateful to Thomas Förster, who gave me the opportunity to publish what I wanted to write about lipids for quite a long time.

REFERENCES

1. Reichert B, Frederichs G, Arends G, Zörnig H. Hargers Handbuch der Pharmazeutischen Praxis: 1. Ergänzungsband, 2. Auflage, Berlin: Springer Verlag, 1949:403.
2. Braun O. DRP N° 22516:1882.
3. Walker GT. Use of lecithin in cosmetics. American Perfume and Cosmetics 1967; 82–10:73–76.
4. Small DM. Handbook of Lipid Research, Vol. 4, The Physical Chemistry of lipids. London: Plenum Press, 1986.
5. Israelachvili JN. Intermolecular and Surface Forces. London: Academic Press, 1991:366–394.

6. Lipowski R, Sackmann E. Structure and dynamics of membranes. Amsterdam: Elsevier, 1995.
7. Seddan JM, Templer RH. Polymorphism of lipid-water systems. In Lipowski R, Sackmann, E, eds. Structure and dynamics of membranes. Amsterdam: Elsevier, 1995:97–160.
8. Thiele OW. Lipide, Isoproenoide mit Steroiden. Stuttgart: Georg Thieme Verlag, 1979:2.
9. Small DM. A classification of biologic lipids based upon their interaction in aqueous systems. Journal of the American Oil Chemists Society 1968, 45:108–119.
10. Häusler C, Führer F. Zur Kristallographie und Kolloidstruktur von Vaseline. Parfümerie und Kosmetik 1992; 73:600–616.
11. Downing DT, Kranz ZH, Lamberton JA, Murray KE, Redcliffe AH. Studies in Waxes XVIII, Beeswax: A spectroscopic and chromatographic examination. Australian Journal of Chemistry 1961; 14:253–263.
12. Peters Rit AW, Puleo SL, Sprenger J. A new beeswax derivative for cosmetic formulations. Cosmetics and Toiletries 1990; 105-8:53–62
13. Barnett G. Lanolin and derivatives. Cosmetics and Toiletries 1986; 101-3:23–44.
14. Motiuk K. Wool wax alcohols: a review. Journal of the American Oil Chemist's Society 1979; 56:651–658.
15. Motiuk K. Wool wax acids: a review. Journal of the American Oil Chemist's Society 1979 b; 56:91–97.
16. Idson B. What is a moisturizer? American Perfumer and Cosmetics 1972; 87-8:33–35.
17. Nachbar F, Korting HC, Plewig G. Zur Bedeutung des positiven Epicutantests auf Lanolin. Dermatosen 1993; 41:227–236.
18. Clark EW, Blondeel A, Cronin E, Oleffe JA, Wilkinson DE. Lanolin of reduced sensitizing potential. Contact Dermatitis 1975; 7:80–83.
19. Pape W, Teichert T. Produktinformationen zur Sicherheit und Verträglichkeit von Eucerinum® anhydricum. Beiersdorf AG: 1999.
20. Kligman AM. Lanolin allergy: crisis or comedy. Contact Dermatitis 1993; 9:99–107.
21. Kligman AM. The myth of lanolin allergy. Contact Dermatitis 1998; 39:103–107.
22. Conrad LI, Maso HF, DeRagon SA. Influence of lanolin derivatives on dispersed systems II. The dispersion of pigments in aqueous media. American Perfumer and Cosmetics 1967; 3:31–35.
23. Silverman HI. Lip barrier investigation, formulation and evaluation. American Perfumer and Cosmetics 1967; 2:29–32.
24. Vlahov G, Angelo CS. The structure of triglycerides of monovarietal olive oils: a 13C-NMR comparative study. Fett/Lipid 1996; 98:203–205.
25. Hedeman H, Bröndsted H, Müllertz A, Frokjaer S. Fat emulsions based on structured lipids (1,3 specific triglycerides): an investigation of the in vivo fate. Pharmaceutical Research 1996; 13:725–728.
26. Merkle RD. Omega-6 Oils—Significance in Cosmetics. Seifen Öle Fette Wachse 1992; 118:991–999.
27. Schäfer L, Kragballe K. Abnormalities in epidermal lipid metabolism in patients with atopic dermatitis. Journal of Investigative Dermatology 1991; 96:10–15.

28. Wertz PW, Schwarzendruber DC, Abraham W, Madison KC, Downing DT. Essential fatty acids and epidermal integrity. Archives of Dermatology 1987; 123:1381–1384.
29. Hartop PJ, Prottey C. Changes in transepidermal water loss and the composition of epidermal lecithin after applications of pure fatty acid triglycerides to the skin of essential fatty acid-deficient rats. British Journal of Dermatology 1976; 95:255–264.
30. Skolnik P, Eaglstein WH, Ziboh VA. Human essential fatty acid deficiency—treatment by topical application of linoleic acid. Archives of Dermatology 1977; 113:939–941.
31. Bat S, Tannert U. Die Aufarbeitung und Charakterisierung von Sanddornölen. Seifen Öle Fette Wachse 1998; 124:19–20.
32. Rasig M, Seith B, Schweiger P. Extraction of the oils of hemp seed and analysis of the content of fatty acids in hemp oil. Seifen Öle Fette Wachse 1999; 125:16–18.
33. Isbell TA. Development of meadowfoam as an industrial crop through novel fatty acid derivatives. Lipid Technology 1997; 11:140–144.
34. Fishman HM. Silicone derivatives in cosmetics. Happi 1992; 2:32, 167, 201.
35. Devant RM. Chemische Totalsynthese von Sphingosin, dem zentralen Baustein der Sphingolipide. Kontakte 1992; 3:11–28.
36. Dietz T. Basic properties of cosmetic oils and their relevance to emulsion preparations. Seifen Öle Fette Wachse 1999; 125:2–9.
37. Tadros ThF. Fundamental principles of emulsion rheology and their applications. Colloids and Surfaces A 1994; 91:39–55.
38. Ansmann A, Kawa R. Cosmetic water-in-oil emulsions—how to formulate elegant skin care products. Seifen Öle Fette Wachse 1991; 117:369–371.
39. Jenni K, Hameyer P. Was Emulsionen wirklich dick macht. Parfümerie und Kosmetik 1998; 79:22–28.
40. Junginger HE. Kristalline Gelstrukturen in Cremes. Deutsche Apotheker Zeitung 1991; 131:1933–1941.
41. Goldemberg RL, de la Rosa CP. Correlation of skin feel of emollients to their chemical structure. Journal of the Society of Cosmetic Chemists 1971; 22:635–654.
42. Ansmann A, Kawa R, Pratt E, Wadle A. Modern cosmetic emulsions—technology and sensory assessment. Seifen Öle Fette Wachse 1994; 120:158–161.
43. Zeidler U. Über das Spreiten von Ölen auf der Haut. Fette-Seifen-Anstrichmitel, 1985; 87:403–408.
44. Zeidler U. Über die taktilen Eigenschften kosmetischer Öle. Seifen Öle Fette Wachse 1992; 118:1001–1007.
45. Friberg SE, Langlois B. Evaporation from emulsions. Journal of Dispersion Science and Technology 1992; 13:223–243.
46. Jung L. Diploma thesis, 1997.
47. Wepf R. Personal communication, 1999.
48. Ghadially R, Halkier-Sorensen L, Elias PM. Effects of petrolatum on stratum corneum structure and function. Journal of the American Academy of Dermatology 1992; 26:387–396.

49. Brown BE, Diembeck W, Hoppe U, Elias PM. Fate of topical hydrocarbons in the skin. Journal of the Society of Cosmetic Chemists 1995; 46:1–9.
50. Wenck H, Hintze U, Pfeiffer S. The impact of product assessment techniques on skin care product technology. Skin Care Forum 1998; 98–4:6–8.
51. Wepierre J, Adrangui M. Factors in the occlusivity of aqueous emulsions. Journal of the Society of Cosmetic Chemists 1982; 33:157–167.
52. Choudhury TH, Marty JP, Orecchioni AM, Seiller M, Wepierre J. Factors in the occlusivity of aqueous emulsions. Influence of humectants. Journal of the Society of Cosmetic Chemists 1985; 36:255–269.
53. Frömder A, Lippold BC. Water vapour transmission and occlusivity in vivo of lipophilic excipients used in ointments. International Journal of Cosmetic Science 1993; 15:113–120.
54. Filho R. Occlusive power evaluation of O/W/O multiple emulsions on gelatin support cells. International Journal of Cosmetic Science 1997; 19:65–73.
55. Van Reeth I, Wilson A. Understanding factors which influence permeability of silicones and their derivatives. Cosmetics and Toiletries 1994; 109–7:87–92.
56. Tiffany JM. Lipid films in water conservation of biological systems. Cell Biochemistry and Function 1995; 13:177–180.
57. Martin-Polo N, Voilley A, Blond G, Colas B, Mesnier M, Floquet N. Hydrophobic films and their efficiency against moisture transfer. 2. Influence of the physical state. Journal of Agricultural and Food Science 1992; 40:413–418.
58. Kester JJ, Fennema O. Resistance of films to water vapor transmission. Journal of the American Oil Chemists' Society 1989; 66:1139–1146.
59. Tsutsumi H, Utsugi T, Hayashi S. Study on the occlusivity of oil films. Journal of the Society of Cosmetic Chemists 1979; 30:345–356.
60. Wilson AA, Rentsch SF. Silicones as skin protectant ingredients. Cosmetics and Toiletries—manufacture 1992:162–165.
61. Glover D. Alkylmethylsiloxanes. SOAP/Cosmetics/Chemical Specialities 1997; 11:54–57.
62. Elias PM, Menon GK. Structural and lipid biochemical correlates of the epidermal permeability barrier. Advances in Lipid Research 1991; 24:1–26.
63. Norlén L, Nicander I, Lundh Rozell B, Ollmar S, Forslind B. Inter- and intraindividual differences in human stratum corneum lipid content related to physical parameters of skin barrier function in vivo. Journal of Investigative Dermatology 1999; 112:72–77.
64. Hamanaka S, Ujihara M, Serizawa S, Nakazawa S, Otsuka F. A case of recessive X-linked ichtyosis: scale-specific abnormalities of lipid composition may explain the pathogenesis of the skin manifestation. Journal of Dermatology 1997; 24:156–160
65. Lavrijsen APM, Bouwstra JA, Gooris GS, Weerheim A, Boddé HE, Ponec M. Reduced skin barrier function parallels abnormal stratum corneum lipid organization in patients with lamellar ichthyosis. Journal of Investigative Dermatology 1995; 105:619–624.
66. Fartasch M. Epidermal disorders of the skin. Microscopy Research and Technology 1997; 38:361–372.
67. Imokawa G, Abe A, Jin K, Kawashima M, Hidano A. Decreased level of ceramides

in stratum corneum of atopic dermatitis: an etiologic factor in atopic dry skin? Journal of Investigative Dermatology 1991; 96:523–526.
68. Motta S, Mionti M, Sesana S, Caputo R, Carelli S, Ghidoni R. Ceramide composition of the psoriatic scale. Biochimica et Biophysica Acta 1993; 1182:47–151.
69. Di Nardo A, Wertz P, Giannetti A, Seidenari S. Ceramide and cholesterol composition of the skin of patients with atopic dermatitis. Acta Dermato-Venerologica 1998; 78:27–30.
70. Bleck O, Abeck D, Ring J, Hoppe U, Wolber R, Brandt O, Schreiner V. Two ceramide subfractions detectable in Cer(AS) position by HPTLC in skin surface lipids of non-lesional skin of atopic eczema. Journal of Investigative Dermatology 1999; 113:894–900.
71. Sauermann G, Schreiner V. The skin caring effect of topical products containing lanolin alcohol. In: Hoppe U, ed. The Lanolin Book. Beiersdorf AG, Hamburg, 1999; 217–235.
72. Schreiner V, Gooris GS, Pfeiffer S, Lanzendörfer G, Wenck H, Diembeck W, Proksch E, Bouwstra J. Barrier characteristics of different human skin types investigated with X-ray diffraction, lipid analysis, and electron microscopy imaging. Journal of Investigative Dermatology 2000; 114:654–660.
73. Ghadially R, Brown BE, Sequeira-Martin SM, Feingold KR, Elias PM. The aged epidermal permeability barrier: structural, functional, and lipid biochemical abnormalities in humans and a senescent murine model. Journal of Clinical Invesigatoion 1995; 95:2281–2290.
74. Yoshikawa N, Imokawa G, Akimoto K, Jin K, Higaki Y, Kawashima M. Regional analysis of ceramides within the stratum corneum in relation to seasonal changes. Dermatology 1994; 188:207–214.
75. Conti A, Rogers J, Verdejo P, Harding CR, Rawlings AV. Seasonal influences on stratum corneum ceramide 1 fatty acids and the influence of topical essential fatty acids. International Journal of Cosmetic Science 1996; 18:1–12.
76. Basketter DA, Griffiths HA, Wang XM, Wilhelm KP, McFadden J. Individual ethnic and seasonal variability in irritant susceptibility of skin: the implications for a predictive human patch test. Contact Dermatitis 1996; 35:208–213.
77. Agner T, Serup J. Seasonal variations of skin resistance to irritants. British Journal of Dermatology 1989; 121:323–328.
78. Denda M, Sato J, Tsuchiya T, Koyama J, Kuramoto M, Elias PM, Feingold KR. Exposure to a dry environment enhances epidermal permeability barrier function. Journal of Investigative Dermatology 1998; 111:858–863.
79. Denda M, Sato J, Tsuchiya T, Koyama J, Elias PM, Feingold KR. Low humidity stimulates epidermal DNA synthesis and amplifies the hyperproliferative response to barrier disruption: implication for seasonal exacerbations of inflammatory dermatoses. Journal of Investigative Dermatology 1998; 111:873–878.
80. Di Nardo A, Sugino K, Wertz P, Ademola J, Maibach HI. Sodium lauryl sulfate (SDS) induced irritant contact dermatitis: a correlation study between ceramides and in vivo parameters of irritation. Contact Dermatitis 1996; 35:86–91.
81. Di Nardo A, Sugino K, Ademola J, Wertz P, Maibach HI. Role of ceramides in proclivity to toluene- and xylene-induced skin irritation in man. Dermatosen 1996; 44:119–125.

82. Lavrijsen APM, Geelen FAMJ, Oestmann E, Hermans J, Boddé HE, Ponec M. Comparision of human back versus arm skin region for its suitability to test weak irritants. Skin Research and Technology 1996; 2:70–77.
83. Berardesca E, Distante F. The modulation of skin irritation. Contact Dermatitis 1994; 31:281–287.
84. Cooper MD, Jardine H, Ferguson J. Seasonal Influence on the occurrence of dry flaking facial skin. In: Marks R, Plewig G, eds. The Environmental Threat to the Skin. Dunitz 1992:159–164.
85. Schreiner V, Sauermann G, Hoppe U. Characterization of the skin surface by ISO-parameters for microtopography. In: Wilhelm KP, Berardesca E, Maibach HI, eds. Bioengineering and the Skin: Skin Surface Imaging and Analysis. Boca Raton, FL: CRC Press, 1997:129–143.
86. Bouwstra JA, Gooris GS, Dubbelaar FER, Werheim AM, Ijzerman AP, Ponec M. Role of ceramide 1 in the molecular organization of the stratum corneum lipids. Journal of Lipid Research 1998; 39:186–196.
87. Bouwstra JA. The skin barrier, a well-organized membrane. Colloids and Surfaces 1997; 123–124:403–413.
88. Bouwstra JA, Gooris GS, van der Spek JA, Lavrijsen S, Bras W. The lipid and protein structure of mouse stratum corneum: a wide and small angle diffraction study. Biochimica et Biophysica Acta 1994; 1212:183–192.
89. Bouwstra JA, Gooris GS, Bras W, Downing DT. Lipid organisation in pig stratum corneum. Journal of Lipid Research 1995; 36:685–695.
90. Francoeur ML, Golden GM, Potts RO. Oleic acid: its effects on stratum corneum in relation to (trans) dermal drug delivery. Pharmaceutical Research 1990; 7:621–627.
91. Lin S-Y, Duan K-J, Lin T-C. Microscopic FT-IR/DSC combined system used to investigate the thermotropic behavior of lipid in porcine stratum corneum after pretreatment with skin penetration enhancers. Skin Research and Technology 1996; 2:186–191.
92. Ongpipattanakul B, Burnette RR, Potts RO, Francoeur ML. Evidence that Oleic Acid Exists in a Separate phase within stratum corneum lipids. Pharmaceutical Research 1991; 8:350–354.
93. Naik A, Pechtold LARM, Potts RO, Guy RH. Mechanism of oleic acid-induced skin penetration enhancement in vivo in humans. Journal of Controlled Release 1995; 37:299–306.
94. Takeuci Y, Yamaoka Y, Fukusjima S, Miyawaki K, Taguchi K, Yasukawa H, Kishimoto S, Suzuki M. Skin penetration enhancing action of cis-unsaturated fatty acids with omega-9, and omega-12-chain lengths. Biological and Pharmaceutical Bulletin 1998; 21:484–491.
95. Taguchi K, Fukusima S, Yamoka Y, Takeuchi Y, Suzuki M. Enhancement of propylene glycol distribution in the skin by high purity cis-unsaturated fatty acids with different alkyl chain lengths having different double bond position. Biological and Pharmaceutical Bulletin 1999; 22:407–411.
96. Morimoto K, Tojima H, Haruta T, Suzuki M, Kakemi M. Enhancing effects of unsaturated fatty acids with various structures on the permeation of indomethacin through rat skin. Journal of Pharmacy and Pharmacology 1996; 48:1133–1137.

97. Tanojo H, Boelsma E, Junginger HE, Ponec M, Boddé HE. In vivo human skin barrier modulation by topical application of fatty acids. Skin Pharmacology Applied Skin Physiology 1998; 11:87–97.
98. Tanojo H, Bouwstra JA, Junginger HE, Boddé HE. In vitro human skin barrier modulation by fatty acids: skin permeation and thermal analysis studies. Pharmaceutical Research 1997; 14:42–49.
99. Schmalfuss U, Neubert R, Wohlrab W. Modification of drug penetration into human skin using microemulsions. Journal of Controlled Release 1997; 46:279–285.
100. Schmalfuss U, Neubert R, Wohlrab W. The influence of a fatty acid, cholesterol and urea on the penetration of a hydrophilic drug from microemulsion systems. Perspectives in Percutaneous Penetration 1997; 5a:109.
101. Walters KA, Walker M, Olejnik O. Non-ionic surfactant effects on hairless mouse skin permeability characteristics. Journal of Pharmacy and Pharmacology 1988; 40:525–529.
102. Kai T, Mak VHW, Potts RO, Guy RH. Mechanism of percutaneous penetration enhancement: effect of n-alkanols on the permeability barrier of hairless mouse skin. Journal of Controlled Release 1990; 12:103–112.
103. Leopold CS, Maibach HI. Effect of lipophilic vehicles on in vivo skin penetration of methyl nicotinate in different races. International Journal of Pharmaceutics 1996; 139:161–167.
104. Leopold CS, Lippold BC. An attempt to clarify the mechanism of the penetration enhancing effects of lipophilic vehicles with differential scanning calorimetry (DSC). Journal of Pharmacy and Pharmacology 1995; 47:276–281.
105. Yang L, Mao-Quiang M, Taljebini M, Elias PM, Feingold KR. Topical stratum corneum lipids accelerate barrier repair after tape stripping, solvent treatment and some but not all types of detergent treatment. British Journal of Dermatology 1995; 133:679–685.
106. Mao-Quiang M, Brown BE, Wu-Pong S, Feingold KR. Exogenous nonphysiologic vs physiologic lipids—divergent mechanisms for correction of permeability barrier dysfunction. Archives of Dermatology 1995; 131:809–816.
107. Mao-Quiang M, Feingold KR, Thornfeldt CR, Elias PM. Optimization of Physiological Lipid Mixtures for Barrier Repair. Journal of Investigative Dermatology 1996; 106:1096–1101.
108. Imokawa G, Akasaki S, Kawamata A, Yano S, Takaishi N. Water-retaining function in the stratum corneum and its recovery properties by synthetic pseudoceramides. Journal of the Society of Cosmetic Chemists 1989; 40:273–285.
109. Möller H, Knörr W, Weuthen M, Guckenbiehl B, Wachter R. Neue Pseudoceramide durch Verknüpfung von speziellen Fettstoffen mit Kohlenhydratderivaten. Fett/Lipid 1997; 99:120–129.
110. Lintner K, Mondon P, Girard F, Gibaud C. The effect of a synthetic ceramide-2 on transepidermal water loss after stripping or sodium laurylsulfate treatment: an in vivo study. International Journal of Cosmetic Science 1997; 19:15–25.
111. Jánossy IM, Raguz JM, Rippke F, Schwanitz HJ. Effekte einer 12,5 %igen Nachtkerzensamenöl-Creme auf hautphysiologische Parameter bei atopischern Diathese. H+G 1995; 70:498–502.

112. Maas-Irslinger R. Fettstoffwechselstörung der Haut. Der deutsche Dermatologe 1996; 44:2–6.
113. Gehring W, Bopp R, Rippke F, Gloor M. Effect of topically applied evening primrose oil on epidermal barrier function in atopic dermatitis as a function of vehicle. Arzneimittel-Forschung/Drug-Research 1999; 49:635–642.
114. Lodén M, Andersson AC. Effect of topically applied lipids on surfactant-irritated skin. British Journal of Dermatology 1996; 134:215–220.
115. Wachter R, Salka B, Magnet A. Phytosterole-pflanzliche Wirkstoffe in der Kosmetik. Parfümerie und Kosmetik 1994; 75:755–761.
116. Chlebarov S. Die kosmetischen Eigenschaften der Phytosterole. TW Dermatologie 1990; 20:228–237.
117. Morrison DS. Petrolatum: a useful classic. Cosmetics and Toiletries 1996; 111-1: 59–69.
118. Wigger-Alberti W, Elsner P. Petrolatum prevents irritation in a human cumulative exposure model in vivo. Dermatology 1997; 194:247–250.
119. Treffel P, Gabard B, Juch R. Evaluation of barrier creams: an in vitro technique on human skin. Acta Dermato-Venerologica 1994; 74:7–11.
120. Alexander P. Putting up the Barriers. Manufacturing Chemist 1992; 2:16–19.
121. Elsner P, Wigger-Alberti W, Pantini G. Perfluorpolyethers in the prevention of irritant contact dermatitis. Dermatology 1998; 197:141–145.
122. Blanken R, van Vilsteren MJT, Tupker RA, Coenraads PJ. Effect of mineral oil and linoleic-acid-containing emulsions on the skin vapour loss of sodium-laurylsulphate-induced irritant skin reactions. Contact Dermatitis 1989; 20:93–97.
123. Lodén M. The increase in skin hydration after application of emollients with different amounts of lipids. Acta Dermato-Venerologica 1992; 72:327–330.
124. Critchley P. European Patent Application EP 556957: 1993.
125. Schreiner V, Gatermann C. Produktinformation Eucerin® Trockene Haut. Beiersdorf AG 1999.
126. Flesh P. Hydration of the skin by fats and emulsions. American Perfumer and Cosmetics 1962; 77-10:77–80.

10
Benefits of Lipidic Refatters in Surfactant Products

Ulrich Issberner
Cognis Deutschland GmbH, Düsseldorf, Germany

I. INTRODUCTION

Soap and synthetic surfactants are commonly used in personal care products to increase the cleansing performance and to create a pleasant foam. They act on the skin surface by dispersing, emulsifying, and removing all types of surface soil. This includes organic material from detached scales and intercellular material (lipids, fatty acids, amines), body secretions, sebum residues, micro-organisms, and a variety of externally applied substances, such as makeup and medical ointments.

Soap, in use in ancient civilizations, helped to maintain the health of the Phoenicians and Romans, and is still an item of daily use. Synthetic surfactants, which became commercially available after World War II, enabled new types of body cleansing products to be created.

However, surfactants can also damage the skin by removing skin lipids and causing corneocytes to swell. Therefore, along with other environmental influences, everyday washing with soap or surfactants can impair the skin barrier, resulting in an increase in transepidermal water loss, roughness, and scaling of the skin.

Mild surfactants with cleansing properties similar to those of aggressive surfactants do not remove lipids to the same extent, and they irritate skin only when used at high concentrations under artificial conditions. The admixture of lipids and lipid compounds to surfactant-based body cleansing products can counteract the negative side effects. Although skin care ingredients have been a feature of personal cleansing products for some time, selected newer formulations are making more substantial claims, which can be proved in controlled consumer studies.

II. SURFACTANTS USED IN CLEANSING PRODUCTS

Surfactants are characterized by a hydrophilic head group and a hydrophobic tail. Modern body wash products usually combine anionic, amphoteric, and nonionic surfactant types to improve cleansing performance, foaming behavior, consistency, and mildness.

A. Anionic Surfactants

1. Soap

Soap (ROO^-M^+) is the alkali salt of a fatty acid. It is generally produced by the saponification of triglycerides, fatty acids, or fatty acid methyl esters. Soaps are subdivided into water-soluble and water-insoluble types. The former are the salts of fatty acids with ammonia, low-molecular-weight amines (especially alkanolamines), and alkali metals (especially sodium and potassium). Water-insoluble fatty acid salts result from reaction with polyvalent metallic cations, such as zinc and aluminum, alkaline earths such as calcium and magnesium, and long-chain fatty amines. The water-soluble salts of fatty acids, derived from alkaline hydrolysis (saponification) of plant or animal fats and oils, are used widely as skin cleansers and in laundry and other cleaning applications.

Because of soap's poor foaming properties in hard water, its application in body wash products is limited; but in cleansing bars, soaps are still widely used.

2. Alkyl Sulfates

Alkyl sulfates ($ROSO_3^-M^+$) are produced by the sulfation of fatty alcohols. Solubility decreases with increasing carbon chain length (>16), concomitant with an increase in the sensitivity to water hardness: lauryl sulfate has been found to yield the best foam and solubility characteristics, whereas longer chain derivatives, such as cetyl/stearyl sulfate, are better tolerated by skin but are less water soluble and foam less copiously [1,2].

3. Alkyl Ether Sulfates

Alkyl ether sulfates ($RO-(CH_2-CH_2-O)_n-SO_3H$) are made primarily by adding ethylene oxide to fatty alcohols, followed by the sulfation of the ethoxylated fatty alcohols and subsequent neutralization with a suitable base (e.g., sodium hydroxide).

The solubility of the ether sulfates is influenced by the hydrophilic polyglycol ether groups and is noticeably higher than that of the corresponding alkyl sulfates. The solubility of calcium salts is very significant; thus, ether sulfates are quite insensitive to water hardness. An increase in the degree of ethoxylation reduces a product's interfacial activity but improves its dermatological properties. On the other hand, an increase in the alkyl chain length (>16) has a negative influence on foaming power and also on cold-water solubility, but skin compatibility is improved. Ether sulfates exhibit synergistic effects in combination with other surfactants—for example, with regard to foaming power, mildness (e.g., combination with betaines), and oil-dispersing ability [1].

4. Sulfosuccinate Esters

Two types of sulfosuccinate esters exist: the sulfosuccinate monoester ($ROOHCCH_2CH(SO_3^-M^+)COO^-M^+$) is more frequently used in skin cleansers than the diester ($ROOHCCH_2CH(SO_2(SO_3^-M^+)COOR$). They are generally produced from maleic anhydride, but maleic acid or fumaric acid may also be used.

$$XO-\overset{\overset{O}{\|}}{C}CH_2CH\overset{\overset{O}{\|}}{C}-OY$$
$$\underset{SO_3^-M^+}{|}$$

Regardless of substitution, the sulfonic acid group is always ionized, and M^+ represents the counterion. In compounds in which X or Y represents an alkyl or substituted alkyl group, the remaining carboxyl function is also present in salt form (e.g., disodium deceth-6 sulfosuccinate). If both X and Y represent alkyl or substituted alkyl groupings, the sulfosuccinate is a diester (e.g., dioctyl sodium sulfosuccinate). Fairly complex multifunctional substituents can be present on the carboxyl groups of this category. Alternately, the XO or the YO groups may be replaced by a substituted N-atom, in which case the resulting sulfosuccinamate is a monoamide (e.g., disodium stearyl sulfosuccinamate).

5. Acylglutamate

Acylglutamates are amides derived from L-glutamic acid and natural higher fatty acids [3]. They have good foaming and detergency behavior, even with hard water and exert a soft feeling to the skin during and after use. An example is sodium lauroyl glutamate [4].

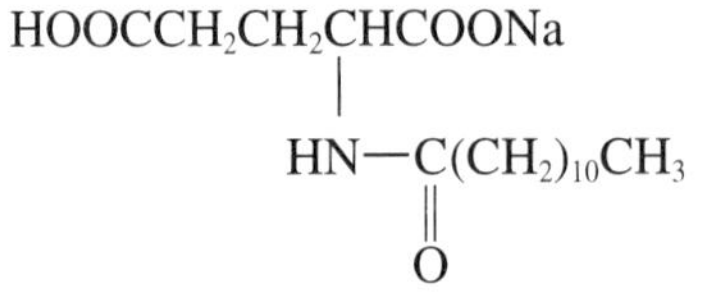

$$\text{HOOCCH}_2\text{CH}_2\text{CHCOONa}$$
$$|$$
$$\text{HN}-\underset{\underset{\text{O}}{\|}}{\text{C}}(\text{CH}_2)_{10}\text{CH}_3$$

B. Amphoteric Surfactants

Amphoteric surfactants are characterized by a molecular structure containing two different functional groups with anionic and cationic character, which change their charge in response to pH.

1. Acyl Derivatives of Ethylene Diamine

These derivatives include monocarboxylates, dicarboxylates, and sulfonates; they differ primarily in their fatty side chain.

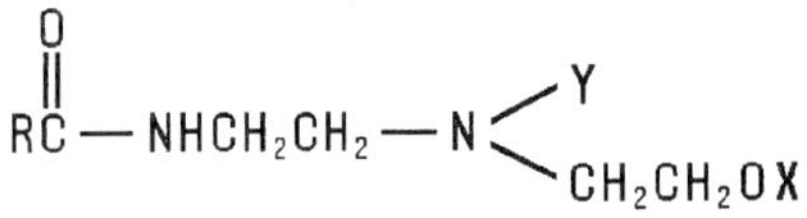

$$\underset{\text{O}}{\overset{\|}{\text{RC}}}-\text{NHCH}_2\text{CH}_2-\text{N}\underset{\text{CH}_2\text{CH}_2\text{OX}}{\overset{Y}{<}}$$

This large group of amphoteric materials exhibits the above structure. They show very good foam stability and hard-water resistance and are compatible with cationic, anionic, and nonionic agents in all preparations [2].

2. Betaines

The betaine group of cosmetic raw materials includes the quaternized alkyl or substituted alkyl derivatives of N,N-dimethyl glycine. They have the following structure,

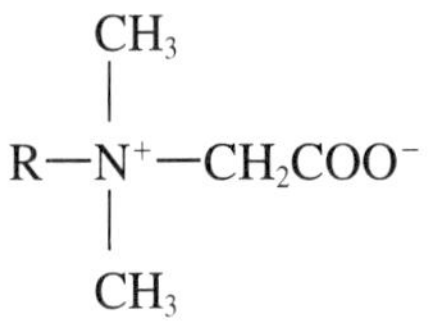

$$\text{R}-\underset{\underset{\text{CH}_3}{|}}{\overset{\overset{\text{CH}_3}{|}}{\text{N}^+}}-\text{CH}_2\text{COO}^-$$

in which the nitrogen atom always carries a positive charge regardless of pH. At the slightly acidic pH values normally encountered in cosmetics, this charge is counterbalanced by a negative charge on the carboxyl group (zwitterion). Betaines are not truly amphoteric compounds, because they do not exhibit exclusively anionic characteristics at high pH. R represents an alkyl group (e.g., $C_{14}H_{29-}$ in myristyl betaine) or an amide-interrupted alkyl group (e.g., $C_{17}H_{35}\text{CONH}(CH_2)_{3-}$) in stearamidopropyl betaine. Related amphoteric compounds are derived from amino propane sulfonic acids (see sulfonic Acids), e.g., lauryl sultaine. The betaines are zwitterionic, surface-active compounds that are employed as emulsifiers, detergents, foam boosters, aqueous viscosity increasing agents, and skin and hair conditioners. They are not affected by water hardness and produce excellent foam with good stability. Exerting very low skin-irritation potential, they have also been found to be effective in reducing the irritation potential of alkyl sulfates, alkyl ether sulfates, and soaps. Alkyl amido betaines are the most common variants of the betaines. Their preparation is carried out in a two-step process, starting with fatty acids or their esters, by condensation with dimethylamino propyl amine and reaction with sodium chloroacetate.

C. Nonionic Surfactants

1. Alkyl Glucosides

Alkyl glucosides are produced by reacting corn starch glucose with fatty alcohol. The resulting nonionic surfactant is used as a cosurfactant in skin cleansers, in which it exerts performance synergy in combination with anionic surfactants.

In addition to the advantage of commercial availability from renewable, vegetable-sourced raw materials, alkyl glucosides display a very high degree of biodegradability.

III. SKIN–SURFACTANT INTERACTIONS

Along with other environmental influences, everyday washing with soap or surfactants can stress the skin and impair the skin barrier. Surfactants wash off not only soil and excess sebum, as required, but also remove some of the skin's intercellular lipids [4–7]. As expected, the strongly cleansing surfactants also have a strong defatting effect on the skin (Fig. 1). If standardized to the same cleansing performance, mild surfactants such as the nonionic surfactant Polysorbate-20 shown in Figure 1 can be identified by their slight defatting action [5].

Even if the amount of removed lipids is only small [8], the selective dissolving out of free fatty acids, fatty acid glycerides, and cholesteryl esters leads to a significant change in the profile of the intercellular lipids [4,9,10]. Moreover,

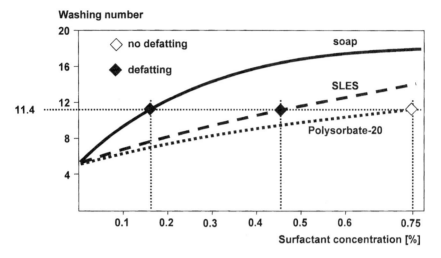

Figure 1 Cleansing performance ("washing number") and defatting by different surfactant solutions. (Adapted from Ref. 5.) (◆ = significant reduction in skin lipids compared to untreated skin, ◊ = no significant changes in skin lipids).

surfactants penetrate into the stratum corneum, where they are adsorbed on the keratin of the corneocytes [11–13] and mix with the intercellular lipids [14]. The consequent impairment of the skin barrier can be demonstrated very sensitively by measurements of the skin's transepidermal water loss (TEWL). As long as the degree of impairment is only slight, the rise in TEWL correlates with a decrease in the lipids' degree of order, thus clearly indicating the important role of the lipid film's morphology in the skin barrier [4].

The corneocytes of the stratum corneum are dead keratinocytes, consisting mainly of the protein keratin, which is present in various forms. Solid keratin microfibrils are embedded in amorphous keratin [9,10]. In addition, dry corneocytes contain about 10% water, which is bound to polar side chains of the keratin [15]. Ionic, polar, and nonpolar bonding forces between the side chains of the keratin polymers are responsible for the mechanical strength of the corneocytes [16].

During washing, water and surfactants penetrate into the corneocytes. Water accumulates at the polar and, above all, the ionic side chains. Due to the high dielectric constant of water, the salt bridges between oppositely charged keratin side chains are weakened and the corneocytes swell [8–10].

Surfactants are adsorbed at nonpolar protein side chains as a result of hydrophobic interactions. Moreover, they can also bond to oppositely charged keratin side chains by electrostatic interaction. At the pH of 6.5 that prevails in many cleansing products, keratin has a slight excess of anionically charged side chains, so anionic surfactants increase the negative charge of the keratin by bonding hydrophobically to nonpolar side chains, thus increasing the degree of swelling (Fig. 2) [11]. At this pH, cationic surfactants neutralize free anionic side chains and reduce the swelling slightly [11]. Like anionic surfactants, nonionic surfactants are adsorbed at nonpolar groups but only enhance the swelling to a small extent.

After excessive exposure to surfactants, TEWL increases sharply and—depending on the disposition and sensitivity of the individual—the skin becomes dry and often scaly, which points to a disruption of the desquamation process [4,8,17]. In this phase, the degree of order of the lipids no longer correlates with the increase in TEWL. This is a further indication that in scaly and irritated skin, other harmful biological mechanisms such as irritation processes are superimposed on the purely physical effects of the surfactants on the lipid film [4,18].

In the cosmetics sector, great efforts are being made to avoid these undesirable effects. Mild surfactants with similar cleansing properties to aggressive surfactants do not attack the intercellular lipids to the same extent [5,6]. The use of refatting oil additives and the application of creams after washing can at least partially restore the lipid film [7,17,19]. A distinction must be drawn at this point between physical effects on the stratum corneum and biological effects on the metabolism of the living epidermis. Petrolatum brings about an immediate partial restoration of the previously impaired skin barrier properties of the stratum corneum, whereas physiological lipid mixtures slowly penetrate into the epidermis and, by naturally accelerating the process of metabolization in the lamellar bodies

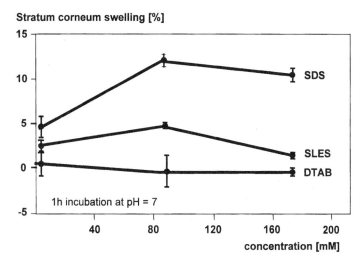

Figure 2 Swelling behavior of human stratum corneum in various surfactant solutions. Error bars indicate SEM. (Adapted from Ref. 11.)

Skin erythema (n=12, % of total scoring)

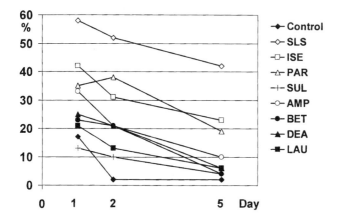

Figure 3 Skin compatibility (erythema) of different surfactant solutions: PAR = sodium C12–15 pareth sulfate (5%; 120 mM); SLS = sodium lauryl sulfate (2%; 65 mM); ISE = sodium cocoyl isethionate (5%; 148 mM); SUL = disodium laureth sulfosuccinate (10%; 181 mM); LAU = lauryl glucoside (10%; 238 mM); DEA = cocamide DEA (10%; 323 mM); AMP = sodium cocoamphoacetate (10%; 274 mM); BET = cocamidopropyl betaine (10%; 286 mM). (Adapted from Ref. 18.)

[19], build up the skin barrier again and help surfactant-induced scaly skin to recover [17].

An overview of the skin compatibility of different surfactants after a human 48-hour patch test was given by Barany et al. [18]. This work is remarkable because of the test concentrations—which were adjusted to concentration *and* molar weight (Fig. 3).

IV. LIPIDIC REFATTERS

The boundaries are merging between skin care and body care: Each skin care ingredient can be used in body care products, in principle. Therefore, the lipids that have been discussed in the preceding chapter (9) can be used as well in personal cleansing products. However, the efficacy of the lipids as active ingredients is highly dependent on the technology of the formulation. In some cases, the use of ready-made compounds can be helpful for the easy production of lipidic surfactant skin cleansers [25].

Benefits of Lipidic Refatters in Surfactant Products

With lipidic refatters, highly sophisticated claims are possible: by delivering proceramides to the upper skin layer, promotion of ceramide production has been asserted, and with high amounts of moisturizers, long-lasting moisturizing is claimed [20,21].

The development of these types of cleansing products can be likened to playing a piano concerto in which many factors come into play: The amount and polarity of the lipids, the combination of synergistically acting surfactants, and the application of the product should be considered to influence the cleansing performance, to balance the foam stability, and to fulfill the promotional claims. A patent disclosure of Unilever comprises a stable cosmetic liquid cleansing composition containing high levels of emollients [22], an example of which is shown in Table 1 [22].

Winder et al. [23] developed a personal cleansing system comprising a diamond mesh bath sponge and a liquid cleanser with moisturizer, which are packed together in a cleansing kit. The following example has been given (Table 2) [23].

Colgate-Palmolive [24] disclosed a skin cleanser comprising surfactants, a silicone fluid, a hydrocarbon, a cationic polymer, a combination of hydroalkyl cellulose and an acrylate copolymer, and water. An example is shown as follows (Table 3) [24].

The above-mentioned examples are composed of at least one lipid ingredient that is claimed to improve not only the mildness of the products but also the efficacy, (e.g., to moisturize the skin). These claims have to be proved in studies using human volunteers. In combination with biophysical measurements which are (explained in detail in Chapter 7), skin care benefits from rinse-off products can be found.

Table 1 Skin Cleansing Lotion

Skin cleansing lotion	Weight %
Sodium laureth sulfate	10.0
Sodium lauroamphoacetate	5.0
Sunflower seed oil	15.0
Lauric acid	2.5
Citric acid	0.8
Magnesium sulfate	1.5
Fragrance	1.0
Water	Add to 100.0

Source: Ref. 22.

Table 2 Moisturizing Liquid Cleanser

Moisturizing liquid cleanser	Weight %
Sodium laureth-3 sulfate	12.0
Cocoamphoacetate/cocoamphodiacetate	6.0
Alkylpolysaccharide	2.0
Coconut monoethanol amide	2.8
Soybean oil	8.0
Maleated soybean oil	2.0
Polymer JR30	0.4
PEG(6) caprylic/caprylglycerate	4.0
Myristic acid	2.0
Glycerine	3.0
Titanium dioxide	0.1
Perfume	1.8
Preservative	0.2
Water	55.7

Source: Ref. 23.

V. ASSESSMENT OF SKIN BENEFITS

A. Single 24-hour Patch Test

The single 24-hour patch test is an epicutaneous plaster test method (COLIPA standard [28]). Applying the test substances to the back under occlusive conditions allows skin compatibility to be studied as a function of application time under various exaggerated conditions. The treatment has been carried out on hu-

Table 3 Skin Cleanser with Improved Stability

Skin cleanser with improved stability	Weight %
Sodium laureth sulfate	7.6
Cocoamidopropylbetaine	2.1
Decylpolyglucoside	0.6
Dimethicone	1.0
Petrolatum	2.0
Polyquaternium-7	0.2
Hydroxypropyl methylcellulose	0.5
Acrylates/C 10–30 alkyl acrylate crosspolymer	0.5
Water	83.5

Source: Ref. 24.

man volunteers by means of epicutaneous plaster for a period of 24 hours. The skin reactions are evaluated at various intervals (6, 24, 48, and 72 hr) under standardized conditions, and the observed reactions are rated in terms of reddening, swelling, scaling, and fissures. For each of these parameters, the volunteers' individual reaction ratings are added together to give a total irritation score.

The skin compatibility of test substances can be determined by direct comparison within the group and by comparison with a positive control (= 100%). The positive control has to exhibit slight to moderate skin irritation under the test conditions.

Shower gel formulations used in the single 24-hour patch test caused no immediate skin reactions when used in dilutions approximating the usual application concentrations (below 5%—i.e., less than 1% surfactant active substance). For this reason, Förster et al. [25] increased the overall concentrations to 30% (i.e., 5% surfactant active substance). Under these conditions, even mild formulations induce skin reactions, whose strength and frequency can be used to differentiate between the products (Fig. 4). The shower gel without added refatting substances (a combination of sodium lauryl ether sulfate with alkyl glucoside and

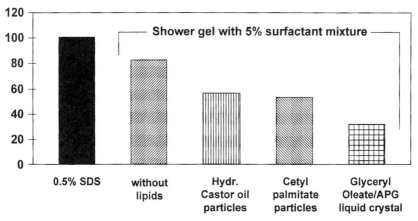

Figure 4 Skin compatibility of surfactant mixtures: Irritation sum scores of shower gel formulations in the 24-hour patch test. The differences between shower gel without lipids and shower gel with hydrogenated castor oil were significant ($P < 0.05$). The differences between shower gel without lipids and shower gel with cetyl palmitate particles and shower gel with glyceryl oleate/APG liquid crystal were highly significant ($P < 0.01$). (From Ref. 25, with kind permission.)

Table 4 Surfactant Formulations Used in a Single 24-hr Patch Test

Cleansing shower gel compositions	1	2	3	4
Sodium laureth sulfate (70% AS)	16.0	16.0	16.0	16.0
Cocamidopropyl betaine (40% AS)	4.0	4.0	4.0	4.0
Coco glucoside (50% AS)	3.75	3.75	3.75	2.7
Coco Glucoside (35% AS) (and) glyceryl oleate	0.0	0.0	0.0	3.0
Hydrogenated castor oil	0.0	1.0	0.0	0.0
Cetyl Palmitate (and) beheneth-10 (and) hydrogenated castor oil (and) glyceryl stearate	2.3	0.0	0.0	0.0
Phenoxyethanol (and) dibromo dicyanobutane	0.1	0.1	0.1	0.1
Water	to 100	to 100	to 100	to 100
pH	5.5	5.5	5.5	5.5

Source: Ref. 25.

cocamidopropylbetaine, formula 1 in Table 4) exhibited a higher irritation potential under these conditions. The strength and frequency of the observed skin reactions were comparable with the positive standard SDS 0.5%. Shower gels containing refatting substance performed much better (formulas 2 to 4 in Table 4). Shower gels with wax particles of hydrogenated castor oil or cetyl palmitate at these concentrations evoked comparably few skin reactions, and the shower gel with added glyceryl oleate and cocoglucoside caused the fewest skin reactions [25].

B. Arm Flex Wash Test

The arm flex wash test is an open epicutaneous test method that is used to study formulations by repeated application in association with mechanical factors (e.g., by rubbing on the skin). This test is suitable for use as a controlled application test (COLIPA) to determine the skin compatibility of two different products by comparing them under conditions of everyday use [26,27].

The inside surface of the elbow of one arm of each volunteer is washed with the test solution. The test design provides for a total application period of 6 minutes. The product is applied, followed by 2 minutes scrubbing and then intensive washing. This procedure is repeated once. After the third and final application, the product is left on the skin for 2 minutes without scrubbing and then washed off. The other arm is treated in exactly the same way with the second test solution. The two test products are used randomly in pair comparisons (right/left). The washing procedure is done twice daily over a total period of 10 washing days. Before each wash and after the final application, skin irritation at the test zones is assessed visually (erythema, scaling, sensory perceptions) on the basis of a scoring system (0–4) and by TEWL measurements.

The arm flex wash test enables the skin compatibility of body cleansing products to be assessed against objective and subjective criteria under near-normal conditions (Fig. 5). In a study of Jackwerth et al. [27], 20 volunteers used a cleansing product twice daily on the sensitive inside surface of the elbow over a period of 2 weeks. Reactions were then assessed in terms of reddening (erythema), scaling, edema (swelling), and fissures (small cracks in the skin). In addition, the degree of skin barrier damage was determined objectively by measuring TEWL. Finally, the volunteers gave their subjective assessment of whether the cleansing product evoked itching, tingling, burning, or a feeling of tautness.

The 10% solution of the anionic surfactant sodium lauryl ether sulfate caused slight but measurable and tangible skin irritation [27]. In comparison, a nonionic alkyl glucoside was considerably gentler to the skin. Erythema and skin scaling were decreased by three-quarters and there was no subjective feeling of skin irritation (see Fig. 5). In this test, too, a combination of the strongly foaming sodium lauryl ether sulfate and the nonionic alkyl glucoside resulted in a clear improvement in skin compatibility. This surfactant basis can thus be used to produce cleansing formulations with tailor-made application properties such as skin compatibility, cleansing performance, foaming properties, and viscosity.

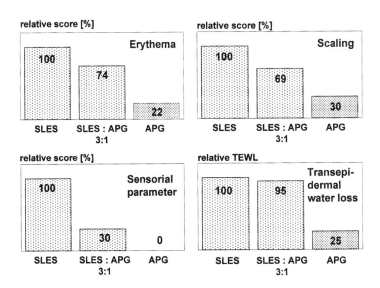

Figure 5 Results of the arm flex wash test with 10% surfactant solutions of sodium lauryl ether sulfate, alkyl glucoside (APG) and a 3:1 mixture. (Adapted from Ref. 27.)

C. Application Study with Lipid Analysis

For the purpose of eluting the refatter glyceryl oleate from the stratum corneum, Förster et al. [25] wetted the inside surface of the lower arm of each of 20 volunteers with warm water (30–34°C); the product was then used for 15 seconds. The skin zone was subsequently rinsed with warm water for 30 seconds, carefully patted dry with paper towels, and allowed to dry for 2 minutes. The lipid film absorbed was finally eluted with a medical cotton pad soaked in ethanol. For the elution, a plastic ring with a defined internal diameter was placed on the treated zone on the lower arm of the volunteer, and the skin within this test zone was wiped three times with the moist cottonwool pad with a circular movement. The pad was then kept in ethanol until analyzed. The refatter in the eluate was quantitatively determined by a gas chromatography/mass spectroscopy method.

Applications of body wash products were made once or over a period of days or weeks to measure short-term and cumulative skin refatting effects. A mild body wash product without lipids was used in a 1-week preconditioning phase and for the duration of the application phase. The use of creams and lotions on the forearms was forbidden during the study.

With this technique, Förster et al. [25] could detect the refatter glyceryl oleate in ethanolic eluates after a single application, after 1 week of use, and after 2 weeks of use (Fig. 6). A statistically significant increase after 1 week in comparison with a single application was detected without an "overfatting" of

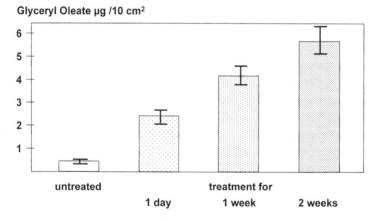

Figure 6 Penetration of glyceryl oleate into the stratum corneum after treatment with a shower gel containing a refatter. The differences after 1-day treatment are highly significant compared to untreated ($P < 0.01$). The differences between 1-day treatment and 1-week treatment are highly significant. (From Ref. 25, with kind permission.)

the skin but as a pleasant skin feel, as has been shown by the volunteers' answers to questions about their sensory perceptions.

D. Forearm Wash Test with Skin Roughness Determination

To assess skin roughness, Förster et al. [25] combined a forearm wash study with fast optical in vivo topometry of the skin (FOITS [29]). The comparative study of shower products was carried out with a group of 30 women; the inside surfaces of the lower arms were the test fields. An image of each test zone was prepared with FOITS before and after 3 weeks of product use. The measuring system consists of a projection unit and a charge coupled device (CCD) camera, fixed at the so-called triangulation angle. This technique involves projecting a sequence of grid lines on the skin. The recorded strip pattern enables the order, phase, and strip quality of each point of the image of the examined zone to be measured.

Changes in the skin profile were quantified on the basis of the parameters Ra and Rz. Ra is the arithmetic mean of the absolute values of all deviations of the roughness profile from the center line within the total path. The rough structure of the test zone is described by the parameter Rz. The product effects were statistically confirmed by ANOVA calculations.

The assessment was made by comparing the initial and final values. During a 3-week study period, the volunteers used the product three times daily. This was done by wetting the test zone and then applying 0.5 ml of product. After a first washing phase of 30 seconds, the product was washed off with warm water and the zone was washed again for 30 seconds. All of the studies were done in a specially climatized laboratory with a constant room temperature of 22 ± 1°C and a relative humidity of 60 ± 5%.

The FOITS results proved the positive effect of the skin-refatting component. Whereas the test sample without refatting components tended to increase the roughness of the skin, the glyceryl oleate–containing formulation resulted in smoother skin after 3 weeks of use. Statistical analysis of the results showed significant differences between the two formulations after all measurement intervals. The skin surface profile was clearly improved by application of the refatting formulation (Fig. 7).

E. Leg Wash Protocol

To assess the dry skin improvement of personal cleansing products, Ertl et al. [30] developed a wash test on dry leg skin. Washes were conducted over a period of days or weeks.

The volunteers had to shave their legs 3 days before the baseline treatment evaluation. A mild cleansing bar was used instead of the usual shower or bath

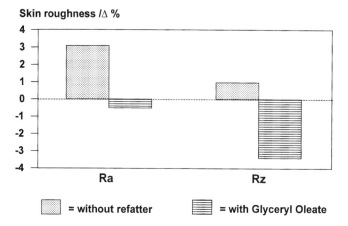

Figure 7 Skin roughness measurement values after a 3-week period of use of a shower gel containing a refatter and a shower gel with no refatter. The differences between the products are highly significant for both Ra and Rz, with $P < 0.01$. (From Ref. 25, with kind permission.)

product for the duration of the study. The subjects were obliged not to use the bar or its lather directly on their legs. Bath oil, cream, or lotion use on the legs was not allowed, as were leg skin–affecting situations (excessive ultraviolet light exposure, swimming in chlorinated pools).

Visual and instrumental evaluations were conducted on each site before the start of the treatment (baseline). Visual dryness was measured using a grading procedure [31]. Only volunteers with a visual dryness score between 2 and 4 on each site at the baseline treatment evaluation continued into the treatment phase. Stratum corneum hydration was assessed via skin capacitance measurements. TEWL was measured with an evaporimeter.

Two regimens of the preconditioning phase were evaluated: Washing the legs twice daily with a supplied commercial soap bar for one week or leaving the legs untreated. In an evaluation study, the changes in clinical parameter values on subjects legs after a 1-week preconditioning phase were compared [30]. The subjects used a syndet bar for cleansing one leg and washed their other leg twice daily in a controlled manner with the soap bar. Greater damage to the soap-washed leg was demonstrated (Table 5) [30].

An analysis of variance (ANOVA) model was used to analyze the data, with baseline skin condition being used as a covariant in the analysis. Data from studies comparing the same clinical protocol were pooled using a meta-analysis technique [32].

Table 5 Skin Properties after Washing with a Syndet or Soap Bar

	Visual dryness	Visual redness	Skin capacitance (pF)	TEWL (g $*$ m^{-2} $*$ hr^{-1})
Soap bar—controlled wash	0.96	0.37	−1.34	0.71
Mild syndet bar—ad lib wash	0.64	0.45	−0.61	0.09

Comparison of clinical parameters after one week of preconditioning phase. Both instrumental end points showed significantly (P < 0.008) greater damage to the soap-washed leg after this phase (n = 24)
Source: Adapted from Ref. 30.

With this protocol, Ertel et al. [30] found that some cleansing products can lead to moisturization of the dry skin, even for periods as long as 24 hours. In a series of 11 leg wash studies, a body wash product yielded significant reduction in visual dryness at all evaluation time points and a significant improvement in stratum corneum hydration at three of the evaluation time points (Table 6).

VI. OUTLOOK

As the result of innovative lipid-surfactant product concepts—and also the development of new kinds of applications (e.g., pore strips, wet wipes), cleansing is one of the fastest growing categories within the body and skin care market. Stud-

Table 6 Changes in Visual Dryness and Skin Capacitance for a Body Wash Product Relative to a No-Treatment Control

Evaluation visit	Dryness change (mean ± SEM)	Capacitance change (mean ± SEM)
3 hours post first wash	−0.6 ± 0.044*	1.2 ± 0.267*
24 hours post first wash	−0.2 ± 0.038*	0.4 ± 0.282†
24 hours post fourth wash	−0.3 ± 0.041*	1.6 ± 0.317*
3 hours post fifth wash	−0.8 ± 0.047*	2.5 ± 0.355*

* P < 0.0001.
† P = 0.10.
Source: Adapted from Ref. 30.

ies with surfactant-lipid mixtures demonstrate that boundaries are merging between skin care and body care.

The compatibility of surfactant mixtures can be increased by selecting mild surfactants and by combining favorable properties. Lipid-containing compounds are the focus of intensive research efforts. Improvement of the skin lipid structure by lipid-containing surfactant products prevents dehydration and scaling of the corneocytes and thus reduces skin roughness and increases skin hydration. In addition to their primary cleansing effect, skin care products must meet consumer expectations of skin-compatible body care ingredients and support other aspects of modern body cleansing such as fun and wellness.

REFERENCES

1. Biermann M, Lange F, Piorr R, Ploog U, Rutzen H, Schindler J, Schmid R. Synthesis of Surfactants. In: Falbe J, ed. Surfactants in Consumer Products. Berlin, Heidelberg: Springer-Verlag, 1987:24–124.
2. Thau P. Surfactants for Skin Cleansers. In: Rieger MM, Rhein LD, eds. Surfactants in Cosmetics. New York, NY: Marcel Dekker, 2nd Ed. 1997:285–306.
3. Acylglutamate Bulletin, Ajinomoto Co., Inc., 1991.
4. Denda M, Koyama J, Namba R, Horii I. Stratum corneum lipid morphology and transepidermal water loss in normal skin and surfactant-induced scaly skin. Arch Dermatol Res 1994; 286:41–46.
5. Gloor M, Munsch K, Friederich HC. Über die Beeinflussung der Hautoberflächenlipide durch Körperreinigungsmittel I. Derm Mschr 1972; 158:576–581.
6. Gloor M, Tretow CW, Friederich HC. Über die Beeinflussung der Hautoberflächenlipide durch Körperreinigungsmittel II. Derm Mschr 1974; 160:291–296.
7. Gloor M. Influence of surface active agents, emollients, and cosmetic applications on skin and hair lipids. Cosmet Toilet 1977; 92:54–62.
8. Froebe CL, Simion FA, Rhein LD, Cagan RH, Kligman A. Stratum corneum lipid removal by surfactants: relation to in vivo irritation. Dermatologica 1990; 181:227–283.
9. Fulmer AW, Kramer GJ. Stratum corneum lipid anormalities in surfactant-induced dry scaly skin. J Invest Dermatol 1986; 86:598–602.
10. Dykes P. Surfactants and the skin. Int J Cosmetic Sci 1998; 20:53–61.
11. Rhein LD, Robbins CR, Fernee K, Cantore R. Surfactant structure effects on swelling of isolated human stratum corneum. J Soc Cosmet Chem 1986; 37:125–139.
12. Robbins CR, Fernee KM. Some observations on the swelling of human epidermal membrane. J Soc Cosm Chem 1983; 34:21–34.
13. Rhein LD. Review of properties of surfactants that determine their interactions with stratum corneum. J Soc Cosmet Chem 1997; 48:253–274.
14. Friberg SE, Goldsmith L, Suhaimi H, Rhein LD. Surfactants and the stratum corneum lipids. Colloids Surfaces 1988; 30:1–12.

15. Potts RO. Stratum corneum hydration: experimental techniques and interpretations of results. J Soc Cosmet Chem 1986; 37:9–33.
16. Speakman JB, Stott E. Trans Faraday Soc 1934; 30:539–548.
17. Imokawa G, Akasaki S, Minematsu Y, Kawai M. Importance of intercellular lipids in water-retention properties of the stratum corneum: induction and recovery study of surfactant dry skin. Arch Dermatol Res 1989; 281:45–51.
18. Barany E, Lindberg M, Loden M. Biophysical characterization of skin damage and recovery after exposure to different surfactants. Contact Dermatitis 1999; 40:98–103.
19. Mao-Qiang M, Brown BE, Wu-Pong S, Feingold KR, Elias PM. Exogenous non-physiologic vs physiologic lipids. Arch Dermatol 1995; 131:809–816.
20. Unilever Plc. Patent No. WO9612469.
21. Procter & Gamble Patent No. GB2297762.
22. Unilever PLC, WO9904751.
23. Procter & Gamble, US5650384.
24. Colgate-Palmolive Co, WO9962493.
25. Förster Th, Issberner U, Hensen H. Lipid/surfactant compounds as a new tool to optimize skin-care porperties of personal-cleansing products. J Surf Det 2000; 3:345–352.
26. Strube DD, Koontz SW, Murahata RI, Theiler RF. The flex wash test: a method for evaluating the mildness of personal washing products. J Soc Cosmet Chem 1989; 40:297–306.
27. Jackwerth B, Krächter H-U, Matthies W. Dermatological test methods for optimising mild tenside preparations. Parfümerie Kosmetik 1993; 74:134–141.
28. Walker AP, Basketter DA, Baverl M, Diembeck W, Matthies W, Mougin D, Paye M, Röthlisberger R, Dupuis J. Test guidelines for the assessment of skin compatibility of cosmetic finished products in man. Food Chem Toxicol 1996; 34:651–660.
29. Rohr M, Schrader K. Fast Optical In Vivo Topometry of Human Skin (FOITS), SÖFW-Journal 1998; 124:52–59.
30. Ertel KD, Neumann PB, Harwig PM, Rains GY, Keswick BH. Leg wash protocol to assess the skin moisturization potential of personal cleansing products. Int J Cosm Sci 1999; 21:383–397.
31. Ertel KD, Keswick BH, Bryant PB. A forearm controlled application technique for estimating the relative mildness of personal cleansing products. J Soc Cosmet Chem 1995; 46:67–76.
32. Boisits EK, Nole GE, Cheney MC. The refined regression method. J Cut Aging Cosmet Dermatol 1989; 3:155–163.

11
Sensory Assessment of Lipids in Leave-On and Rinse-Off Products

Thomas Gassenmeier and Peter Busch*
Henkel KGaA, Düsseldorf, Germany

I. SURVEY OF SENSORY MEASUREMENT TECHNIQUES

The sensory assessment of all kinds of consumer goods, foods, or objects of daily life is doubtless as old as mankind itself. Standardization of testing procedures was first introduced for the assessment of certain foods and drinks—for example, wine, spirits, coffee, fish, and meat. At the start of the 20th century, the developing food industry required a scientific sensory assessment of its products. It is not surprising, therefore, that the generally accepted basic principles of a modern, formalized, and systematic methodology of sensory assessment were established in the food industry.

The present chapter is designed to give a short overview of these generally accepted basic principles. A detailed description of sensory techniques can be found in *Sensory Evaluation Techniques* [1].

A. Measurable Properties in Sensory Assessment

Which properties can be measured by sensory assessment? In principle, everything that we can perceive with our five senses—for example:

Appearance (e.g., color, luster, shape, size, transparency)
Taste, aroma
Odor, smell
Consistency, texture (e.g., viscosity, cohesiveness, softness, wetness, dryness)
Sounds (e.g., pitch, loudness)

---

* *Current affiliation*: Cognis Deutschland GmbH, Düsseldorf, Germany.

Obviously, relevant quantities vary from product to product. Thus, taste plays a role in all foods but is of no importance in cosmetics. Among cosmetic products themselves, different parameters are likewise of interest for different categories: the quantity and characteristics of the foam produced are of elementary importance for surfactant-based agents for washing the body and the hair but are not significant for other skin care products.

B. Influences on the Assessment Procedure

The various sensory characteristics such as external appearance, smell, aroma, flavor, consistency, and sounds are perceived primarily through the sense organs, namely the eyes, skin, nose, tongue, and ears: chemical or physical stimuli are registered and transmitted (Fig. 1). Both physiological and psychological factors have an influence on the primary registration of these stimuli. The physiological factors include the phenomena of adaptation or potentiation. For example, if, when assessing the sweetness of a particular test material, the subject has been given a sweet material immediately beforehand, the test material is perceived to be significantly less sweet. The reason: the reduction in the sensitivity of the sense of taste as a result of the repeated administration.

Psychological factors play a particular role in the registration and processing of stimuli. These influences include a particular expectation, errors due to adaptation, logical errors, so-called halo effects (i.e., where one perception influences a succeeding one), the order in which samples are presented, influence by third parties, or personal motivation at the time.

What is really decisive for the measurement of effects, in addition to the actual registration of the stimuli by the sense organs, is their—practically simultaneous—interpretation by the mental process.

Both the sensory perception and the mental interpretation of the registered stimuli depend on a large number of particular factors. Thus, the anatomic realities are different for every human being; moreover, they change during each person's lifetime. The psyche, which plays a role in mental interpretation, likewise differs from person to person, depending on the individual's culture, membership in particular social groups, age, education, religion, and upbringing. These differences, which can hardly be systematized in any way, are responsible for the fact that the typical findings of a sensory assessment study show a general tendency toward vagueness.

The world of individual experience is reflected particularly strongly in language. Language, indeed, is what opens our door to experience: the greater our language resources, the more we experience. The proof can be adduced by means of "open interviews" [3]. If different panels are made up of subjects with different personal experience histories or training levels, and they are asked to describe

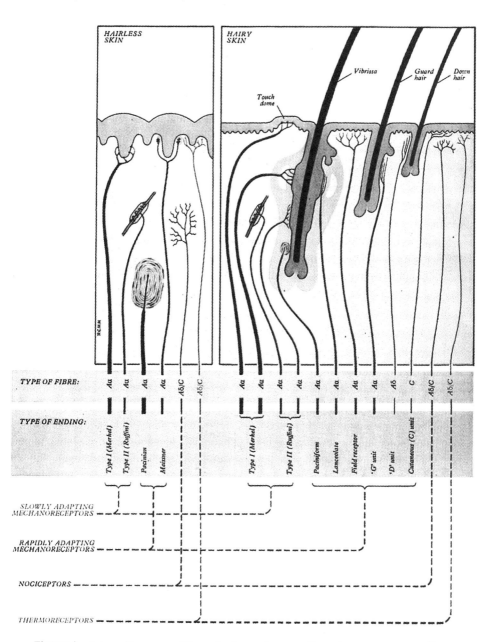

Figure 1 Schematic survey of the major types of mammalian cutaneous sensory endings and their afferent fibers. (From Ref. 2.)

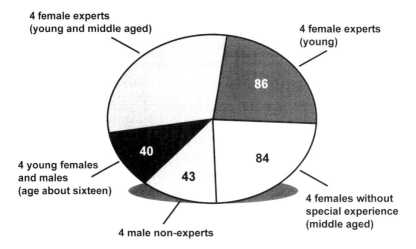

Figure 2 Panel test for generation of words.

a face-care cream in words, the great differences in linguistic ability become apparent (Fig. 2).

C. Measures to Reduce Extraneous Influences in Sensory Assessment

1. Test Controls

If usable test results are to be obtained, the practical implementation of sensory tests requires the choice of a careful test design that keeps the effect of the influences described above as small as possible. One basic precondition is that sensory assessment be carried out under strictly controlled conditions. First of all, this means that the test venue must be arranged in such a way that the panelists are able to perform as well as possible. Some of the design requirements for a sensory assessment laboratory are listed in a guideline issued by the American Society for Testing Materials (ASTM) [4]. In addition, controlled conditions must also be maintained with respect to the test products (manufacture, presentation) and the panelists (training).

2. Test Setup

A skillfully setup test is a further precondition for keeping the effect of extraneous influences as small as possible. Various techniques will be described below.

3. Tests for Sameness/Difference

Triangle test. In the triangle test, three samples are administered, of which two are identical and one different. The subjects must identify the different product.

Two-out-of-five test. This test is also designed to establish differences between two samples. Here, two or three (as the case may be) of the five samples are identical. The advantage of this test lies in the fact that the probability of a result that is correct by chance is small (1 in 10). The disadvantage is that the numerous samples lead to adaptation and fatigue effects, which make the test really only suitable for visual and tactile applications.

Duo-trio test. In contrast to the triangle test, subjects in the duo-trio test are required to compare two different samples with a reference sample (one of the samples is the same as the reference, the other is not). The disadvantage of this setup is the high probability of a chance correct response (50%); the advantage is that it is easy to understand.

Simple difference test. This form of test requires subjects to compare two different or—as a control—two identical test products.

Difference-from-control test. The difference-from-control test provides information about a possible difference between two samples and—unlike the test setups mentioned hitherto—also about the degree of difference. In actual practice, the subjects are given one or more test products and a reference sample. What is measured is the degree of difference between each test product and the reference.

Sequential tests. Sequential tests are carried out in order to save time and costs by keeping down the number of tests needed. In the test procedures described above, with a fixed number of judgments and resulting type I error (α = probability of stating that a difference occurs when it does not), the type II error (β = probability of stating that no difference occurs when it does) is minimized. This means that the possibility of false positive results is accepted in order to avoid false negative results and to make do with a fixed number of judgments. In sequential tests, by contrast, α and β are laid down in advance, and the required number of judgments is calculated from test results as they are obtained. Depending on the course taken by the test, more or fewer assessments are required in order to ensure a particular degree of statistical certainty. Sequential testing in practice often uses triangle or duo-trio tests.

4. Tests to Determine Differences in Particular Characteristics

Paired comparison and multisample difference tests. Distinctions between two or more products in respect to a particular characteristic can be made in paired-comparison or multisample-difference tests. Here, the subjects are given two or more samples and required to assess whether the products are different. If more than two products are assessed, the intensity of the relevant characteristic is also assessed for the various test products. Depending on the number of samples, different procedures are required for statistical evaluation (e.g., analysis of variance, Friedman analysis, or other specific statistical procedures).

5. Descriptive Analysis Technique

In descriptive analysis, the panelists must be able to perceive and describe the sensory characteristics of a sample. This methodology thus includes a qualitative aspect, defining the product, as well as the quantification of the respective qualitatively discovered parameters. A detailed description of the method can be found in *Sensory Evaluation Techniques* [1].

An important precondition of descriptive analysis is for the judges to receive considerably more elaborate training. This is necessary in order to explain to them the nomenclature of the sensory parameters used, and to make sure that the words for the quantitative assessment of intensity differences are used in a uniform way. In addition, defined reference patterns are used to exemplify the terminology.

6. Panel Management

A further important measure to ensure good results by sensory analysis is the selection and training of the test panel. General guidelines on this topic can be found in the literature [5,6]. The first stage in the setting-up of a test panel is the selection of suitable judges. This can be done by means of preliminary tests in which, for example, potential panelists are required to determine the sameness of, or the differences and the size of the differences between, various samples. In the case of panels for descriptive testing, the greater demands make it highly desirable to give particular care to the recruitment of suitable panelists. The usual method is to draw up a short list by means of a questionnaire, and to make the final selection by means of acuity and ranking tests using known samples. Once selected, the judges have to be trained. For example, the panelists must be taught about correct behavior before the test, about the use of the samples, about ways of reducing adaptation, and about the product characteristics that are to be assessed. Where descriptive analysis is concerned, the panelists must first of all learn the necessary parameters for the product category in question, and the precise expressions used to assess them. This is followed by practical training with exercises of increasing difficulty. The usual total training time required ranges from 40 to 100 hours, depending on product category.

D. Test to Establish Product Acceptance

To establish product acceptance, it is usual practice to carry out consumer tests or in-house panel tests. The advantage of consumer tests lies in their high degree of realism; in other words, the panelists consist exclusively of "normal" consumers. In practice, consumer tests are a commonly used instrument to assess the acceptance of newly introduced or modified market products or to assess possible market potential. What all consumer tests have in common, however, is that they need a large number of subjects (approx. 100–400) in order to achieve reliable results. This is associated with high costs, and, at several weeks, a great deal of time.

II. SENSORY ASSESSMENT OF LIPIDS IN COSMETICS

Following the general explanation of sensory techniques, we shall now, with the help of numerous examples, give an overview of the sensory assessment of lipid-containing cosmetic preparations.

A. Measurement of Cosmetic Effects

There are two basic ways of obtaining information about the effects of cosmetic products. On the one hand, there are the various biophysical or dermatological measuring techniques, such as the measurement of transepidermal water loss, sebumetry, corneometry, patch tests, and so forth. These procedures are described in detail in other chapters of this book. The advantage of these objective test methods is that they provide exact, comprehensible, and easy-to-communicate (i.e., intersubjective) results. The disadvantage of many physical or physicochemical procedures is that their practical relevance is not always certain. Transepidermal water loss (TEWL) measurements, for example, can give very precise information on the barrier characteristics of the stratum corneum—but whether consumers can really notice any significant measurable difference in the barrier characteristics of their own stratum corneum following cosmetic treatment is questionable, in view of the fact that there is no organ to provide direct information about them.

Another possibility for obtaining information on the effects of cosmetic preparations is to carry out subjective tests. The advantage of these methods is that they are, without question, highly realistic. The disadvantage is that the results are usually imprecise. The reasons for the imprecision have already been mentioned, and these include variations in skin status, variations in the functional efficiency of the sense organs, and, of course, differences due to cultural, social, or religious influences in the mental processing of identical stimuli. In sensory

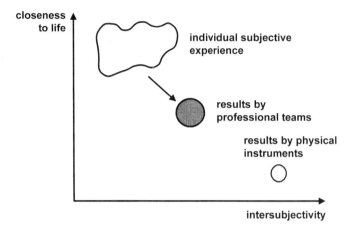

Figure 3 Intersubjectivity box.

assessment, the assessment of products by trained professional teams leads to a significant increase in intersubjectivity while maintaining a high degree of realism (Fig. 3).

B. Sensory Assessment Practice in Cosmetics

For the subjective assessment of cosmetic effects, all the above-mentioned technologies and procedures could in principle be used. In actual practice, for the sensory assessment of cosmetic preparations, it is primarily difference tests and quantitative descriptive analysis that are suitable and described in the literature [7–10].

The usual test methods all have in common this requirement: to reduce the influence of extraneous factors that could vitiate the precision of the method, the test conditions must be strictly controlled. Training the panelists—which, for a descriptive analysis in cosmetic matters, takes at least 30 hours per person—is also expensive and time-consuming.

C. Quantitative Testing

The requirements described above regarding enhancing the precision of sensory assessment, in particular the extensive training of the panelists, are, as stated, both expensive and time-consuming. For this reason, a simplified methodology has been developed that builds on two basic principles: the use of precisely defined sensory parameters and the carrying out of a comparative test using reference products [3].

1. Test Parameters

Table 1 sets out test parameters for skin care products and surfactant-based body-cleansing preparations. These parameters have been established in preliminary linguistic tests. Each of the given parameters is precisely defined (Figs. 4 and 5).

2. Reference Products

The use of reference products to facilitate assessment is a standard procedure, being a component, for example, of the duo-trio and difference-from-control tests outlined above. Normally, the reference products are selected from products on the market. As the composition and characteristics of products on the market are changing with increasing speed as a result of diminishing product-cycle times, it is advisable to use one's own standard products whose composition is known and constant. In a recent article [3], the use of reference samples derived from the cremor basalis of DAC (German Pharmaceutical Codex) is suggested for the sensory assessment of skin care products (Figs. 6 and 7).

This aqueous hydrophilic cream can be varied easily in respect to its content of different modules such as emulsifiers, lipids, or rheological modifiers [11]. For every sensory parameter to be assessed (Table 1), three reference compositions are used, which respectively evince the relevant characteristic strongly, weakly, or to an intermediate degree.

In practice, these reference samples are used as follows. The test product (i.e., the one to be assessed) is first tested for the relevant sensory characteristic against the intermediate reference sample. If judged identical, it is given a score of 4 on a 7-point scale. If judged to be different, the test-product is tested against

Table 1 Sensory Parameters of Skin Care and Surfactant-Based Products

Skin care products	Surfactant-based products
Pick-up	Distribution
Consistency	Slipperiness
Peaking	Flash foam
Cushion	Amount of foam
Distribution	Foam bubble size
Absorption	Creaminess
Smoothness	Foam stability
Stickiness	Foam cushion
Oiliness (residues)	Smoothness of the skin
Waxiness (residues)	Skin dryness

Cushion/Body

150 mg of substance are placed on the fingertips of both index fingers. The index finger and thumb are rubbed against each other. Evaluate the amount of product perceived between index finger and thumb during the rubbing procedure. The more product is perceived, the higher the cushion effect.

Figure 4 Definition of Cushion/Body.

the reference sample that deviates in the same direction from the standard. At this point at the latest, the final score on the 7-point scale can be given, whereby scores 2, 4, and 6 represent identity with a standard product, 3 and 5 a position between the reference samples, and 1 and 7 extreme values (Fig. 8).

3. Validation

In order to test the extent to which the method provides comparable results between different laboratories, a validation procedure was carried out for the sensory assessment of cream products.

The experiment was set up in the following way: one test product was assessed in five laboratory panels each with 12 judges, and the results then statistically evaluated. Those who carried out the test in the respective laboratories had previously taken part in a one-off training session. The 12 panelists were familiarized with the test procedure for each of the sensory parameters to be assessed,

Sensory Assessment of Lipids in Skin Care Products

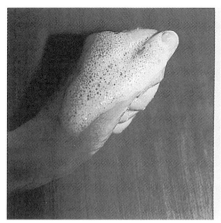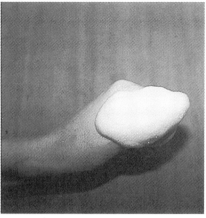

Flash Foam

Both hands are immersed in warm water (38°C) before the test. Then 0.5 g of the product to be tested is placed in the palm of the hand together with 2.5 g water. The panelist is asked to start the washing procedure by using the second hand, which is moved with circle movements with slight pressure and a frequency of two movements/sec. for 10 sec. over the area where the product has been applied. After that, again 2.5 g of water are applied and the procedure is repeated for another 10 sec. The speed of foam formation is evaluated in comparison to the reference product.

Figure 5 Definition of Flash Foam.

Ingredient	Amount [% (w/w)]
Glyceryl Stearate	4.0
Cetearyl Alcohol	6.0
Caprylic/Capric Triglyceride	7.5
Petrolatum	25.5
PEG-20 Glyceryl Stearate	7.0
Propylene Glycol	10.0
Aqua	40.0
	100.0

Figure 6 Cremor Basalis DAC.

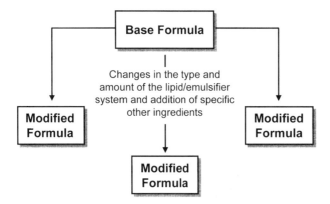

Figure 7 Strategy for generation of reference products.

and they were subsequently in a position to assess the test product immediately. Ten sensory parameters were covered: pick-up, consistency, peaking, cushion, distribution, absorption, smoothness, stickiness, oiliness, and waxiness. Each of the parameters was precisely defined in a test protocol according to the kind and quantity of the application of the test or reference product and the characteristic to be assessed by the subject.

The mathematical-statistical evaluation of the ring experiment was performed according to the following procedure [12]. First, the arithmetic mean,

Sensory Parameters			Ref. 1		Ref. 2		Ref.3		
Evaluation*		1	2	3	4	5	6	7	
Pick-up	good								less good
Consistency	low								high
Peaking	little								high
Cushion	little								much
Distribution	easy								difficult
Absorption	fast								slow
Smoothness	little								much
Stickiness	little								much
Oiliness	little								much
Waxiness	little								much

1 = Testproduct < Ref. 1
2 = Testproduct = Ref. 1
3 = Ref. 1 < Testproduct < Ref. 2
4 = Testproduct = Ref. 2
5 = Ref. 2 < Testproduct < Ref. 3
6 = Testproduct = Ref. 3
7 = Testproduct > Ref. 3

Figure 8 Sensory profile sheet.

Table 2 Raw Data Ring-Test

Parameter	mean DGK mean of all members	Laboratory 1			Laboratory 2			Laboratory 3			Laboratory 4			Laboratory 5		
		a	b	N	a	b	N	a	b	N	a	b	N	a	b	N
Pick-up	5.15	3.91	1.76	11	5.08	1.56	12	5.57	1.28	14	6.08	0.67	12	4.92	2.07	12
Consistency	4.56	4.82	0.60	11	4.67	0.65	12	4.29	0.91	14	4.50	0.67	12	4.58	0.79	12
Peaking	3.18	3.36	1.50	11	3.25	1.21	12	2.93	0.62	14	2.50	0.67	12	3.92	1.38	12
Cushion	5.28	4.36	1.86	11	5.42	0.90	12	5.79	1.12	14	5.25	1.14	12	5.42	0.79	12
Distribution	4.38	3.73	1.56	11	4.42	1.56	12	4.29	0.99	14	5.00	0.43	12	4.42	0.90	12
Absorption	4.10	4.28	1.42	11	3.92	1.62	12	3.57	1.83	14	5.00	0.43	12	3.83	1.70	12
Smoothness	4.46	5.00	1.73	11	4.42	1.51	12	4.50	1.61	14	4.75	1.22	12	3.67	0.89	12
Stickiness	2.31	3.09	2.07	11	2.17	1.19	12	3.43	2.06	14	1.25	0.45	12	1.50	0.67	12
Oiliness	3.87	3.00	1.55	11	4.08	0.90	12	3.07	1.86	14	4.92	0.29	12	4.33	0.89	12
Waxiness	2.38	2.45	0.69	11	1.83	0.72	12	2.79	0.58	14	2.42	0.67	12	2.33	0.49	12

a = mean; b = standard deviation

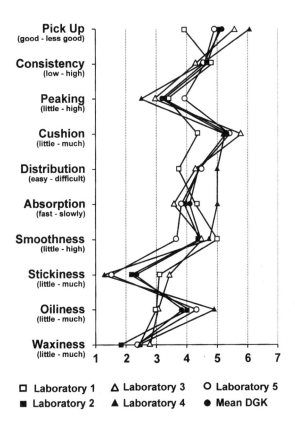

Figure 9 Sensory profiles of a test cream (ring-test).

the standard deviation, and the total number of scores were established for each laboratory and each parameter. The test to reject extreme scores, recommended in the quoted literature at this point, was not employed in the current investigation, as we did not consider that the distribution of scores by the judges pointed to any gross errors of measurement. The next steps in the evaluation of this experiment were the Bartlett test and the F-test, which provide a measure of the homogeneity of the standard deviations and arithmetic means. Iterations of these two tests were continued until no further significances were obtained. Finally, all the individual data for each parameter were summarized as a single overall arithmetic mean, overall standard deviation, and a total number of scores. If one looks at the experiment in conjunction with the mathematical evaluation as described, it can be seen that the results from the different laboratories are largely comparable, and produced very similar qualitative sensory profiles (see Table 2 and Fig. 9).

A detailed examination shows that in the case of nine parameters, all the laboratories showed no significant differences concerning arithmetic means. As for the standard deviations, in four parameters one laboratory in each case stood out by reason of a lower standard deviation. The circumstance that in three of these cases the laboratory in question was the one where the method had originally been developed can be explained by the far higher level of training its judges had received. As far as the parameters stickiness and oiliness are concerned, two laboratories stand out by reason of a high standard deviation. A possible explanation for this is that these parameters, in the case of the test cream, were assessed differently depending on the time when they were assessed. Presumably these two laboratories carried out part of the assessment somewhat earlier than laid down in the test regulations.

Summarizing the results of this experiment, it is apparent that very similar sensory profiles were obtained for the same test cream in the different laboratories. From a statistical point of view as well, the results from the different laboratories are comparable. In view of the brief, one-off training received by the judges, and the high scatter usual in subjective testing methods, this result provides confirmation that the method presented here really can achieve replicable results with a relatively inexpensive and short period of training.

III. SENSORY ASSESSMENT OF GALENIC CONCEPTS IN COSMETICS

When developing cosmetic preparations, it is often important to design products with very specific characteristics using appropriate galenic formulations. The method of sensory assessment outlined above can be used to very good effect to test the effect of galenic concepts quickly and without undue expense [13]. In addition, it will reveal any unintended "side effects".

When various galenic concepts are implemented in cosmetics, a role is often played not only by particular active constituents but also by lipids and their derivatives—for example, emollients, consistency-giving factors, emulsifiers, or refatting agents. In the following section, with the help of several examples, we shall discuss what influences systematic galenic modifications of a cream formulation have on its sensory characteristics.

We chose as our starting formulation an oil-in-water (O/W) cream, based on ethoxylated cetylstearylalcohol as emulsifier and cetylstearylalcohol and glycerine monostearate as co-emulsifiers, and an oil phase consisting of dicaprylyl ether, hexyldecanol, hexyldecyl laurate, cocoglycerides, and almond oil. Glycerine was chosen as moisturizer. The emulsifier/thickening system accounted for approximately 8%, the oil phase 20%, and the moisturizer 5% of the formulation. Classification of the emulsion thus produced showed that it could most appropri-

Function	Ingredients	Amount [% (w/w)]
Emulsifier	Ceteareth-20	2.2
Consistency giving factors	Cetearylalkohol Glyceryl Stearate	4.0 2.0
Emollients	Dicaprylyl Ether Hexyldecanol, Hexyldecyl Laurate Cocoglycerides Almond Oil	5.0 8.0 4.0 3.0
Moisturizer	Glycerin Aqua	5.0 ad 100.0

Figure 10 Base emulsions for proof of galenic concept.

ately be used as intensive skin care or night care cream. The thickening system, the emulsifier, the oil phase, and the moisturizer were now systematically varied and the sensory changes recorded using the sensory profile sheets shown already in Figure 7 (Fig. 10).

A. Modification of the Consistency-Giving Factors

Where emulsions are concerned, a whole range of different compounds are used as consistency-giving factors. The fatty alcohols and partial glycerides used in the initial formulation exercise their consistency-giving effect largely by forming a coherent lamellar gel network in the aqueous phase together with the hydrophilic emulsifiers. This process also takes place in the absence of an oil phase, and leads then to so-called complex emulsifiers or surfactant gels. If an oil phase is present, this can in itself lead to increased viscosity if the proportions are appropriate.

Another group of frequently-used thickeners are polymer compounds such as polyacrylates or various polysaccharides. In the case of polyacrylates, the thickening effect comes from the way the molecules arrange themselves following the neutralization of the free acrylic acid side groups in the aqueous phase of the emulsion. The polyacrylate gels thus formed are, in contrast to the hydrophilic gels formed by polysaccharides, highly shear- and electrolyte-sensitive.

In the case of W/O emulsions, the consistency-giving factors here described are ineffectual, because the consistency is determined first and foremost by the nature, amount, and viscosity of the emulsifiers and the oil phase, and

Sensory Assessment of Lipids in Skin Care Products

also by physical properties, such as particle size and the proportion of dispersed aqueous phase.

In order, by way of example, to register the sensorily relevant changes when a consistency-giving system is modified, formulations were produced and subjected to sensory assessment; the quantities of consistency-giving factors present in the initial formulation were doubled or else largely replaced by a polysaccharide or a polyacrylate. The results (Fig. 11) show that when the amount of consistency-giving factor is doubled, the consistency of the product increases as expected, and along with this, it is more difficult to spread; also, peaking and cushion are somewhat increased, and it is harder to get the cream out of the pot. The higher proportion of fatty alcohol and partial glyceride also leads to greater waxiness. All in all, however, the sensory profile does not undergo any fundamental change. The relevant properties of the initial formulation, which in any case

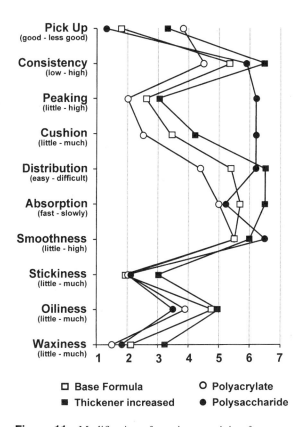

Figure 11 Modification of consistency-giving factors.

already has a high proportion of the factor, are increased, but without any additional effects.

If the initial thickening system is strongly reduced, and a polyacrylate used instead, the sensory characteristics change markedly. First of all, the formulation is more difficult to remove from the container. This is because the polyacrylate, besides its thickening properties, is both shear-sensitive and sensitive to electrolytes, which means that skin contact alone causes the formulation to partially break down when it is taken from the container. The reduction in peaking and cushion can be explained in the same way. As the polymer leaves residues on the skin, stickiness is increased while the smoothness of the skin is somewhat reduced. The sensory changes resulting from these changes in formulation would be assessed as negative if we were talking, for example, about a face cream. If, however, the preparation were intended to be applied to a large area of skin, as a body lotion for example, the use of these polymer thickeners could be entirely positive.

Finally, the use of a polysaccharide in combination with a reduced quantity of fatty alcohol leads to a product that shows a marked increase in peaking and cushion effects, while at the same time the polysaccharide increases skin smoothness and facilitates the removal from the container. The effects are due to the fact that while this polymer only thickened the emulsion slightly, it did not itself show any noticeable shear-sensitivity or sensitivity to electrolytes.

B. Modification of the Emulsifier

The choice of emulsifier system is of fundamental importance for the sensory characteristics of the emulsion. First, it is the emulsifier that determines whether the oil or the water forms the outer phase of the emulsion. Second, the differences in chemical structure, even within one type of emulsion where the formulation is otherwise identical, lead to a marked change in sensory characteristics when the emulsifier is changed. Among the reasons are the different physical properties and the different propensities of different emulsifiers to form liquid-crystalline associates. Among the O/W (oil-in-water) emulsifiers commonly used in cosmetics are such different compounds as ethoxylated fatty alcohols, partial glycerides or sorbitan fatty acid esters, alkyl polyglucosides, polyglycerol esters, alkyl phosphates, fatty alcohol sulfates, alkali salts of fatty acids, and amphiphile polymer compounds, to name but a few.

To illustrate this, first of all an ethoxylated cetylstearylalcohol in the model formulation was exchanged for a cetylstearyl glucoside. The basic properties of the cream remained unchanged (Fig. 12). Because of the particularly strongly marked propensity of alkyl glucoside to form liquid-crystalline gel phases, the consistency and the cushion effect both increase. The remaining parameters are largely unchanged.

When the complete emulsifier/thickener system is replaced by polyglyc-

Sensory Assessment of Lipids in Skin Care Products

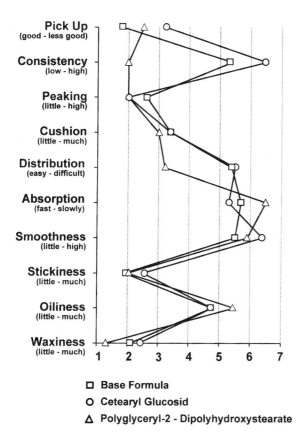

Figure 12 Modification of the emulsifier-system.

eryl-2-dipolyhydroxystearate, the result is a thin W/O emulsion. This can be spread easily and, despite its low consistency, shows a relatively high cushion effect. This is typical of W/O creams, which mostly have a high cushion score. In their absorption score, as in the other parameters, there are only small deviations from that of the initial O/W formulation. As a result of the change described here, therefore, merely changing the emulsifier system, while keeping the same oil phase, can result in a rich, semi-solid cream being turned into a low-viscous, easily spread lotion that leaves behind a strongly "caring" impression.

C. Modification of the Oil Phase

Cosmetic oils are often used specifically to implement certain galenic concepts. There is a range of emollients to choose from, a whole range of different chemical classes, such as ester oils, glycerides, hydrocarbons, ether, guerbet alcohols, or silicone derivatives. Within the various groups, in turn, substances can be chosen with quite different viscosity or spreadability or other physical properties. In general, increasing the quantity of emollient will lead to richer formulations that take longer to absorb. If the amount of oil-phase is kept constant, the physicochemical properties of their composition will in turn affect the sensory characteristics.

In order to illustrate these correspondences, the relatively high proportion of cosmetic oils in the initial formulation was first reduced in the model formulation. The sensory assessment (Fig. 13) shows two conspicuous sensory changes: first, the formulation is absorbed much more quickly and is much less oily; second, the cream has a much higher consistency, and a larger quantity of waxy

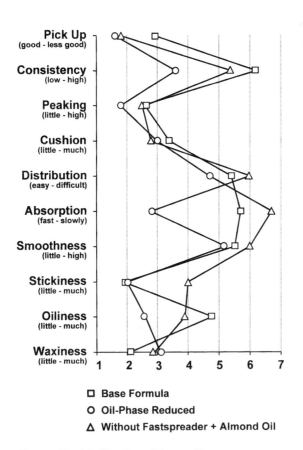

Figure 13 Modification of the emollients.

residue could be detected. These effects are due to a relative disproportion of consistency-giving factors. Altogether, the result was a more compact cream that left a less "caring" impression. For particular applications, as in a warm humid climate, a formulation of this kind might well be positive.

A formulation that omits high-spreading dicaprylyl ether and replaces it by low-spreading almond oil results in some surprising changes: in spite of the higher viscosity of the almond oil, the consistency of the cream is reduced. This result can be explained by the fact that the oil phase in total is more polar and thus dissolves more fatty alcohols, which are then no longer available as consistency-giving factors to form the liquid-crystalline surfactant phases that determine viscosity. The increased subjective stickiness, the slower absorption, and the greater waxiness that result when the high-spreading factor is omitted are, by contrast, all to be expected. The experiment shows that the implementation of galenic concepts by a change in the oil phase must always be accompanied by an adjustment in the emulsifier/co-emulsifier system if the desired effect is to be achieved without unintended side effects.

D. Variation of the Moisturizers

One of the most important tasks of skin-care emulsions is to supply moisture. To retain moisture in the external layers of the skin, moisturizers are normally employed, typically various polyols. These substances really do improve the moisture of the skin, but their maximum concentration has an upper limit. Using sensory assessment, for example, it can be proved that an increase in the glycerol content to 20% results in a very marked increase in stickiness (Fig. 14). Because of the co-solvent effect of glycerol, there is also a reduction in the liquid-crystalline surfactant phases, and thus a reduction in the consistency of the cream.

It is interesting that glycerol carbonate, which can also be used as a moisturizer, does not result in either a marked change in consistency or in the high stickiness associated with glycerol, even when used in concentrations of 20%. Compounds with properties of this kind allow the formulation of elegant cosmetic emulsions with a very long-lasting moisturizing effect.

E. Multiple Emulsions

Multiple emulsions are characterized by the presence of one or more additional inner phases. In the case of W/O/W emulsions, oil droplets are distributed within a coherent aqueous phase; the oil droplets in their turn contain small droplets of water.

In the investigation described below, the question to be answered was whether there are any sensory characteristics associated specifically with multiple emulsions. For this purpose, three different products were subjected to sensory

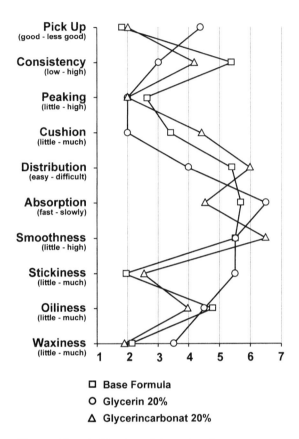

Figure 14 Modification of the moisturizer.

assessment. One product was based on a polymer emulsifier with a polyacrylate as thickening agent; a second on polyglycerol ester and ethoxylated fatty alcohols; and a third product on ethoxylated fatty acids and partial glycerides.

The sensory profiles of the creams investigated (Fig. 15) are different. The formulation thickened with polyacrylates rapidly broke down after being applied to the skin, and the cream felt quite watery. Then the formulation was quickly absorbed, evincing a certain stickiness. The other emulsions show sensory profiles that are markedly different in several parameters.

These investigations show that there is no set of sensory characteristics specific to all W/O/W creams. Only in the case of one product can the multiple emulsion be directly detected by the senses, in that it feels very watery when first applied, and then producing, if anything, an oily and sticky impression.

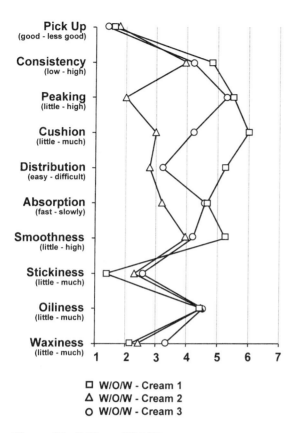

Figure 15 Different W/O/W creams.

The actual usefulness of multiple emulsions lies in the fact that they make it theoretically possible to keep certain normally incompatible ingredients separate during storage. In this way, it has been possible to prepare unstable formulations that allow novel or long-term effects.

F. Sensory Assessment of Water-Free Emollients

Cosmetic lipids are usually one component of formulations and are used together with numerous other substances, mostly emulsifiers. As shown above, the sensory characteristics of the finished product are always the result of a complicated interaction of all the ingredients. In certain cases, such as in skin oils or other formulations with a high lipid content, the emollients are the only, or at least the clearly dominant, factor as far as the formulation's sensory characteristics are concerned.

Cosmetic emollients are the subjects of detailed description elsewhere in this book. In principle, the emollients differ along the following parameters:

Polarity: nonpolar emollients—paraffins, iso-paraffins; polar emollients—triglycerides, straight-chained, branched and unsaturated esters
Molecular weight
Degree of branching
Viscosity
Surface tension
Functional groups

An important parameter of emollients is their spreadability, as it makes an important contribution to absorbability and to the time taken for fatting. Zeidler [14] has examined the connections between the spreading speed of different emollients on the skin and their physicochemical and chemical properties. The results showed that within functionally similar product series, the spreading values increased with increasing molecular weight. In addition, moderately hydrophobic lipids with ester groups (i.e., fats and wax esters) spread—although molecular

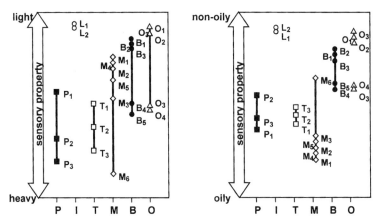

Figure 16 Comparative evaluation of various emollients. (From Ref. 14.)

weights were comparable—more strongly than nonpolar hydrocarbons; and branched lipids spread more strongly than nonbranched.

Further connections between the sensory characteristics of emollients and their chemical and physicochemical properties are described elsewhere [15,16] (Fig. 16). Nonpolar emollients are, compared with polar substances, heavy and draggy. Branched substances are if anything light and smooth, as are substances with low viscosity and low molecular weight. Nacht et al. [17] describe an inverse proportionality between perceived greasiness and the skin friction coefficient.

The connections described are confirmed by the following sensory profile,

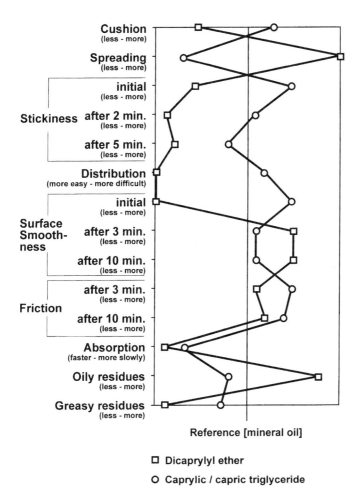

Figure 17 Sensory profile of dicaprylyl ether and capric/caprylic triglyceride.

in which dicaprylyl ether (low viscous, nonpolar) and caprylic/capric triglycerides (low viscous, polar) were compared in pairs with a mineral oil as reference (Fig. 17). It can clearly be seen that the low-viscous or polar oil components were, overall, assessed as being the lighter, more quickly absorbed oils. This sensory assessment was carried out as a pair comparison, whereby the sensory parameters were defined in similar fashion to those for the emulsions described above.

G. Sensory Assessment of Lipids in Surfactant-Based Rinse-Off Products

Sensory assessment can also be applied to other cosmetic product categories. Depending on the product category concerned, individual sensory parameters and

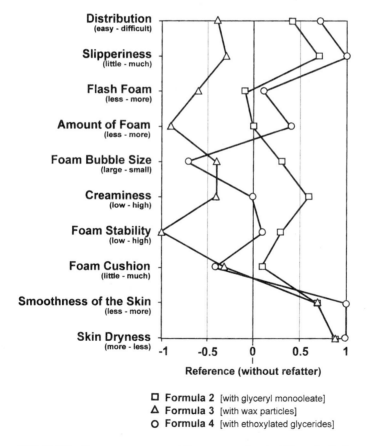

Figure 18 Sensory assessment (direct comparison) of formula 2–4 versus reference.

reference products must be developed. For surfactant-based body-cleansing products for example, foaming and post-use skin-feel are the chief parameters; in the case of shampoos, other parameters are more important (e.g., volume, care, combability, gloss, or stylability).

In the field of surfactant-based body-cleansing preparations, in the early 1980s more surfactants that are gentler to the skin and hair came on the market. This concept, however, is no longer regarded as original. The topic of refatting or moisturizing is interesting in this connection. This is due on the one hand to changed consumer habits: today, people take showers and wash their hair as a rule several times a week, which leads to repeated removal of skin lipids with the result that the barrier function of the stratum corneum can be gradually weakened, resulting in turn in increasing dryness of the skin. Second, demographic shift means that the older section of the community is growing markedly. This consumer group suffers particularly severely from dry, itchy, and taut skin. Particularly important for them are not only gentle but, above all, refatting surfactant preparations.

Sensory assessment has been employed to determine the efficacy of novel refatting agents [18]. A compound of glycerol monooleate and APG and a wax particle dispersion were tested. In both cases, a clear improvement in skin softness and skin moisture was registered after treatment with refatting shower-baths (Fig.

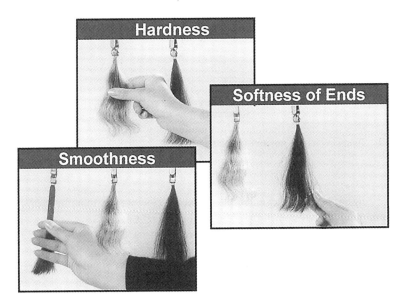

Figure 19 Sensory assessment of hair.

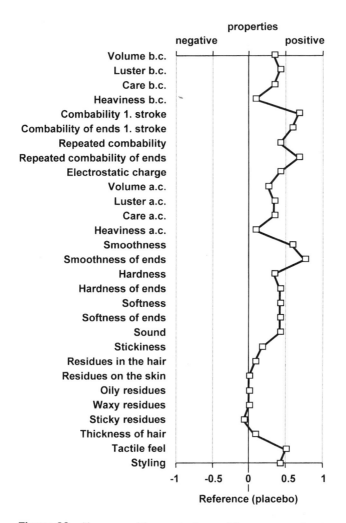

Figure 20 Shampoo with a nonionic conditioner versus placebo.

18). Interesting in this connection are the marked differences in foaming characteristics that were discovered in this sensory investigation.

The other big surfactant-based product category is shampoos, which may contain functional ingredients designed, for example, to increase hair gloss or to develop a conditioning effect. Typically, alongside various polymers, silicone derivatives or other conditioning lipids are employed. The effects on the hair are usually determined by half-side tests in hair studios; the sensory assessment is carried out by hairdressers.

Sensory Assessment of Lipids in Skin Care Products

Another possibility consists in treating strands of hair and having a panel of experts assess these strands along precisely defined parameters. In Figure 19, the test procedures for the parameters hardness, smoothness, and softness of ends are shown by way of example. The effects following application of a shampoo containing a conditioning lipid are shown in Figure 20. Positive effects were detected above all for combability, smoothness, and tactile feel.

IV. CORRELATION OF SENSORY ASSESSMENT AND CONSUMER TESTS

What all the forms of practical implementation of sensory assessment described above have in common is the fact that the panelists are more or less highly trained experts. If sensory assessment is also to result in statements on the potential market acceptance of a newly developed or modified product, therefore, there is a definite need to know about any correlations between the results of sensory assessment and consumer tests.

Below, the results of an investigation are presented in which three moisturizing creams of type O/W (creams A, B, C) were examined in a sensory assessment (paired comparison test) and in a typical consumer test.

A. Sensory Assessment

The three products were compared in pairs in a paired comparison test (A versus B, A versus C, and B versus C) along certain sensory parameters. The sensory profiles resulting from this investigation showed that product A was markedly different from the other products in respect to the majority of sensory parameters. Samples B and C showed only minor differences in the sensory parameters investigated.

B. Consumer Test

The three products were subjected in parallel to consumer tests in Germany and France, each using 60 participants. These consumer tests also showed up differences between product A and samples B and C. But it transpired that not all the established differences (e.g., between samples A and C) were perceived. Table 3 lists the respective parameters for which differences were noted in the sensory assessment and in the consumer test.

Overall, the comparison shows that the consumer tests, which cost much more in terms of money and, in particular, time, came up with fewer differences in the descriptive analysis of the three samples than the sensory assessment. It is interesting that this survey was also able to register the influence of language/

Table 3 Product Comparison A versus C, Distinguishable Parameters

Sensory assessment	German consumer panel	French consumer panel
Appearance	Appearance	Appearance
Color	Color	
Pick-up	Consistency	Consistency
Peaking		
Cushion		
Consistency	Consistency	Consistency
Distribution		Distribution
Absorption		Absorption
Smoothness		
Stickiness		
Oiliness		
Waxiness		
Fragrance		
Shine	Skin gloss	Skin gloss

cultural background on the product assessment, as revealed in the varying ability of the consumers to register differences.

The advantage of the consumer tests is that they provide a direct statement on whether consumers like the product or not. In the present case, the different consumer panels agreed in the ranking of the three test products as far as consumer acceptance was concerned: cream C > cream B ≫ cream A.

Only exact knowledge of the correlations between the results of expert assessment and consumer acceptance of the product enables sensory assessment to be used to make a forecast on product acceptance. The exact evaluation of such parallel tests could therefore be used to dispense with consumer tests in future.

V. RHEOLOGY AND SENSORY CHARACTERISTICS

In view of the relative expenditure of time and money, sensory assessment has the advantage over the consumer test. However, it still involves considerable expense in view of the relatively large number of people involved. For this reason, attempts have been made to establish test systems that can come up with statements on sensory characteristics of consumer goods without involving human subjects at all. In this connection, it is above all the rheological characteristics that have been subjected to more detailed investigation.

In the literature, there has been controversial debate on the question of whether the sensory characteristics of lipid-containing topical emulsions correlate with their rheological characteristics. Moscowitz, for example, describes the dependence of the sensory attributes of skin-care fluids on their rheological characteristics [19]; Brummer [20] finds a correlation of primary skin feeling with the shear stress at the onset of flow τ_F and the dynamic viscosity η_{dyn} and secondary skin feeling with the value of stationary viscosity η for the rate of shear prevailing at the end of application to the skin.

S. Wang measured TEWL and skin capacitance using several emulsions that contained the same oil but had different rheological properties and found no effect from rheology on moisturizing efficacy [21]. This study, as well as another [22], could detect no clear correlation with various sensory characteristics.

Basically, the rheological characteristics of lipid-containing emulsions in W/O systems are determined by the phase/volume relationship between oil and aqueous phase, whereas in O/W formulations the dominant influence on the flow behavior comes from the co-emulsifiers and thickeners. Examples of this can be found in Förster [23] (Figs. 21 and 22). This source also provides an example of the correlation between particular sensory and rheological parameters. Thus, it could be shown that the consistency determined by sensory assessment correlates with viscosity at low shear rates, and the dispersability (or distribution) of the product accords well with the viscosity at higher shear rates (Fig. 23).

Taking a general view, it seems to be the case that rheological measurements may be suited for making forecasts of certain simple sensory parameters,

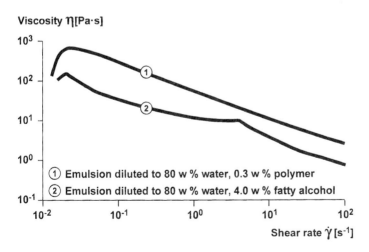

Figure 21 Viscosity curves of differently thickened emulsions.

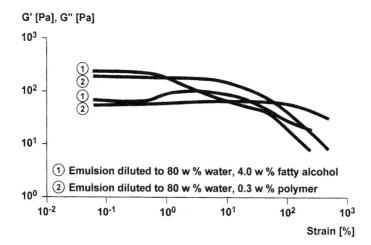

Figure 22 Strain-sweep curves of differently thickened emulsions: upper set of curves shows G′, lower set G″.

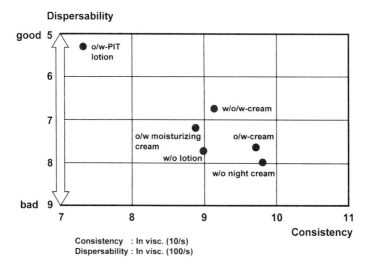

Figure 23 Estimation of sensory properties of creams.

depending on the formulation. In the case of certain complex parameters, such as skin feel, absorption, or skin smoothness, rheological measurements are not capable of delivering a reliable forecast.

VI. SUMMARY

In this chapter, we have discussed the basics and various test methods for the sensory investigation of lipid-containing cosmetic preparations. The sensory assessment of lipids in leave-on and rinse-off products such as emulsions and surfactant-based formulations for skin and hair care has been presented with numerous examples. In addition, we have pointed out correspondences between the physicochemical and galenic characteristics of cosmetics on the one hand, and the subjectively perceptible characteristics profile on the other.

REFERENCES

1. Meilgaard M, Civille GV, Carr BT. Sensory Evaluation Techniques. Boca Raton, FL: CRC Press, 1991.
2. Williams PL, Warwick R, Dyson M, Bannister LH (eds.). Gray's Anatomy, 37th ed. Edinburgh: Churchill Livingstone, 1989.
3. Busch P, Gassenmeier T. Sensory assessment in the cosmetic field. Parfümerie und Kosmetik 1997; 78:16–21.
4. American Society for Testing and Materials. Physical requirement guidelines for sensory evaluation laboratories. In: Eggert J, Zook K, eds., ASTM Special Technical Publication 913, Philadelphia, 1986.
5. ASTM, Commitee E-18. Guidelines for the selection and training of sensory panel members. ASTM Special Technical Publication 758, Philadelphia, 1981.
6. International Organization for Standardization, TC34/SC12. Sensory analysis—choosing and training assessors. Draft International Standard ISO/DP 8586, Tour Europe, Cedex 7, 92080 Paris La Défense, 1985.
7. Stone H, Sidel JL. Sensory evaluation for skin care products. Cosmet Toil 1986; 101(3):45–50.
8. Stone H, Sidel JL. Sensory evaluation practices. San Diego, CA: Academic Press, 1993.
9. Civille GV, Dus CA. Evaluating tactile properties of skincare products: a descriptive analysis technique. Cosmet Toil 1991; 106(5):83–88.
10. Aust LB, et al. The descriptive analysis of skin care products by a trained panel of judges. J Soc Cosmet Chem 1987; 38:443–449.
11. Muckenschnabel R. Hydrophilic und ambiphile Cremesysteme—Beitrag zur Kenntnis von Struktur und Galenik. Dissertation Erlangen 1983.
12. Gottschalk G, Kaiser RE. Einführung in die Varianzanalyse und Ringversuche. Mannheim, BI-Taschenbuch Nr. 775.

13. Gassenmeier T, Busch P. Sensory Assessment zur Prüfung der Wirkung galenischer Konzepte. Lecture held at the DGK Symposium Bad Neuenahr, 1999.
14. Zeidler U. Über das Spreiten der Lipide auf der Haut. Fette, Seifen, Anstrichmittel 1985; 87:403–408.
15. Kameshwari V, Mistry ND. Sensory properties of emollients. Cosmet Toil 1999; 114:45–51.
16. Alexander P. Assessing emolliency. Manufacturing Chemist 1992; 26–27.
17. Nacht S, Close, JA, Yeung D, Gans EH. Skin friction coefficient: changes induced by skin hydration and emollient application and correlation with perceived skin feel. J Soc Cosmet Chem 1982; 32:55–65.
18. Gassenmeier T, Busch P, Hensen H, Seipel W. Some aspects of refatting the skin. Cosmet Toil 1998; 113:89–92.
19. Moskowitz HR, Fishken D. Rheological characteristics and consumer acceptance of emulsion products. Cosmet Toil 1978; 93:31–44.
20. Brummer R, Godersky S. Rheological studies to objectify sensations occurring when cosmetic emulsions are applied to the skin. Colloids Surfaces A 1999; 152:89–94.
21. Wang S, Kislalioglu MS, Breuer M. The effect of rheological properties of experimental moisturizing creams/lotions on their efficacy and perceptual attributes. Int J Cosmetic Sci 1999; 21:167–188.
22. Wiechers JW. Comparing instrumental and sensory measurements of skin moisturization. Cosmet Toil 1999; 114:29–34.
23. Förster T, von Rybinski W. Applications of emulsions. In: Binks BP, ed. Modern Aspects of Emulsion Science. Royal Society of Chemistry; 1998:395–426.

Index

Acylglutamate, 301
Alkyl chain
  all-trans conformation of, 81, 85
  conformation, 107
  trans fraction of, 81, 90
Alkyl ether sulfate, 300, 309
Alkyl glucoside, 303, 309, 311
Alkyl sulfate, 300
Amino acid content, 106
Area per lipid, 81, 86
Arm flex wash test, 310, 311
Ascorbate (see Ascorbic acid)
Ascorbic acid
  cofactor in lipogenesis, 138
  and lipid profile, 140

Barrier (see Skin barrier)
Betaine, 302, 310
Blister technique, 156, 157
Body wash product (see Surfactant)

Ceramide, 1, 6, 8, 14, 42, 79, 92, 228, 270, 271, 280, 307
  analogue, 15, 16, 17
  and cholesterol and fatty acid (see Skin barrier lipid)
  composition and aging, 110
  glycosyl, 5, 8, 14, 28, 101, 104, 137, 281
    in skin models, 137, 138
  syntase, 103

[Ceramide]
  nature-identical, 15
  precursor, 14
  synthesis of, 103, 111
  synthetic, 13, 42
  type-1, 3, 7, 12, 15, 279, 280
  type-2, 7, 12, 77, 281
  type-3, 3, 7, 17, 18, 29, 30, 42
  type-4, 3, 7, 138, 281
  type-5, 3, 7, 12, 138, 140, 279, 281
  type-6, 3, 7, 138, 140
  type-7, 138, 140
Cleansing (see Washing)
  milk, 248
Cholesterogenesis, 102
Cholesterol, 5, 12, 42, 43, 45, 51, 52, 79, 84, 85, 92, 102, 109, 138, 173, 283, 286
  addition of, 240, 288
  and ceramide and fatty acid (see Skin barrier lipid)
  effect on area per lipid, 86
  synthesis of, 111
Cholesterol sulfate, 101, 104, 166, 172, 228, 246
Chromameter, 215
Collagen
  type I, 133
  type III, 133
  type V, 133

[Collagen]
  type VII, 133
  type XII, 128
  type XIV, 128
Confocal laser scanning microscopy (CLSM), 56, 58, 215, 275
Consumer test, 325, 347, 348
Corneometer, 188
  and capacitance, 189
  and conductance, 189
  and phase angle, 189
Corneosurfametry, 209
Cornified envelope, 37, 107, 112, 113
Cream, 228, 275, 286
  absorption, 330
  cold, 256
  consistency, 330, 335, 338, 339
  cushion, 330, 336
  lamellar, 249, 250
  multiple, 339
  oiliness, 330, 333
  oil-in-water, 273, 285, 333, 347
  peaking, 330, 335, 336
  pick-up, 330
  and restoration of lipid film, 305
  spreadability, 273, 274, 338
  stickiness, 330, 333, 336, 339, 340
  water-in-oil, 248, 249, 273, 285, 334, 337
  waxiness, 330, 335, 338, 339
Culture
  fibroblast, 122
  keratinocyte, 122
Cutometer, 204

Defatting, 303
Dermal-epidermal junction, 128, 130
Desmosome, 207
Desquamation, 206
  degree of, 207
  process, 139, 160
  rate of, 207
Differentiation
  marker, 133, 160
  and proliferation, 136
DSC (*see* Thermal analysis)

Elastin, 128
Emollient, 188, 202, 214, 307, 341, 342
  absorbability, 342
  spreadability, 343
Emulsifier, 214, 257, 258, 303, 327, 333, 334, 336, 341
Emulsion (*see* Cream)
Epidermal
  barrier, 3
  cell division, 14
  lipids, 210

Fat (*see* Triacylglyceride)
Fatty acid/ester ratio, 237, 246
Fibrillin-1, 133
Fibronectin, 133
Filaggrin, 133
Foam, 299, 301, 302, 311, 345
Fourier transform infrared spectroscopy (FTIR) (*see* Infrared spectroscopy)
Frictionmeter, 216
Free fatty acid, 8, 43, 45, 102, 160, 167, 171, 172
  addition of, 288
  in aged skin, 279
  area per lipid, 86
  and ceramide and cholesterol (*see* Skin barrier lipid)
  removal of, 246, 303
  in skin models, 138, 140
  synthesis of, 103, 111
Freeze fracture electron microscopy (FFEM), 38, 47, 56

Gas-bearing electrodynamometer, 204
Gas chromatography, 165
Gas-liquid chromatography, 171
  and extracting solvent, 171
Greasiness, 237

Hydrolipidic film, 210

Infrared spectroscopy, 38, 43, 58, 76, 77, 85, 107, 137, 138, 189, 228, 229, 232, 282

Index

[Infrared spectroscopy]
  and assignment of bands, 235
  and ATR technique, 230
Interlayer distance, 10, 39, 45

Keratin, 10, 14, 130, 133
Keratinocytes
  differentiation of, 128
    and retinoids, 130
  joined by desmosomes, 129

Lamellar body, 5, 99, 100, 101, 104, 136, 137, 138
Lamellar granules, 129
Laminin, 133
Lanolin, 256, 265, 287
Laser-Doppler-flowmeter, 282, 283
Lecithin, 257
Lipid, 327
  alkyl chain order, 47, 238, 246
  barrier composition
    and age, 150
    and anatomical site, 150
    and dietary influences, 151
    and exposure to UV, 152, 154
    and hormonal effects, 151
    and location in the epidermis, 163
    and race, 151
    and seasonal variations, 152
    and sex, 150
    and winter xerosis, 150, 152
  bilayer, 4, 10, 12, 76, 78, 80, 99, 106, 228
    protective function of, 100
  lamellae, 38
    arrangement of, 84
    swelling of, 47
    water content of, 47
  lamellar structure of, 4, 12, 13, 37, 43
  lateral packing of, 41
  layer (*see* Bilayer)
  metabolism
    and calcium, 112
    and enzymes, 111
    and magnesium, 112
    and potassium, 112

[Lipid]
  packing order, 244, 249, 282
  profile, 163, 164, 173
Liposome, 43, 48, 49, 55
  gel-state, 49, 51
  liquid-state, 49, 51
  of skin lipids, 51

Magnetic resonance imaging, 215
Model
  collagen gel, 127
    addition of adipocytes to, 128
    support of hair follicles by, 128
  dermal analogue (*see* Dermal substrate)
  dermal equivalent, 121
  dermal substrate
    based on collagen, 126
    based on collagen-glycosaminoglycan-chitosan, 127, 128, 133
    noncollagenous, 127
  epidermis, 121–124, 140, 159, 242
  horny layer lipid, 9, 76
  membrane system, 42
  skin barrier, 77
  skin equivalent, 121–124, 242
    inclusion of endothelial cells in, 124
    inclusion of Langerhans cells in, 124
    inclusion of melanocytes in, 124
    and lipids, 159
  stratum corneum (SC) lipid (*see* horny layer lipid)
Moisturizer, 258, 307, 333
Molecular modeling, 75

Natural moisturizing factor (NMF), 188
Nuclear magnetic resonance spectroscopy (NMR), 76, 92, 189, 228

Occlusivity, 276, 277, 278, 280, 287, 288, 289
Oil
  ester, 266, 278
  Jojoba, 266

Optical coherence tomography, 215
Order parameter, 87, 88, 92

Packing parameter, 76
Paraffin, 256, 262, 264
Patch test, 308
Periodicity (*see* Interlayer distance)
Penetration, 281
  enhancer, 52, 55, 241, 281, 282, 283, 284
  of lipids, 54, 275, 276, 288
  of surfactants, 304
  of water, 275
Percutaneous absorption (*see* Penetration)
Permeation study, 49, 137
Petrolatum, 256, 278, 286, 287, 305
Phase
  behavior, 45
    of stratum corneum lipids, 238
  diagram, 10
  gel, 84, 240
  lamellar, 39, 41, 45
  separation, 42, 43, 59
Phosphatidylcholine, 84
  bilayer, 79
Phospholipid, 6, 8, 101, 174
  membrane, 47
  saturated, 50
  unsaturated, 50
  vesicle, 51
Phytosterol, 286
Polymer, 262, 307, 336, 340, 346
Profilometer, 198
Pseudoceramide, 15, 17, 18, 19, 20–24, 29, 30, 270, 271
Psoriasis, 279

Raman spectroscopy, 84, 228, 231, 232
  and assignment of bands, 235
Refatting, 305, 312, 345
Rheological modifier (*see* Thickener)
Rheology, 272, 349

Scanning electron microscopy (SEM), 199

Sebum, 5, 99, 100, 102, 107, 108–110, 210,
  excretion rate of, 211, 247
  effect of age on, 211
  effect of sex on, 211
  effect of body site on, 211
  wash off by surfactants, 303
Sebumeter, 212
Sebutape, 211
Second messenger, 14
Sensory assessment, 273, 275, 319, 335, 345
  comparative, 326
  correlation with consumer test, 347
  correlation with rheology, 348
  guidelines for, 324
  and language, 320
  and psychological factors, 320
  test set-up for, 322
Sensory characteristic (*see* Sensory assessment)
Sensory profile
  of cleansing products, 345
  of creams, 332, 334, 340
  of emollients, 341, 343
Shower gel (*see* Surfactant)
Silicone, 268, 275, 278, 307, 338, 346
Skin
  absorption, 273, 274, 351
  atopic, 279, 285
  color, 215
  compatibility, 188
    of cleansing agent, 188, 189, 306, 309, 310, 311
  dry, 105, 106, 110, 152, 154, 187, 194, 207, 243, 244, 248, 279, 280, 285, 313, 315, 345
  and aging, 109
  elasticity, 202
  extensibility, 202
    and aging, 205
  feel, 273, 351
  flexibility, 202
  greasy, 244
  hydration (*see* Moisture)
  irritation, 283, 303, 305
    of cleansing products, 209, 310

[Skin]
   moisture, 14, 106, 193, 237, 243, 246, 249, 250, 339
   moisturization, 187, 202, 284, 286, 288, 289, 307, 345
      and humectants, 188, 191
      and occlusive products, 193
   pH, 216
   redness, 195, 311
   replica, 197
   rheometer, 205
   roughness, 14, 152, 197, 200, 280, 284, 289, 313
   scaly, 106, 110, 152, 207, 208
      surfactant induced, 107, 305, 306, 311
   smoothness, 286, 330, 336, 351
   softness, 203, 286
   surface lipids, 210
   swelling, 311
   temperature, 216
   texture
      and age, 197
      and weather, 197
      and topical application, 197
   thickness, 201, 215, 280
   type, 212, 244, 248
   xerosis, 15, 110, 194
Skin barrier
   damage, 13, 99, 194, 195, 303
      prevention of, 196
   disruption, 109
   formation
      and nuclear hormone receptor, 113
   function, 14, 37, 97, 102, 105, 137, 193
   and histamine receptor, 113
   homeostasis, 98, 99, 114
   lipid (SBL), 7, 9, 17, 25, 38, 77, 89, 101, 136, 166, 193, 278, 279, 285
   recovery, 286
      and circadian rhythm, 98
      by creams, 196
      after psychological stress, 98
      by replacement of lipids, 284, 285

[Skin barrier]
   and sex, 150
   after topical application, 108
   repair, 13, 107, 111, 112
   strengthening of, 191, 194, 248, 249
Soap, 299, 300
Solvent mixture, 160
Sphingolipids, 160, 228
   signaling role of, 14
   sphinganine, 1
   sphingosine, 1, 15, 16
   sphingomyeline, 101
Squalene, 5, 12, 102, 107, 108, 166, 210
Squamometry, 209
Stratum corneum
   barrier (*see* Skin barrier)
   lipid (*see* Skin barrier lipid)
   thickness, 97, 129, 215
   turnover, 207
Sulfosuccinate ester, 301
Surface area (*see* Area per lipid)
Surface tension, 273, 274, 275
Surfactant, 188, 228, 247, 299, 307, 312, 315, 345
   amphoteric, 302
   damaging effect of, 195, 196, 277, 285, 309, 311

Tackiness, 275
Tape stripping, 155, 156
   by adhesive tape, 155
   and barrier disruption, 107, 156
   by cyanoacrylate resin, 156, 162
   and lipid gradient, 163
   and scaling, 110, 208
Thermal analysis, 77, 85, 228, 239, 282, 283
Thickener, 257, 327, 333, 334, 340
Thin-layer chromatography, 137, 138, 165
   and ceramide profiles, 165
   and eluting solvent, 167
   and pure lipid standards, 166
   and total lipid determination, 165
Tocopherol, 108, 138, 268

Topical application
  of mevalonic acid, 111
  of petrolatum, 108
  of retinoic acid, 113
  of stratum corneum lipids, 108
Topometry, in vivo, 199, 313
Torquemeter, 204
Transepidermal water loss (TEWL), 13, 107, 249, 280, 281, 283, 285, 305, 349
  in dry environment, 97
  in the elderly, 110
  and film formation, 277
  and healthy skin, 19
  and impairment of skin barrier, 304
  inhibiting of, 265
  normalization of, 196
  and occlusivity, 278, 287
  and skin irritation, 311
  in skin models, 137
Transfersome, 52, 54, 55, 58
Transglutaminase, 133
Transmission electron microscopy (TEM), 59, 76, 133, 137
Triacylglyceride (*see* Triglyceride)
Triacylglycerol (*see* Triglyceride)
Triglyceride, 6, 7, 110, 167, 173, 210, 342
  addition of, 288
  in creams, 256, 267
  in skin models, 137, 138, 140

Two-photon excitation microscopy, 59, 61

Ultrasound measurement, 214

Vaseline, 262
Vesicle, 59
  nonionic surfactant, 50
Viscosity, 273, 274, 275, 303, 311, 334, 338, 339, 349
Vitamin C (*see* Ascorbic acid)
Vitamin E (*see* Tocopherol)

Washing, 189, 246, 247, 299, 304, 307, 311
  dehydrating effect of, 189
Water-binding capacity, 13
Water domains, 48
Wax, 256, 259, 264, 270, 278, 286, 310, 342
Wrinkles, 197, 200

X-linked ichthyosis, 109, 279
X-ray scattering, 38, 43, 47, 280
  small-angle (SAXS), 10, 39, 44, 76, 137, 138, 139, 228
  wide-angle (WAXS), 39, 41, 44, 76, 105